Reading Essentials
An Interactive Student Workbook

ips.msscience.com

 **Glencoe**

New York, New York   Columbus, Ohio   Chicago, Illinois   Peoria, Illinois   Woodland Hills, California

To the Student

In today's world, knowing science is important for thinking critically, solving problems, and making decisions. But understanding science sometimes can be a challenge.

Reading Essentials takes the stress out of reading, learning, and understanding science. This book covers important concepts in science, offers ideas for how to learn the information, and helps you review what you have learned.

In each chapter:

- **Before You Read** sparks your interest in what you'll learn and relates it to your world.
- **Read to Learn** describes important science concepts with words and graphics. Next to the text you can find a variety of study tips and ideas for organizing and learning information:
  - The **Study Coach** offers tips for getting the main ideas out of the text.
  - **Foldables™ Study Organizers** help you divide the information into smaller, easier-to-remember concepts.
  - **Reading Checks** ask questions about key concepts. The questions are placed so you know whether you understand the material.
  - **Think It Over** elements help you consider the material in-depth, giving you an opportunity to use your critical-thinking skills.
  - **Picture This** questions specifically relate to the art and graphics used with the text. You'll find questions to get you actively involved in illustrating the concepts you read about.
  - **Applying Math** reinforces the connection between math and science.
- Use **After You Read** to review key terms and answer questions about what you have learned. The **Mini Glossary** can assist you with science vocabulary. Review questions focus on the key concepts to help you evaluate your learning.

See for yourself. *Reading Essentials* makes science easy to understand and enjoyable.

Glencoe

The McGraw-Hill Companies

Send all inquiries to:
Glencoe/McGraw-Hill
8787 Orion Place
Columbus, OH 43240

ISBN 0-07-867089-6
Printed in the United States of America
6 7 8 9 10 11 024 10 09 08

Table of Contents

The Nature of Science

section ❶ What is science?

● Before You Read

Have you ever wondered how something works? On the lines below, describe a time that you wondered how something worked. Did you find out how it worked? Explain how.

_____

_____

● Read to Learn

Learning About the World

When you think of a scientist, do you think of someone in a laboratory with charts, graphs, and bubbling test tubes? Anyone who tries to learn about the natural world is a scientist—even you. <u>Science</u> is a way of learning more about the natural world. Scientists want to know why, how, or when something happened. Learning usually begins by keeping your eyes open and asking questions about what you see.

What kinds of questions can science answer?

Scientists ask many questions. How do things work? What are they made of? Why does something take place? Some questions cannot be answered by science. Science cannot help you find the meaning of a poem or decide what your favorite color is. Science cannot tell you what is right, wrong, good, or bad.

What are possible explanations?

Learning about your world begins with asking questions. Science tries to find answers to these questions. However, science can answer questions only with the information that exists at the time. Any answer found by science could be wrong or could change because people can never know everything about the world around them.

What You'll Learn

- what science is and what science cannot answer
- what theories and laws are
- identify a system and its parts
- the three main branches of science

Study Coach

Outlining As you read the section, create an outline using each heading from the text. Under each heading, write the main points or ideas that you read.

FOLDABLES

Ⓐ Build Vocabulary
Make the following Foldable to help you define and learn the vocabulary terms in this section.

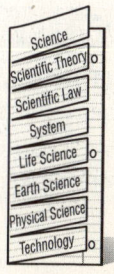

How does new information affect old explanations?

As time passes, people learn more about the world around them. As you can see in the diagram below, new information might make scientists look at old ideas and think of new explanations. Science finds only possible explanations. For example, people once thought Earth was the center of the solar system. Through the years, new information about the solar system showed this is not true.

Picture This

1. **Explain** Look at the diagram to the right. How can new information affect an old explanation for something?

_____

_____

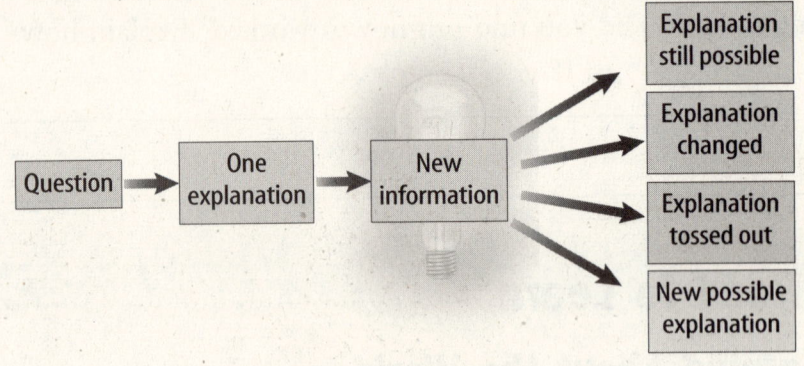

Possible outcomes

Question → One explanation → New information →
- Explanation still possible
- Explanation changed
- Explanation tossed out
- New possible explanation

What are scientific theories?

A <u>scientific theory</u> is an attempt to explain a pattern seen repeatedly in the natural world. Theories are not just guesses or opinions. Theories in science must have observations and results from many investigations to back them up. They are the best explanations that have been found so far. Theories can change. As new data are found, scientists decide how the new data fit the theory. Sometimes the new data do not support the theory. Then scientists can change the theory to fit the new data better.

What are scientific laws?

A <u>scientific law</u> is a rule that describes a pattern in nature. For an observation to become a scientific law, it must be observed happening over and over again. The law is what scientists use until someone makes observations that do not follow the law. A law helps you predict what will happen. If you hold an apple above the ground and drop it, it always will fall to Earth. The law tells you the apple will fall, but the law does not explain why the apple will fall. A law is different from a theory. It does not try to explain why something happens. It simply describes a pattern. ☑

✔ Reading Check

2. **Determine** Which describes a pattern in nature, a scientific theory or a scientific law?

_____

Systems in Science

Scientists can study many different things in nature. Some scientists study how the human body works. Others might study how planets move around the Sun. Still others might study the energy in a lightning bolt. What do all of these things have in common? All of them are systems. A **system** is a group of structures, cycles, and processes that are related to each other and work together. Your stomach is a structure, or one part of, your digestive system. ☑

Where are systems found?

You can find systems in other places besides science. Your school is a system. It has structures like school buildings, furniture, students, teachers, and many other objects. Your school day also has cycles. Your daily class schedule and the school calendar are examples of cycles. Many processes are at work during the school day. Your teacher may have a process for test taking. Before a test, the teacher might ask you to put your books away and get out a pencil. When the test is over, the teacher might ask you to put down your pencil and pass the test to the front of the room.

In a system, structures, cycles, and processes work together, or interact. What you do and what time you do it depends on your daily schedule. A clock shows your teacher that it is time for your lunch break. So, you go to lunch.

How are parts of a system related to a whole system?

All systems are made up of other systems. For example, the human body is a system. Within the human body are many other systems. You are part of your school. Your school is probably part of a larger district, state, or national system. Scientists often solve problems by studying just one part of a system. A scientist might want to know how the construction of buildings affects the ecosystem. Because an ecosystem has many parts, the scientist might study one particular animal in the ecosystem. Another might study the effect on plant life.

The Branches of Science

Science is often divided into three main parts, or branches. These branches are life science, Earth science, and physical science. Each branch asks questions about different kinds of systems.

✔ **Reading Check**

3. **List** What are the three parts of a system?

_____

_____

_____

💡 **Think it Over**

4. **Describe** Buildings usually have a heating system. Write each of the following by the part of the system it best represents. *turning on and off, thermostat, spreading heat*

Structure:

_____

Process:

_____

Cycle:

_____

What is life science?

<u>Life science</u> is the study of living systems and the ways in which they interact. Life scientists try to answer questions like "How do whales know where they are swimming in the ocean?" and "How do vaccines prevent disease?" Life scientists can study living things, where they live, and how they act together.

People who work in the health field, like doctors and nurses, know a lot about life science. They work with systems of the human body. Some other people that use life science are biologists, zookeepers, farmers, and beekeepers.

What is Earth science?

<u>Earth science</u> is the study of Earth systems and systems in space. It includes the study of nonliving things such as rocks, soil, clouds, rivers, oceans, planets, stars, meteors, and black holes. Earth science also includes the weather and climate systems on Earth. Earth scientists ask questions like "How do you know how strong an earthquake is?" and "Is water found on other planets?" They make maps and study how Earth's crust formed. Geologists study rocks and Earth's features. Meteorologists study weather and climate. There are even volcanologists who study volcanoes. ☑

What is physical science?

<u>Physical science</u> is the study of matter and energy. Matter is anything that takes up space and has mass. Energy is the ability to cause matter to change. All systems—living and nonliving—are made of matter.

Chemistry and physics are the two areas of physical science. Chemistry is the study of matter and the way it interacts. Chemists ask questions like "What can I do to make aspirin work better?" and "How can I make plastic stronger?" Physics is the study of energy and its ability to change matter. Physicists ask questions like "How does light travel through glass?" and "How can humans use sunlight to power objects?"

How are science and technology related?

Learning the answers to scientific questions is important. However, these answers do not help people unless they can be used in some way. <u>Technology</u> is the practical use of science in our everyday lives. Engineers use science to create technology. The study of how to use the energy of sunlight is science. Using this knowledge to create solar panels is an example of technology.

✔ **Reading Check**

5. **Apply** What might an Earth scientist study that is not on Earth?

_____

_____

💡 **Think it Over**

6. **Apply** Which of the following is an example of technology? Circle the correct answer.
   a. finding out how light travels
   b. creating solar-powered cars
   c. deciding which rock is the hardest
   d. making strong plastic

● After You Read

Mini Glossary

Earth science: the study of Earth systems and systems in space

life science: the study of living systems and the ways in which they interact

physical science: the study of matter and energy

science: a way of learning more about the natural world

scientific law: a rule that describes a pattern in nature

scientific theory: an attempt to explain a pattern seen repeatedly in the natural world

system: a group of structures, cycles, and processes that are related to each other and work together

technology: is the practical use of science in our everyday lives

1. Review the terms and their definitions in the Mini Glossary. When you see lightning strike, you probably will hear thunder soon. Is this statement a scientific theory or a scientific law? Explain.

_____

_____

2. Fill in the graphic organizer below with explanations of science, each branch of science, and technology.

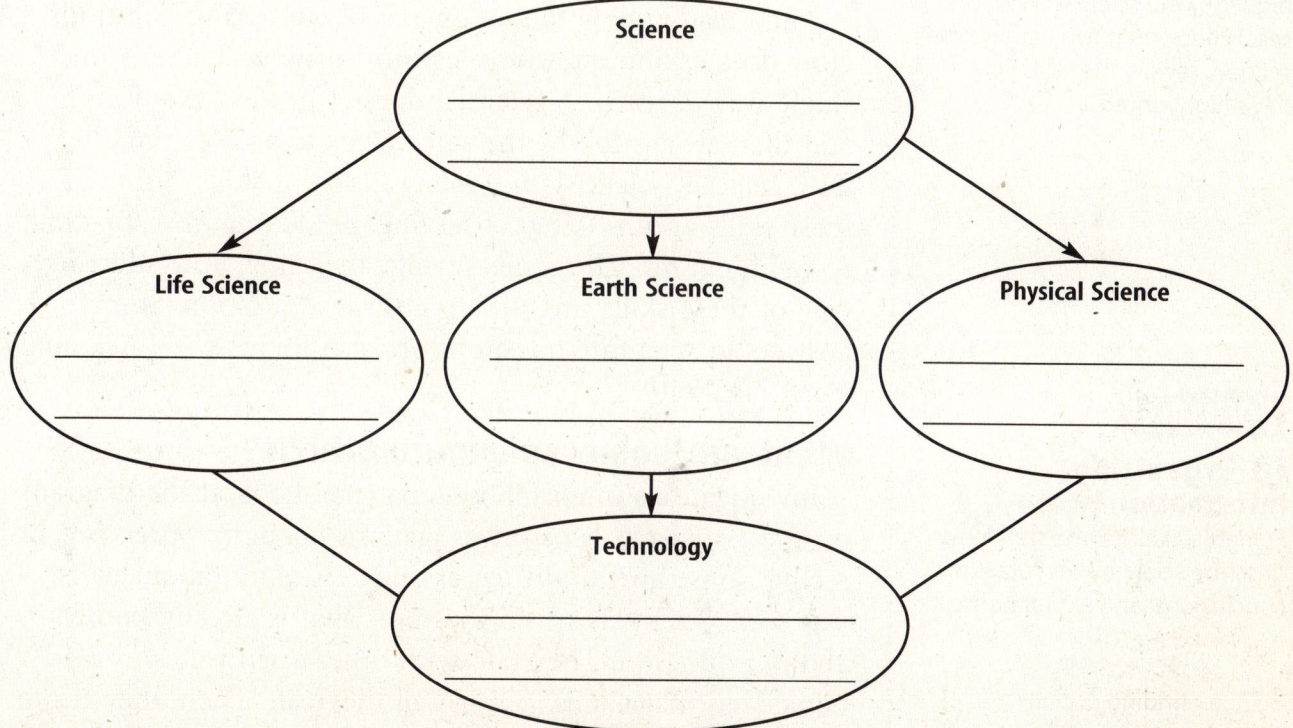

Science●nline Visit **ips.msscience.com** to access your textbook, interactive games, and projects to help you learn more about what science is.

End of Section

Reading Essentials **5**

chapter 1 The Nature of Science

section 2 Science in Action

What You'll Learn

- skills that scientists use
- the meaning of hypothesis
- the difference between observation and inference

Before You Read

Think of some skills that you have. You may be good at basketball. Of all the skills you have, which do you think is your best? How did you learn that skill?

Mark the Text

Highlighting As you read the text under each heading, highlight the science skills you see. When you finish reading the section, review the skills you have highlighted.

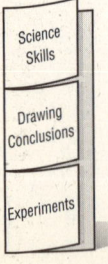

FOLDABLES

B Organizing Information Make a Foldable like the one shown to describe science skills, drawing conclusions, and experiments.

Science Skills

Drawing Conclusions

Experiments

Read to Learn

Science Skills

You already know that science is about asking questions. How does asking questions lead to learning? There is no single way to learn. A scientist doesn't just ask a question and then always follow the same steps to answer the question. Instead, scientists use many different skills. Some of these skills are thinking, observing, predicting, investigating, researching, modeling, measuring, analyzing, and inferring. Any of these skills might help answer a question. Some answers to scientific questions are also found with luck and using creativity.

What are some science methods?

Investigations often follow a pattern. Look at the diagram on the next page. Most investigations begin by observing, or seeing, something and then asking a question about what was seen. Scientists try to find out what is already known about a question. They talk with other scientists, and read books and magazines to learn all they can. Then, they try to find a possible explanation. To collect even more information, scientists usually make more observations. They might build a model, do experiments, or both.

A Scientific Method

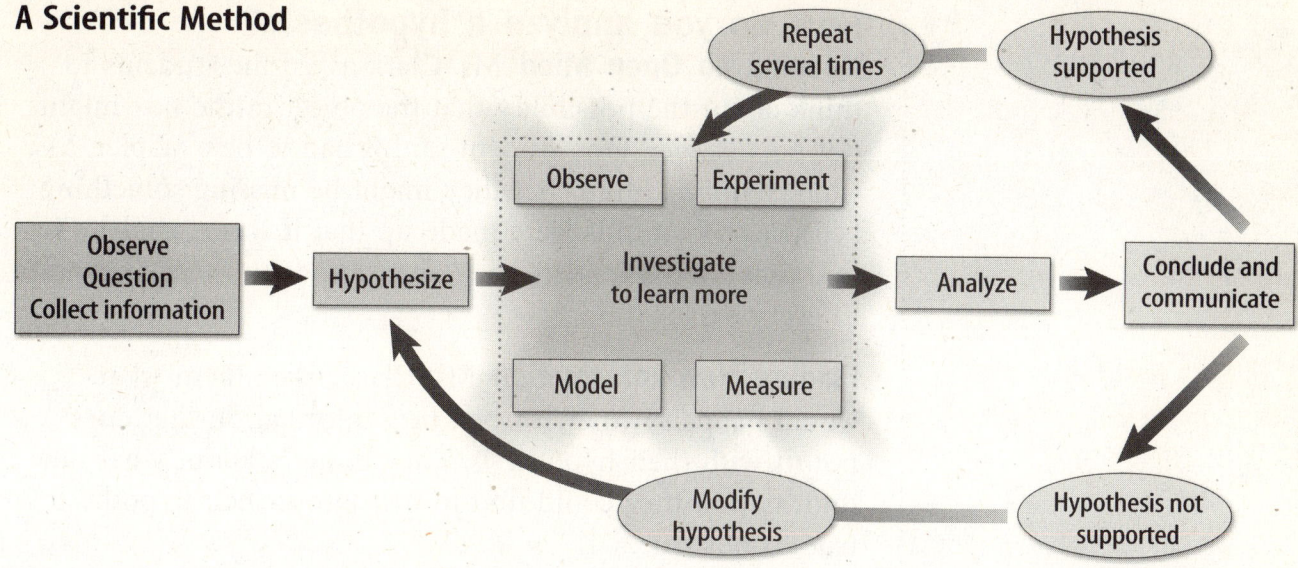

How do you question and observe?

You ask questions and make observations all the time. For example, Ms. Clark, a middle school science teacher, placed a sealed shoebox on the table at the front of the classroom. Everyone in the class began to ask questions about what was inside the box. Some students picked up the box. There was something loose inside it. Some students said it sounded like metal when it hit the sides of the box. It wasn't very heavy. Ms. Clark passed the box around. Each student made observations and wrote them in his or her journal. These students were using science skills without even knowing it.

How is guessing part of science?

Ms. Clark asked her students to guess what was in the box. One student thought it was a pair of scissors. Another student thought it was too heavy to be scissors. All agreed it was made of metal. The students then guessed that it was probably a stapler in the box. Ms. Clark told them that by guessing a stapler was in the box, they had made a hypothesis.

What is a hypothesis?

A **hypothesis** is a reasonable answer based on what you know and what you observe. The students observed that whatever was in the box was small, heavier than a pair of scissors, and made of metal. The students knew that a stapler is small, heavier than a pair of scissors, and made of metal. They used this information to make the hypothesis that there was a stapler in the box.

Picture This

1. **Interpret a Diagram** Why are there two arrows going in different directions at the right end of the diagram?

_____

_____

_____

Think it Over

2. **Compare and Contrast** How is a hypothesis different than a guess?

_____

_____

_____

How do you analyze a hypothesis?

Keeping an Open Mind Ms. Clark asked the students to think about the possibility that the object in the box might not be a stapler. One student said it had to be a stapler. Ms. Clark reminded him that they might be missing something because their minds were made up that it was a stapler. She said that good scientists keep open minds to every idea and to every explanation.

Finding New Information Ms. Clark asked them what would happen if they learned new information that does not fit with their hypothesis. They thought about what new information they could find to help prove their hypothesis true or not true.

The students decided they could get another shoebox and put a stapler in it. Then they could shake it to see whether it felt and sounded the same as the other box. By getting more information to find out if their hypothesis was correct, the students could analyze, or examine, their hypothesis.

How do you make a prediction?

Ms. Clark asked the students to predict what would happen if their hypothesis was correct. A prediction is what you think will happen. The students all agreed the second box should feel and sound like the mystery box.

How do you test a hypothesis?

Sometimes a hypothesis is tested by making observations. Sometimes building a model is the best way to test a hypothesis. To test their hypothesis, the students made a model. The new shoebox with the stapler inside of it was the model.

After making a model, you must test it to make sure it is the same as the original. So the students needed to test their model to see if it was the same as the original shoebox. When they picked up the model box, they found it was a little heavier than the first box. Also, when they shook the model box, it did not make the same sound as the first box.

The students decided to find the mass of each box. Then they would know how much more mass a stapler has compared to the object in the first box. The students used a balance to find that the box with the stapler had a mass of 410 g. The mystery box had a mass of 270 g.

Think it Over

3. **Explain** What will the box sound and feel like if the students' hypothesis is correct?

_____

_____

_____

How do you organize your findings?

Ms. Clark suggested that the students organize the information they had before making any new conclusions. By organizing their observations, they had a summary to look at while making conclusions. The students put their information into a table like the one below.

Observation Chart		
Questions	**Mystery Box**	**Our Box**
Does it roll or slide?	It slides and appears to be flat.	It slides and appears to be flat.
Does it make any sounds?	It makes a metallic sound when it strikes the sides of the box.	The stapler makes a thudding sound when it strikes the sides of the box.
What is the mass of the box?		

How do you draw conclusions?

When the students looked at the observations in their table, they decided that their hypothesis was not correct. Ms. Clark asked them if that meant that there was not a stapler in the mystery box. One student said that there could be a different kind of stapler in the mystery box. The students were inferring that the object in the mystery box was not like the stapler in their test box. To **infer** means to make a conclusion based on what you observe. ☑

Another student suggested that they were right back where they started with the mystery box. Ms. Clark pointed out that even though their observations did not support their hypothesis, they knew more information than when they started.

How do you continue to learn?

A student asked if she could open the box to see what was in it. Ms. Clark explained that scientists do not get to "open the box," to find answers to their questions. Some scientists spend their entire lives looking for the answer to one question. When your investigation does not support your first hypothesis, you try again. You gather new information, make new observations, and form a new hypothesis.

Picture This
4. Complete a Table
Using information from the previous page, complete the last row of the table.

✔ **Reading Check**

5. Determine What did the students in Ms. Clark's class infer?

_____

_____

_____

_____

6. Draw Conclusions
Suppose scientists did not show their methods and results to other scientists. How would this affect scientific discoveries?

_____

_____

_____

How do you communicate your findings?

Sometimes scientists try to continue or repeat the work of other scientists. It is important for scientists to explain the results of their investigations and how they did their investigations. Scientists write about their work in journals, books, and on the Internet. They also often go to meetings and give speeches about their work. An important part of doing science is showing methods and results to others.

Experiments

Different types of questions need different types of investigations. Ms. Clark's class needed to answer the question "What is inside the mystery box?" To answer the question, they built a model to learn more about the mystery box. Some scientific questions are answered by doing a type of investigation called a controlled experiment. A **controlled experiment** involves changing only one part in an experiment and observing what that change does to another part of the experiment.

What are variables?

A **variable** is a part of an experiment that can change. Imagine an experiment that tests three fertilizers to see which one makes plants grow tallest. This experiment has two variables. One variable is the fertilizer used. Since three different fertilizers are used, each one can have a different outcome in the experiment. The second variable is the height of the plants. The different fertilizers can affect the height of the plants.

Independent Variables The **independent variable** is a variable that is changed in an experiment. The fertilizer is changed by the scientist. So, the fertilizer is an independent variable. In an experiment, there should be only one independent variable.

Dependent Variables The **dependent variable** is a variable that depends on what happens in the experiment when the independent variable is used. The dependent variable is also the variable measured at the end of the experiment. The height of the plants is the dependent variable. The height of each plant may be different, depending on which fertilizer is used. The scientist will measure the height of each plant at the end of the experiment to see what fertilizer affects the height the most. ✔

✔ Reading Check

7. Explain what a dependent variable is in an experiment.

_____

_____

_____

What are constants?

A <u>constant</u> is a part of an experiment that is not changed. There can be more than one constant. In the fertilizer experiment, the constants could be the type of plant, the amount of water or sunlight the plants get, or the kind of soil the plants are planted in. The scientist keeps all of these constants the same for all the types of fertilizer that are tested.

Laboratory Safety

In your science class, you will perform many kinds of investigations. Performing investigations involves more than just following steps. You must learn how to keep yourself and those around you safe. Always obey the safety symbol warnings shown below.

Safety Symbols

 Eye Safety

 Clothing Protection

 Disposal

 Biological

 Extreme Temperature

 Sharp Object

 Fume

 Irritant

 Toxic

 Animal Safety

 Flammable

 Electrical

 Chemical

 Open Flame

 Handwashing

How do you practice safety in the lab?

When scientists work in a lab, they take many safety precautions. You must also take safety precautions in the science lab. The most important safety advice is to think before you act. You should always check with your teacher during the planning stage of your investigation. Make sure you know where the safety equipment is in your lab or classroom. You also need to make sure you know how to use the safety equipment. Safety equipment includes eyewashes, thermal mitts, and the fire extinguisher. ☑

Think it Over

11. Infer If you are doing a science experiment in the lab or in the field, what is the one thing that you should always wear?

_____

What are some good safety habits?

Good safety habits include the following suggestions:

- Find and follow all safety symbols before you begin an investigation.

- Always wear an apron and goggles to protect yourself from chemicals, flames, and pointed objects.

- Keep goggles on until activity, cleanup, and handwashing are complete.

- Always slant test tubes away from yourself and others.

- Never eat, drink, or put on makeup in the lab.

- Report all accidents to your teacher.

- Always wash your hands after working in the lab.

How do you practice safety in the field?

Investigations are also done outside the lab. You can do investigations in streams, farm fields, and other places. Scientists call this working in the field. Scientists must follow safety regulations in the field as well as in the lab. Always wear eye goggles and other safety equipment that you need. Never reach into holes or under rocks. Always wash your hands after you have finished your work in the field or in the lab.

Why have safety rules?

Doing science in the lab or in the field can be much more interesting that just reading about it. But doing experiments can be dangerous and accidents can happen. If you follow safety rules closely, an accident is less likely to happen. Still, you cannot predict when something will go wrong.

Think of a person taking a trip in a car. Most of the time the person is not in a car accident. However, to be safe, drivers and passengers must wear their safety belts. Wearing safety gear in the lab is like wearing a safety belt in a car. It can keep you from being hurt in an accident. You should wear safety gear even if you are just watching an experiment. Always keep safety in mind when conducting an experiment.

● After You Read

Mini Glossary

constant: a part of an experiment that is not changed

controlled experiment: involves changing only one part of an experiment and observing what that change does to another part of the experiment

dependent variable: a variable that depends on what happens in the experiment when the independent variable is used

hypothesis: a reasonable answer based on what you know and what you observe

independent variable: a variable that is changed in an experiment

infer: to make a conclusion based on what you observe

variable: a part of an experiment that can change

1. Review the terms and their definitions in the Mini Glossary. In this section, what was an example of a controlled experiment?

_____

_____

2. In the flowchart below, complete the steps that a scientist might take when conducting a scientific investigation. Use these words or group of words to complete the chart:
   conclude and communicate
   hypothesize
   experiment, investigate, or model

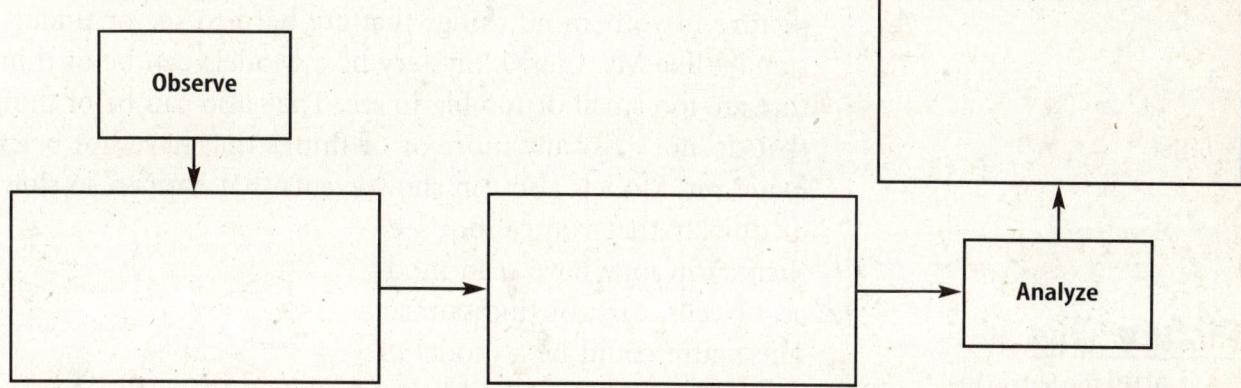

3. At the beginning of the section, you were asked to highlight science skills in the section. What is another method you could have used to learn about science skills?

_____

 **Science**●**nline** Visit **ips.msscience.com** to access your textbook, interactive games, and projects to help you learn more about action.

End of Section

The Nature of Science

section ❸ **Models in Science**

What You'll Learn

- to describe different types of models
- the uses of models

● **Before You Read**

Have you ever built a model? Why did you build the model? Tell about a model you built or want to build.

_____

_____

Study Coach

Identifying the Main Point When you read a paragraph, look for the main idea and write it down on a piece of paper or in your notebook.

Picture This

1. **Label** the Sun in the model of the solar system.

● **Read to Learn**

Why are models necessary?

There are many ways to test a hypothesis. In the last section, Ms. Clark's class tested their hypothesis with a model of the mystery box. A **model** is something that represents an object, event, or idea in the natural world. Models can help you picture in your mind things that are hard to see or understand—like Ms. Clark's mystery box. Models can be of things that are too small or too big to see. They also can be of things that do not exist any more or of things that have not been made yet. Models also can show events that happen so slowly or quickly that you cannot see them. You may have seen models of cells, cars, or dinosaurs. The figure could be a model of the solar system. It could help you understand which planets are next to each other.

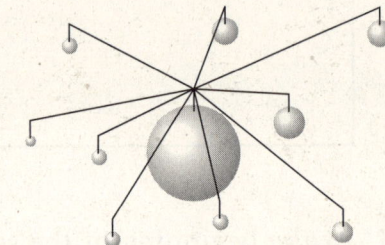

Types of Models

There are three main types of models—physical models, computer models, and idea models. Scientists can use one or more types of models to help them answer questions. Different models are used for different reasons.

What are physical models?

Models that you can see and touch are physical models. The figure on the previous page is a physical model. A globe of Earth is also a physical model. Models show how parts relate to each other. They also can show how things look when they change position or how things react when a force is put on them.

What are computer models?

Computer models are built using computer software. You can't touch them, but you can look at them on a computer screen. Computer models can show events that happen too quickly or too slowly to see. For example, a computer can show how large plates in Earth move. They also can be used to predict when earthquakes might happen.

Computers also can model movements and positions of things that might take hours or days to do by hand, or even using a calculator. They also can predict changes caused by different systems or forces. For example, computer models help predict the weather. They use the movement of air currents in the atmosphere to make these predictions.

What are idea models?

Some models are ideas that describe what someone thinks about something in the natural world. Idea models cannot be built like physical models because they are just ideas. A famous idea model is Albert Einstein's theory of relativity. One model for this theory is the mathematical equation $E = mc^2$. This explains that mass, m, can be changed into energy, E. ☑

Making Models

Have you ever seen a sketch artist at work? The artist tries to draw a picture of someone from a description given by someone else. The more detailed the description is, the better the picture will be. Sometimes the artist uses descriptions from more than one person. If the descriptions have enough information, the sketch should look realistic. Scientific models are much the same way. The more information the scientist finds, the more accurate the model will be.

FOLDABLES™

C Organize Information
Use a half sheet of paper to help you organize information about models in science.

Models in Science

☑ **Reading Check**

2. **Determine** Which type of model can you see and touch: physical, computer, or idea model?

_____

_____

💡 **Think it Over**

3. **Apply** Suppose you want to make a model of a plant cell for your science project. What will help you make the most accurate model?

_____

_____

Using Models

When you think of a model, you might think of a model airplane or a model of a building. Not all models are for scientific uses. You may even use models and not know it. Drawings, maps, recipes, and globes are all models.

How are models used?

Communicate Some models communicate ideas. Have you ever drawn a map to show someone how to get to your home? If so, you used a model to communicate. It is sometimes easier to show ideas than to tell them.

Test Predictions Other models test predictions. Engineers often use models of airplanes or cars in wind tunnels to test predictions about how air affects them.

Save Lives, Money, and Time Models are often used because it is safer and less expensive than using the real thing. For example, crash test dummies are used instead of people in automobile crash tests. NASA has built a special airplane that models the conditions in space. It creates freefall for 20 to 25 seconds. Astronauts can practice freefall in the airplane instead of in space. Making many trips in the airplane is easier, safer, and less expensive than a trip into space.

Limitations of Models

The solar system is too big to see all at once. So, scientists have built models. The first solar system models had the planets and the Sun revolving around Earth. Later, as scientists learned new information, they changed their models. A new model explained the solar system in a different way, but Earth was still the center. Still later, after more observations, scientists discovered that the Sun is the center of the solar system. A new model was made to show this. Even though the first solar system models were incorrect, the models gave scientists information to build upon. Models are not always perfect, but they are a tool that scientists can see and learn from. ✓

💡 Think it Over

4. Describe What is another model, besides a map, you could use to communicate with?

_____

_____

_____

✔ Reading Check

5. Explain Why are models that have been proven to be wrong still helpful?

_____

_____

● After You Read

Mini Glossary

model: something that represents an object, event, or idea in
the natural world

1. Review the term and its definition in the Mini Glossary. Which of the three types of
   models can you touch? Explain why you cannot touch the other types of models.

   _____

   _____

   _____

2. Complete the graphic organizer below to describe the three types of models and their uses.

Types of Models
Physical models are models you can see and touch.

Uses of Models
to save money, time, and lives

3. How do you think making a physical model can help you learn more about how
   models work?

   _____

   _____

   _____

Science●nline Visit **ips.msscience.com** to access your textbook, interactive
games, and projects to help you learn more about models.

End of
Section

The Nature of Science

section ➍ Evaluating Scientific Explanation

What You'll Learn

- to evaluate scientific explanations
- how to evaluate promotional claims

● Before You Read

Have you ever played the game where you whisper a message into a person's ear, and then that person repeats the message to another person, and so on? What usually happens by the time the message gets to the last person?

_____

_____

Study Coach

Asking Questions As you read the section, write down any questions you might have about what you read.

● Read to Learn

Believe it or not?

Think of something someone told you that you didn't believe. Why didn't you believe it? You probably decided there was not enough proof. What you did was evaluate, or judge, the reliability of what you heard. You can evaluate the reliability of a statement by asking "How do you know?" If what you are told seems reliable, you can believe it.

What is critical thinking?

When you decide to believe information you read or hear, you use critical thinking. **Critical thinking** means using what you already know and new facts to decide if you agree with something. You can decide if information is true by breaking it down into two parts. Based on what you know, are the observations correct? Do the conclusions make sense?

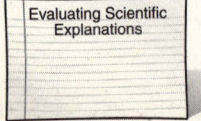 FOLDABLES™

ⓓ Finding Main Ideas
Use a half sheet of paper to help you list the main ideas about how to evaluate scientific explanations.

> Evaluating Scientific
> Explanations

Evaluating the Data

Data are observations made in a scientific investigation. Data are gathered and recorded in tables, graphs, or drawings during a scientific investigation. Always look at the data when you evaluate a scientific explanation. Be careful about believing any explanation that is not supported by data.

Are the data specific?

The data given to back up a statement should be specific, or exact. Suppose a friend tells you that many people like pizza better than they like hamburgers. What do you need to know before you agree with your friend? You need some specific data. How many people were asked which food they like more? Specific data makes a statement more reliable and you are more likely to believe it.

How do you take good notes?

In this class you will keep a science journal. You will write down what you see and do in your investigations. Instead of writing "the stuff changed color," write "the clear liquid turned to a bright red when I added a drop of food coloring." It is important to record your observations when they happen. Important details can be forgotten when you wait. ☑

Evaluating Conclusions

When you think about a conclusion that someone has made, you can ask yourself two questions. First, does the conclusion make sense? Second, are there any other possible explanations? Suppose you hear that school will be starting two hours late because of bad weather. A friend decides that the bad weather is snow. You look outside. There is no snow on the roads. The conclusion does not make sense. Are there any other possible explanations? Maybe the roads are icy. The first conclusion is not reliable unless other possible explanations are proven to be wrong.

Evaluating Promotional Materials

MIRACLE CREAM MAKES WRINKLES DISAPPEAR OVERNIGHT

Look at the newspaper advertisement. It seems unbelievable. You should hear some of the scientific data before you believe it. The purpose of an ad is to get you to buy something. Always keep this in mind when you read an ad. Before you believe ads like this one, evaluate the data and conclusions. Is the scientific evidence from a good, independent laboratory? An independent laboratory is not related to the company selling the product. Always evaluate data and ask questions before you spend your money.

☑ **Reading Check**

1. **Explain** Why should you record your observations when they happen?

_____

_____

💡 **Think it Over**

2. **Describe** Think of an advertisement that you have seen that you did not believe. Explain why you did not believe the advertisement.

_____

_____

_____

_____

● After You Read

Mini Glossary

critical thinking: using what you already know and new facts to decide if you should agree with something

data: observations made in a scientific investigation

1. Review the terms and their definitions in the Mini Glossary. Write one sentence using both terms.

_____

_____

2. Use the graphic organizer below to record some of the questions you should ask when you read the results of a scientific investigation.

Questions to ask when you evaluate a scientific investigation
Is the data specific?

3. How could you teach an elementary science class about how to use critical thinking?

_____

_____

_____

End of Section

 Visit **ips.msscience.com** to access your textbook, interactive games, and projects to help you learn more about evaluating scientific explanations.

20   The Nature of Science

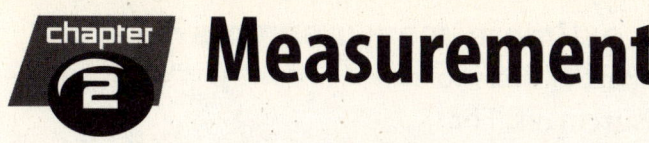

Measurement

section ❶ Description and Measurement

● Before You Read

Weight, height, and length are common measurements. List at least five things you can measure.

What You'll Learn
■ how to estimate
■ how to round a number
■ the difference between precision and accuracy

● Read to Learn

Measurement

A **measurement** is a way to describe objects and events with numbers. You use measurements to answer questions like how much, how long, or how far. You can measure how much sugar to use in a recipe, how long a snake is, and how far it is from home to school. You also can measure height, weight, time, temperature, volume, and speed. Every measurement has a number and a unit of measure. There are many different units of measure. Some are meter, liter, and gram.

How do measurements describe events?

Events like races can be described with measurements. In the 2000 summer Olympics, Marion Jones of the United States won the women's 100-m dash in a time of 10.75 s. In this example, measurements tell about the year of the race, its length, and the runner's time. The name of the event, the runner's name, and her country are not measurements.

Estimation

What happens if you want to know the size of something but it is too large to measure or you don't have a ruler? You can use **estimation** to make a rough guess about the size of an object. You can use the size of one object you know to help you estimate the size of another object. Estimation can help when you are in a hurry or don't need an exact measurement. You will get better at estimation with practice.

Study Coach

Make an Outline Make an outline of the main ideas in this section. Use the headings in the section as major headings. Be sure to include all words in bold.

FOLDABLES

A Organize Make a Foldable, as shown below, and label the tabs Measurement, Estimation, Precision, and Accuracy.

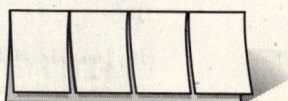

1. Draw and Compare

You can show how the tree is about twice as tall as the person. Draw another person the same height that is standing on the shoulders of the person in the figure.

_____

_____

_____

✔ Reading Check

2. Identify What word tells you that a measurement is an estimate?

_____

_____

💡 Think it Over

3. Apply Give a time when estimating might be helpful.

_____

_____

_____

How do you estimate measurements?

You can compare objects to estimate a measurement. For example, the tree in the figure is too tall to measure. You can estimate the height of the tree by comparing it to the height of the person. The tree is about twice the height of the person. If the person is about 1.5 m tall, the tree must be 2 × 1.5 m, or about 3 m tall.

Estimated measurements often use the word *about*. For example, a soccer ball weighs about 400 g. You can walk about 5 km in an hour. When you see or hear the word *about* with a measurement, the measurement is an estimation. ☑

When is estimation used?

Estimation is not only used when an exact measurement cannot be made. It is also used to check that an answer is reasonable. A reasonable answer makes sense. Suppose you calculate a friend's running speed as 47 m/s. Does it make sense that your friend can run that fast? Think about how far 47 m is. That's almost a 50-m dash. That means your friend could run a 50-m dash in about 1 s. Can your friend do that? No, he can't. So, 47 m/s is too fast and is not a reasonable answer. You need to check your work.

Precision

Precision describes how close measurements are to each other. Some measurements are more precise than others. Suppose you measure the distance from home to school four times. Each time you get 2.7 km. Your neighbor measures the same distance four times. He measures 2.9 km two times and 2.7 km two times. Your measurements are closer to each other than your neighbor's measurements. So, your measurements are more precise.

The timing for Olympic events has become very precise. Years ago, events were measured to tenths of a second. Now they are measured to hundredths of a second. The instruments we use to measure today are more precise than those used years ago.

Accuracy

<u>Accuracy</u> is the closeness of a measurement to the true value. Suppose you measured the length of your shoelace two times. One time you measured 12.5 cm and the other time you measured 12.3 cm. Your measurements are precise because they are close together. However, if the shoelace is actually 13.5 cm long, the measurements are not accurate.

What makes a good measurement?

A good measurement must be both precise and accurate. A precise measurement is not always a good measurement. A watch that has a second hand is more precise than a watch without one. But, the watch with the second hand could be set 1 hour earlier or later than the real time. In that case, the watch would not be accurate at all. Since the time on the watch is precise but not accurate, the time on the watch is not a good measurement.

How do you round a measurement?

Suppose you need to measure the length of the sidewalk outside your school. You could measure to the nearest millimeter. But, you probably only need to know the length to the nearest meter or tenth of a meter. Suppose you find that the length of the sidewalk is 135.481 m. How do you round this number to the nearest tenth of a meter? Follow these two steps:

1. Look at the digit to the right of the place being rounded to.
   • If the digit is 0, 1, 2, 3, or 4, the digit being rounded to stays the *same*.
   • If the digit is 5, 6, 7, 8, or 9, the digit being rounded to *increases by one*.

So, to round to the nearest tenth of a meter, look for the digit in the tenths place, 4. Find the digit to the right of 4. The 4 increases to 5 because the digit to the right of 4 is 8.

2. Look at the digit being rounded to. Then look at the digits to its right. If those digits are to the right of a decimal, they are removed. If they are to the left of a decimal, change them to zeros. For example, 432.9 rounded to the nearest hundred is 400. Since the 9 is to the right of the decimal, it is removed.

So, 135.481 rounded to the nearest tenth of a meter is 135.5 m.

💡 **Think it Over**

4. **Apply** Give another example of a measurement that is precise but not accurate.

_____

_____

_____

_____

Applying Math

5. **Rounding Values**
   What is 135.481 m rounded to the nearest ten meters?

_____

Some measurements are not precise.

If a measurement doesn't need to be precise, you can round your measurement. For example, suppose you want to divide a 2-L bottle of soda equally among seven people. When you use a calculator to divide 2 by 7, you get 0.285 714 28. You can round this number to 0.3. This is a little less than 0.333..., or $\frac{1}{3}$. You pour a little less than $\frac{1}{3}$ L of soda for each person.

What are significant digits?

Significant digits are the number of digits that show the precision of a number. For example, 18 cm has two significant digits: 1 and 8. There are four significant digits in 19.32 cm: 1, 9, 3, and 2. All non-zero digits are significant digits. Sometimes zeros are significant and sometimes they are not. Use these rules to decide if a zero is a significant digit.

6. Counting Significant Digits How many significant digits are in 28.070?

	Rule	Example	Number of Significant Digits
Always significant	Final zeros after a decimal point	4.5300	5
	Zeros between other digits	502.0301	7
Not significant	Zeros at the beginning of a number	0.00059	2
May be significant	Zeros in whole numbers	16,500	3 or 5

How do you calculate with significant digits?

There are rules to follow when deciding the number of significant digits in the answer to a problem.

Multiplying or Dividing First, count the significant digits in each number in your problem. Then, multiply or divide. Your answer must have the same number of significant digits as the number with fewer significant digits in the problem. If it does not, you must round your answer.

$$6.14 \quad \times \quad 5.6 \quad = \quad \boxed{34}.384$$

3 digits   2 digits   round to 2 digits

7. Calculate What is 13.2 × 4.628, rounded to the nearest significant digit? Show your work.

Adding or Subtracting When you add or subtract, find the least precise number in the problem (the number with the fewest decimal places). Then add or subtract. Your answer can show only as many decimal places as the least precise number. If it does not, you must round your answer.

$$\begin{array}{r} 6.14 \quad \text{(hundredths)} \\ + \ 5.6 \quad \text{(tenths)} \\ \hline \boxed{11.7}4 \quad \text{(round to the tenths)} \end{array}$$

● After You Read

Mini Glossary

accuracy: the closeness of a measurement to the actual measurement or value

estimation: a method used to guess the size of an object

measurement: a way to describe objects and events with numbers

precision: describes how close measurements are to each other

1. Review the terms and their definitions in the Mini Glossary. How is precision different from accuracy?

   _____

   _____

2. Complete the chart that shows how to round a number.

   ┌───┐
   │ **Look at digit to the** _____   │
   │ **of the place being rounded to.**   │
   └───┘

   ┌──────────────────────────────────┐   ┌──────────────────────────────────┐
   │ **If the digit is less than 5, the digit being rounded to** │   │ **If the digit is 5 or greater, the digit being** │
   │   │   │   │
   │ _____ . │   │ **rounded to** _____ . │
   └──────────────────────────────────┘   └──────────────────────────────────┘

   ┌──┐
   │ **If digit(s) to the right of the digit being rounded are to the *right* of a decimal,** │
   │   │
   │ **they are** _____ . **If they are to the *left* of a decimal,** │
   │   │
   │ **they are** _____ .   │
   └──┘

3. You were asked to make an outline as you read this section. Did your outline help you learn about measurement? What did you do that seemed most helpful?

   _____

   _____

   _____

Science Online Visit **ips.msscience.com** to access your textbook, interactive games, and projects to help you learn more about description and measurement.

End of Section

Measurement

section ② SI Units

What You'll Learn

- what the SI is and why it is used
- the SI units of length, volume, mass, temperature, time, and rate

Identify Definitions As you read, highlight the definition of each word that appears in bold.

Applying Math

1. **Converting Measurements** What do you multiply by to change a hectometer measurement into meters?

● Before You Read

Some people use metric units such as meters, grams, and liters. Others use units such as feet, pounds, and gallons. Explain why using different units might cause a problem.

_____

_____

_____

● Read to Learn

The International System

Scientists created the International System of Units, or SI. The <u>SI</u> is a system of standard measurement that is used worldwide. Some of the units in the SI system are meter (m) for length, kilogram (kg) for mass, kelvin (K) for temperature, and second (s) for time.

Units in the SI system represent multiples of ten. You know the multiple of ten by looking at the prefix before the unit. For example, *kilo-* means 1,000. So, one *kilo*meter is equal to 1,000 meters.

To change a smaller unit to a larger SI unit, multiply by a power of 10 as shown in the table. To rewrite a *deci*meter measurement in meters multiply by 0.1.

$$10 \text{ dm} \times 0.1 = 1 \text{ m}$$

1 dm and 0.1 m describe the same length. So do 5 km and 5,000 m.

Prefix	Meaning	Multiply by:
giga-	one billion	1,000,000,000
mega-	one million	1,000,000
kilo-	one thousand	1,000
hecto-	one hundred	100
deka-	ten	10
[Unit]		1
deci-	one-tenth	0.1
centi-	one-hundredth	0.01
milli-	one-thousandth	0.001
micro-	one-millionth	0.0001
nano-	one-billionth	0.000 000 001

Length

Length is the distance between two points. Metric rulers and metersticks are used to measure length. A **meter** (m) is the SI unit of length. One meter is about as long as a baseball bat. A meter is used to measure distances such as the length and height of a building.

Some units are used to measure short distances. One millimeter (mm) is about the thickness of a dime. A millimeter is used to measure very small objects, such as the length of a word on this page. One centimeter (cm) is about the width of a large paper clip. A centimeter also can be used to measure small objects, such as the length of a pencil.

A kilometer is used to measure long distances, such as distances between cities. A kilometer (km) is a little over half of a mile.

Volume

Volume is the amount of space an object fills. To find the volume of a rectangular object like a brick, measure its length, width, and height. Then multiply them together.

$$\textbf{Volume} = \textbf{length} \times \textbf{width} \times \textbf{height}$$
$$V = l \times w \times h$$

The volume of a cube with side lengths of 10 cm is $10 \times 10 \times 10 = 1,000$ cubic centimeter (cm^3). You probably have seen water in 1-L bottles. A liter is a measurement of liquid volume. One liter is equal to $1,000\ cm^3$. So, a cube with side lengths of 10 cm has the same volume of a 1-L water bottle. A cube with side lengths of 1 cm has the volume of $1\ cm^3$. It can hold 1 mL ($1\ cm^3$) of water.

How can you find the volume of ice cubes in water?

What happens when you add ice cubes to a glass of water? The height of the water increases, but the amount of water does not change. The ice cubes take up space because they have volume, too. So, the glass contains both the volume of the water and the volume of the ice.

Not all objects have a regular shape like a brick. A rock has an irregular shape. You cannot find the volume of a rock by multiplying its length, width, and height. When you want to measure the volume of an irregular object, you can find the volume by immersion.

FOLDABLES

B Compare and Contrast
Make the Foldable below to help you understand the different types of measurement using SI units. Take notes on each type of measurement as you read.

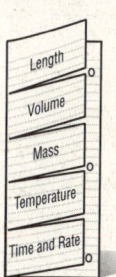

Think it Over

2. Infer The volume formula can be used to find the volume of

a. a tissue box.

b. a lamp.

c. a computer mouse.

Applying Math

3. Calculating Volume
Find the volume of a cube that measures 2 cm on each side. Show your work.

_____

_____

What is immersion?

Start with a known volume of water and place the object in it. Then find the volume of the water with the object in it. The difference between the two volumes is equal to the volume of the object.

For example, to find the volume of a rock, first find the volume of water in a container. Next, place the rock in the water, making sure the water covers all of the rock. Then find the volume of the water and rock together. Subtract the smaller volume from the larger one. The difference is the volume of the rock.

Mass

The **mass** of an object measures the amount of matter in the object. The **kilogram** (kg) is the SI unit for mass. One liter of water has the mass of about 1 kg. Smaller masses are measured in grams (g). The mass of a paper clip is about 1 g.

Mass versus Weight Why use the word *mass* instead of *weight?* Mass and weight are not the same. Mass depends only on the amount of matter in an object. Your mass on the moon would be the same as it is on Earth. **Weight** is a measure of the gravitational force, or pull, on the matter in an object. In other words, weight depends on gravity. You would weigh much less on the moon because the gravitational force on the moon is less than on Earth. ☑

How much would you weigh on other planets?

Keep in mind that weight is a measure of force. The SI unit for force is the newton (N). Suppose you weigh 332 N on Earth. That would be a mass of about 75 pounds, or 34 kg. Remember that the gravitational force is different on different planets. On Mars, you would weigh 126 N. On Jupiter, you would weigh much more—782 N. Your mass would still be the same on all the planets because the amount of matter in your body has not changed.

Time

Time tells how long it takes an event to happen. The SI unit for time is the second (s). Time is also measured in minutes (min) and hours (h). Time is usually measured with a clock or a stopwatch.

Reading Check

4. **Determine** What does weight depend on?

_____

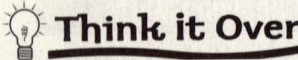

Think it Over

5. **Infer** Which has more gravity, Mars or Jupiter?

_____

Rate

A **rate** is the amount of change in one measurement that takes place in a given amount of time. To find a rate, a measurement is divided by an amount of time. Speed is a common rate that tells how fast an object is moving. Speed is the distance traveled in a given time. The formula for speed is distance (length) divided by time, such as 80 kilometers per hour (km/h). The unit that is changing does not have to be an SI unit. You can use rate to tell how many cars pass through an intersection in an hour (cars/h). ☑

Temperature

Temperature is how hot or cold an object is. The **kelvin** (K) is the SI unit for measuring temperature. The Fahrenheit (°F) and Celsius (°C) temperature scales are the scales used on most thermometers. Compare the scales in the figure. The kelvin scale starts at 0 K, absolute zero, the coldest possible temperature. One degree of change on the kelvin scale is the same as one degree of change on the Celsius scale.

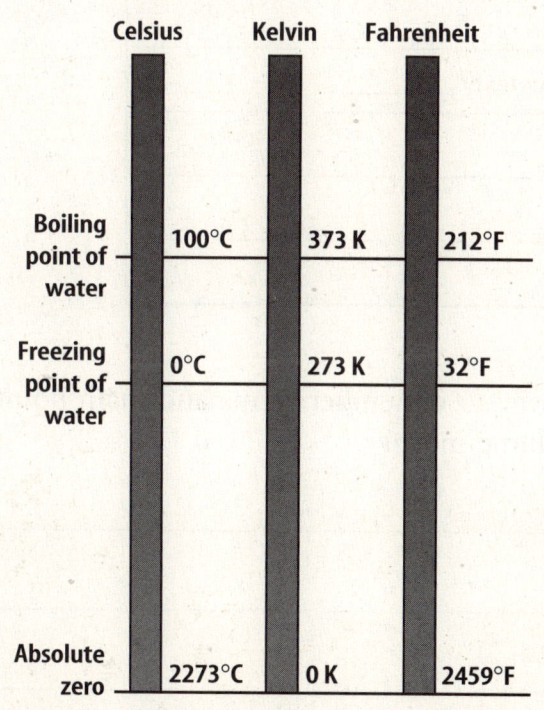

	Celsius	Kelvin	Fahrenheit
Boiling point of water	100°C	373 K	212°F
Freezing point of water	0°C	273 K	32°F
Absolute zero	2273°C	0 K	2459°F

✔ Reading Check

6. **Explain** How do you calculate a rate?

_____

_____

_____

Picture This

7. **Interpreting Graphs** What is the boiling point of water on each of the temperature scales?

Celsius:

_____

Kelvin:

_____

Fahrenheit:

_____

● After You Read

Mini Glossary

kelvin: SI unit for measuring temperature (K)

kilogram: SI unit for measuring mass (kg)

mass: amount of matter in an object

meter: SI unit of length (m)

rate: the amount of change in one measurement that takes place in a given amount of time

SI: a system of standard measurement that is used worldwide

volume: amount of space that fills an object

weight: a measure of the gravitational force, or pull, on the matter in an object

1. Review the terms and their definitions in the Mini Glossary. What are four examples of rate?

   _____

   _____

2. Complete the table to identify common SI base units.

Quantity	SI Unit
Length	
Volume	cubic centimeter or _____
Mass	
Temperature	
Time	

3. Sometimes you need to find your own way to remember terms and main points. How could you remember what mass and volume mean?

   _____

   _____

   _____

End of Section

Science**nline** Visit **ips.msscience.com** to access your textbook, interactive games, and projects to help you learn more about SI units.

chapter 2

Measurement

section 3 Drawings, Tables, and Graphs

● Before You Read

A common saying is "A picture is worth a thousand words." Use the lines below to explain what this means.

_____

_____

● Read to Learn

Scientific Illustrations

Most science books include pictures. These pictures can be drawings or photographs. Drawings and photographs can often explain new information better than words can.

What does a drawing show?

Drawings are helpful because they can show details. Drawings can show things that you cannot see. The drawing below shows details of the water cycle that can't be shown in a photograph. You can also use drawings to help solve problems. For example, you could draw the outline of two continents to show how they might have fit together at one time.

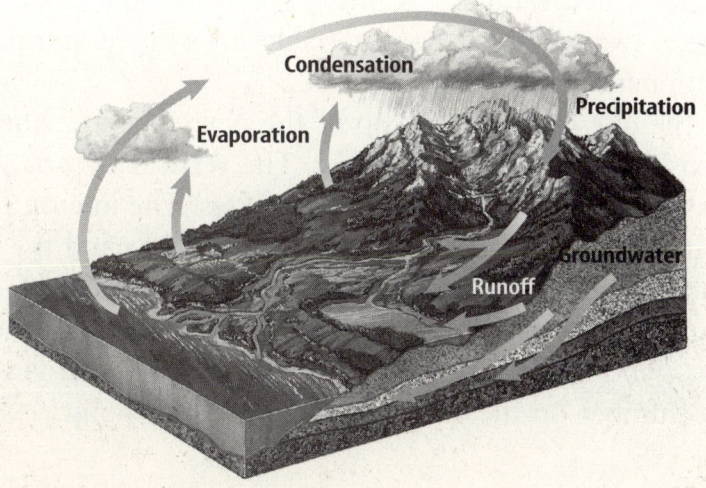

Condensation

Precipitation

Evaporation

Groundwater

Runoff

Picture This
1. **Interpret a Drawing**
   What parts of the water cycle can be shown in a drawing but not in a photograph?

   _____

   _____

   _____

⑥ Compare and Contrast
Make a Foldable as shown. Use the sections to draw a sample of a scientific illustration, a line graphs, a bar graph, and a circle graph.

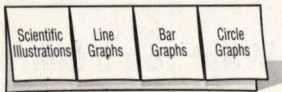

Scientific Illustrations | Line Graphs | Bar Graphs | Circle Graphs

Applying Math

2. Interpret Data How many animal species were endangered in 1992?

a. 192

b. 284

c. 321

d. 389

✔ Reading Check

3. Describe How many variables are shown on a line graph?

What does a photograph show?

A photograph shows an object at one moment in time. A video or movie is made of a series of photographs. A movie shows how an object moves. It can be slowed down or sped up to show interesting things about an object.

Tables and Graphs

Science books contain tables and graphs as well as drawings and photographs. Tables are a good way to organize information. A **table** lists information in columns and rows so that it is easy to read and understand. Columns are vertical, or go up and down. Rows are horizontal, or go across from left to right. This table shows the number of endangered animal species for each year from 1984 to 2002.

A **graph** is a drawing that shows data, or information. Sometimes it is easier to see the relationships when the data is shown in a graph. The three most common types of graphs are line graphs, bar graphs, and circle graphs.

Endangered Animal Species in the United States	
Year	Number of Endangered Animal Species
1984	192
1986	213
1988	245
1990	263
1992	284
1994	321
1996	324
1998	357
2000	379
2002	389

What does a line graph show?

A **line graph** shows changes in data over time. Things that change, like the number of endangered animals and the year, are called variables. A line graph shows the relationship between two variables. Both variables in a line graph must be numbers. ☑

In the line graph at the top of the next page, the horizontal axis, or *x*-axis, shows the year. The vertical axis, or *y*-axis, shows the number of endangered species. The line on the graph shows the relationship between the year and the number of endangered species. To find the number of endangered species in a given year, find that year on the *x*-axis. Find the point on the line that is above that year. See what number on the *y*-axis lines up with the point.

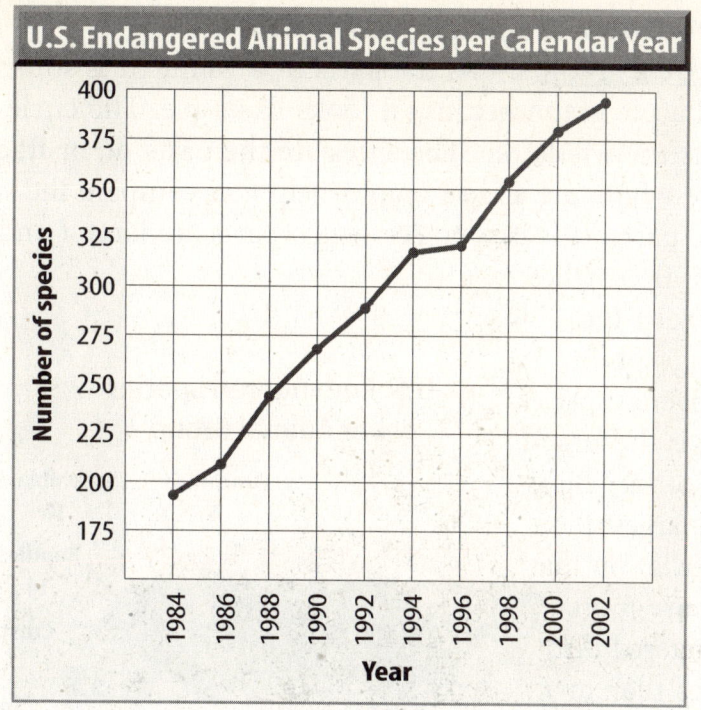

U.S. Endangered Animal Species per Calendar Year

Picture This

4. Interpret Data Look at the line graph. Between which two years did the number of endangered species increase the least? Circle your answer.

a. 1986–1988

b. 1992–1994

c. 1994–1996

d. 2000–2002

What does a bar graph show?

A **bar graph** uses rectangular blocks, or bars, to show the relationships among variables. One variable must be a number. This variable is divided into parts. The other variable can be a category, like kinds of animals, or it could be another number. In the bar graph on endangered species, the height of each bar shows the number of endangered species in each animal group. For example, there are about 30 species of endangered insects. Bar graphs make it easy to compare data.

Picture This

5. Describe Look at the bar graph. How many animal groups have fewer than 30 endangered species?

_____

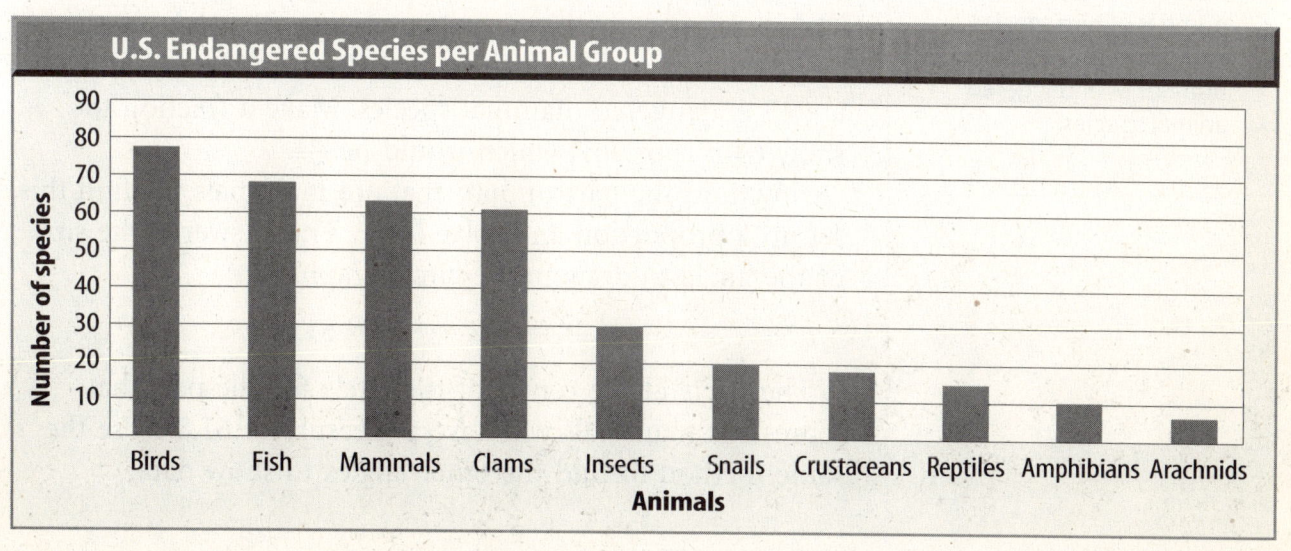

U.S. Endangered Species per Animal Group

What does a circle graph show?

A <u>circle graph</u> shows the parts of a whole. It is sometimes called a pie graph because it looks like a pie. The circle represents the whole pie. The slices are the parts of, or fractions of, the whole pie. Circle graphs help you compare the sizes of the parts. It is easy to see which parts are largest and which are smallest.

Look at the circle graph on endangered species. What is the largest piece of pie? Birds is the largest piece. This tells you that there are more endangered bird species than any other animal group. What is the smallest piece of pie? Arachnids is the smallest. There are fewer endangered arachnids than any other animal group.

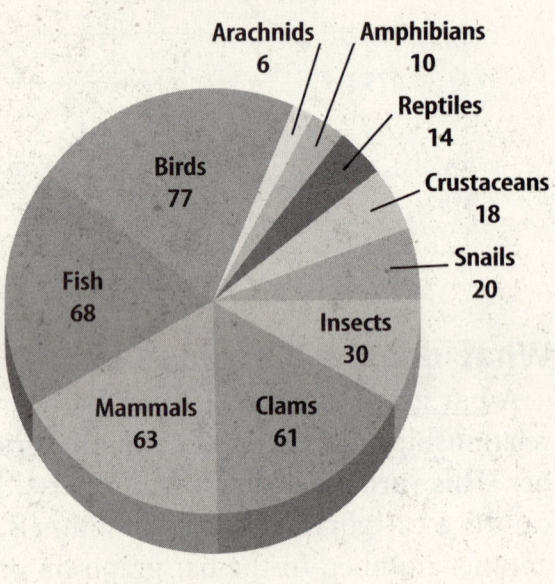

U.S. Endangered Species per Animal Group

How do you make a circle graph?

A circle has 360°. To make a circle graph, find what fraction of 360 each part is. First, find the total number of parts in the whole. The total number of endangered species is 367. Then, compare the number of each animal species with the total. Look at the part of the circle labeled Mammals. There are 63 endangered mammal species. Make a fraction to show 63 out of 367, which would be $\frac{63}{367}$.

Now find the part of 360° that are mammals. To find this, set up a proportion and solve for x. The answer is the size of the angle to draw in the circle graph.

$$\frac{63}{367} = \frac{x}{360°} \qquad x = 61.8°$$

The angle at the center of the circle for the part that represents mammals will have a measure of 61.8°. Use the same method to find the other angles to show data.

Picture This

6. Interpret Data Use the circle graph. Which animal group has the second largest number of endangered species? How many endangered species does this group have?

_____

_____

Applying Math

7. Find the Ratio Look at the circle graph. Write a fraction to show the number of bird species compared to the total number of endangered animal species.

_____

Why is the scale of a graph important?

The two graphs below do not look the same, but they both show the same data. Why do the graphs look different? The graphs look different because their scales are different. Look at the scale on the *y*-axis in each graph. Look at the lowest number. The first graph begins with 310 instead of 0. It appears as though there was a great increase in the number of endangered species from 1996 to 2002. The second graph does begin at 0. It shows that there was a slight increase in endangered species during these years.

A scale that does not start at 0 is called a broken scale. A broken scale makes it easier to see small changes in the data. However, you must read the graph carefully to see if there is a broken scale. When you see a bar graph or a line graph, look at the data carefully. If the graph seems odd, take a closer look at the scale. ☑

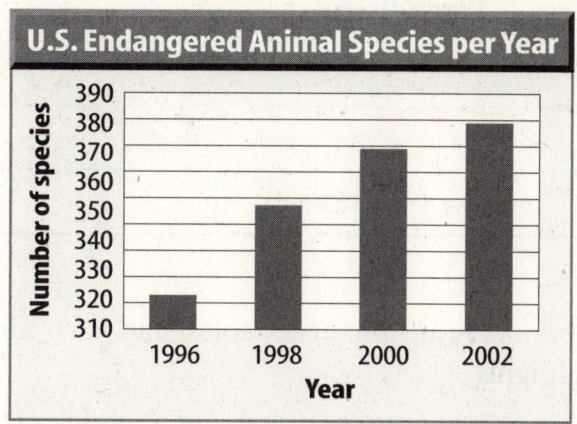

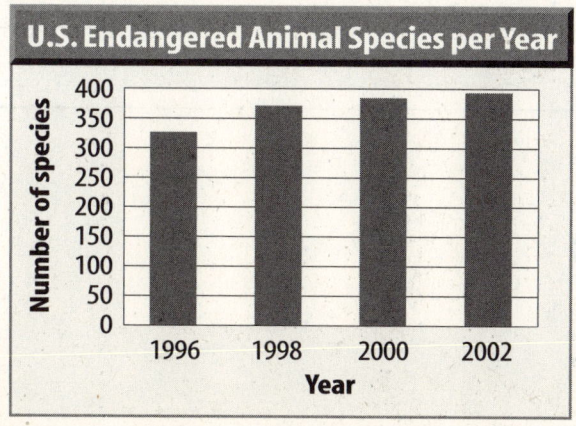

☑ **Reading Check**

8. **Explain** What is a broken scale?

_____

_____

_____

Picture This

9. **Describe** In your own words, tell when a broken scale in a bar graph or a line graph is useful.

_____

_____

_____

● After You Read

Mini Glossary

bar graph: a graph that uses bars to show the relationships between variables.

circle graph: a graph that shows the parts of a whole

graph: a kind of drawing that shows data, or information

line graph: a graph that shows changes in data over time

table: lists information in columns and rows so that it is easy to read and understand

1. Review the terms and definitions in the Mini Glossary. Write a sentence that tells how a graph helps you understand data.

_____

_____

2. Complete the chart below to describe when each graph is most useful.

Graph Type	When Useful
Line graph	
Bar graph	
Circle graph	compare parts to a whole

3. You were asked to highlight two main points about pictures, tables, and each kind of graph. How did you decide what to highlight?

_____

_____

_____

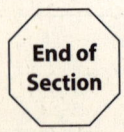
End of Section

Science Online Visit **ips.msscience.com** to access your textbook, interactive games, and projects to help you learn more about drawings, tables, and graphs.

Atoms, Elements, and the Periodic Table

section ❶ Structure of Matter

● Before You Read

Take a deep breath. What fills your lungs? Can you see it or hold it in your hand?

● Read to Learn

What is matter?

Is a glass with some water in it half empty or half full? Neither is correct. The glass is completely full. It is half full of water and half full of air. What is air? Air is a mixture of several gases, including nitrogen and oxygen. Nitrogen and oxygen are kinds of matter. <u>Matter</u> is anything that has mass and takes up space. So, even though you cannot see it or hold it, air is matter. Water also is matter. Most of the things you can see, taste, smell, and touch are made of matter.

What isn't matter?

You could not read the words on this page without light. Light has no mass and does not take up space. So, light is not matter. Is heat matter? Heat has no mass and does not take up space. So, heat is not matter. Your thoughts, feelings, and ideas are not matter, either. ☑

What makes up matter?

Could you cut a piece of wood small enough so it no longer looks like wood? What is the smallest piece of wood you can cut? People have asked questions like these for hundreds of years. They wondered what matter is made of.

✔ Reading Check

1. **List** three things that are not matter.

_____

_____

_____

What was Democritus's idea of matter?

A Greek philosopher named Democritus lived from about 460 B.C. to 370 B.C. He thought the universe was made of empty space and tiny bits of stuff that he called atoms. The word *atom* comes from a Greek word that means "cannot be divided." Democritus believed atoms could not be divided into smaller pieces. Today, we define an **atom** as a particle that makes up most types of matter. The table below shows what Democritus thought about atoms. ☑

2. Summarize What is an atom?

_____

_____

Democritus's Ideas About Atoms
1. All matter is made of atoms.
2. There are empty spaces between atoms.
3. Atoms are complete solids.
4. Atoms do not have anything inside them.
5. Atoms are different in size, shape, and weight.

Democritus also thought that different types of atoms existed for every type of matter. He thought the different atoms explained the different characteristics of each type of matter. Democritus's ideas about atoms were a first step toward understanding matter. In the early 1800s, scientists started building on the concept of atoms to form the current atomic theory of matter.

Can matter be made or destroyed?

For many years, people thought matter disappeared when it burned or rusted. Seeing objects grow, like trees, also made them think that matter could be made. A French chemist named Lavoisier (la VWAH see ay) lived about 2,000 years after Democritus. Lavoisier studied wood fires very carefully. Lavoisier showed that wood and the oxygen it combines with during a fire have the same mass as the ash, gases, and water vapor that are produced by the fire. So, matter is not destroyed when wood burns. It just changes into a different form.

Applying Math

3. Apply Suppose you increase the mass of wood you are burning in a fireplace. What will happen to the total mass of ash, gases, and water vapor?

_____

_____

$$\frac{\text{total mass of}}{\text{wood + oxygen}} = \frac{\text{total mass of}}{\text{ash + gases + water vapor}}$$

From Lavoisier's work came the law of conservation of matter. The **law of conservation of matter** states that matter is not created or destroyed—matter only changes form.

Models of the Atom

Models often are used for things that are too small or too large to be observed. Models also are used for things that are difficult to understand.

Smaller Models One way to make a model is to make a smaller version of something that is large. If you want to design a new sailboat, would you build a full-sized sailboat and hope it would float? It would be safer and cheaper to build and test a smaller version first. Then, if it didn't float, you could change your design and just build another model, not another full-sized sailboat. You could keep trying until the model worked.

Larger Models Scientists sometimes make models that are larger than the actual objects. Atoms are too small to see. So, scientists use large models of atoms to explain data or facts that are found during experiments. This means these models are also theories.

What was Dalton's model of an atom?

John Dalton was an English chemist. In the early 1800s, he made an atomic model that explained the results of the experiments of Lavoisier and others.

Dalton's atomic model was a set of ideas instead of an object. He believed matter was made of atoms that were too small to see. He also thought that each type of matter was made of only one kind of atom. For example, gold rings were made of gold atoms. Iron atoms made up an iron bar. Dalton also thought gold atoms are different from iron atoms. The different types of atoms explain why gold and iron are different. Other scientists made experiments and gathered data based on Dalton's model. Dalton's model became known as the atomic theory of matter. ☑

How small is an atom?

Atoms are so small it would take about 1 million of them lined up in a row to be about as thick as one human hair. Or, imagine you are holding an orange. If you want to see the atoms on the orange's skin, the orange would need to be as big as Earth. Then, imagine the Earth-sized orange covered with billions of marbles. Each marble would represent an atom on the skin of the orange.

FOLDABLES

Ⓐ Compare and Contrast
Use two half-sheets of notebook paper to compare the past atomic model and the present atomic model.

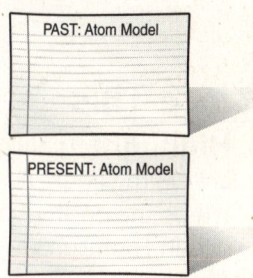

Reading Check

4. **Explain** Dalton's atomic model was not an object. What was it?

_____

What is an electron?

An English scientist named J.J. Thomson discovered the electron in the early 1900s. He experimented using a glass tube with a metal plate at each end, like the one in the figure below. Thomson connected the metal plates to electricity. One plate, called the anode, had a positive charge. The other plate, called the cathode, had a negative charge.

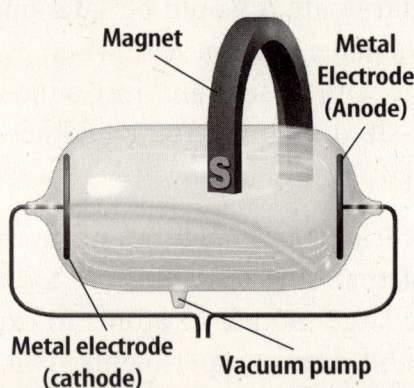

Magnet

Metal Electrode (Anode)

S

Metal electrode (cathode)

Vacuum pump

Picture This

5. Highlight Highlight the area in the figure to show where the positive charge was in Thomson's experiment.

_____

During his experiments, Thomson watched rays travel from the cathode to the anode. Then, he used a magnet to bend the rays. Since the rays could be bent, they were made of particles that had mass and charge. He knew that like charges repel each other and opposite charges attract each other. Since the rays were traveling to the positive plate (the anode), Thomson decided the rays must be made of particles with negative charges. These invisible particles with negative charges are **electrons**. Thomson showed that atoms can be divided into smaller particles.

What was Thomson's model of the atom?

Think it Over

6. Explain Why was Thomson's discovery important?

_____

_____

_____

Matter that has an equal amount of positive and negative charge is neutral. Most matter is neutral. So, Thomson thought an atom was made of a ball of positive charge with negatively charged electrons in it. ☑

Thomson's model of an atom was like a ball of chocolate chip cookie dough. The dough was positively charged. The chocolate chips were the negatively charged electrons.

What was Rutherford's model of the atom?

✔ Reading Check

7. Locate Information What did Thompson think an atom was made of?

_____

_____

_____

Scientists still had questions about how the atom was arranged and about particles with positive charge. Around the year 1910, an English scientist named Ernest Rutherford and his team of scientists tried to answer these questions.

Rutherford's experiment Rutherford's team shot tiny, high-energy, positively charged particles, or alpha particles, at a very thin piece of gold foil. Rutherford thought that the alpha particles would pass easily through the foil. Most of the alpha particles did pass straight through. But, other alpha particles changed direction. A few of them even bounced back.

Since most particles passed straight through the gold, Rutherford thought that the gold atoms must be made of mostly empty space. But, because a few particles bounced off something, the gold atoms must have some positively charged object within the empty space. He called this positively charged object the nucleus. The **nucleus** (NEW klee us) is the positively charged, central part of an atom. Rutherford named the positively charged particles in the nucleus of an atom **protons.** He also suggested that negatively charged electrons were scattered in the empty space around the nucleus. Rutherford's model is shown in the figure below.

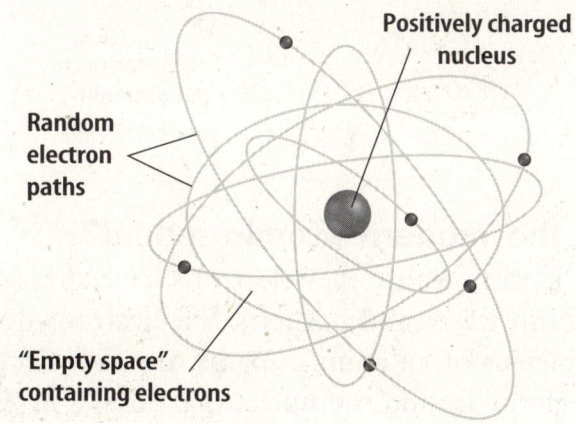

Positively charged nucleus

Random electron paths

"Empty space" containing electrons

Picture This

8. **Draw Conclusions** In Rutherford's model, what is an atom *mostly* made of?

_____

_____

How was the neutron discovered?

Rutherford was puzzled by one observation in his experiment with alpha particles. The nucleus of an atom seemed to be heavier after the experiment. He did not know where this extra mass came from. James Chadwick, one of Rutherford's students, answered the question: The nuclei were not getting heavier. But, the atoms had given off new particles. He found that the path of the new particles was not affected by an electric field. This meant the new particles were neutral—had no charge. Chadwick called these new particles neutrons. A **neutron** (NEW trahn) is a neutral particle in the nucleus of an atom. His proton-neutron model of the nucleus of an atom is still accepted today. ☑

✔ **Reading Check**

9. **Identify** What type of charge do neutrons have?

_____

_____

Improving the Atomic Model

A scientist named Niels Bohr found that electrons are arranged in energy levels in an atom. The figure shows his model. The lowest energy level is closest to the nucleus. It can have only two electrons. Higher energy levels are farther from the nucleus. They can have more than two electrons. To explain these energy levels, some scientists thought that electrons might orbit, or travel, around the atom's nucleus. The electrons were thought to travel in paths that are specific distances from the nucleus. This is similar to how the planets travel around the Sun.

Bohr's Model of an Atom

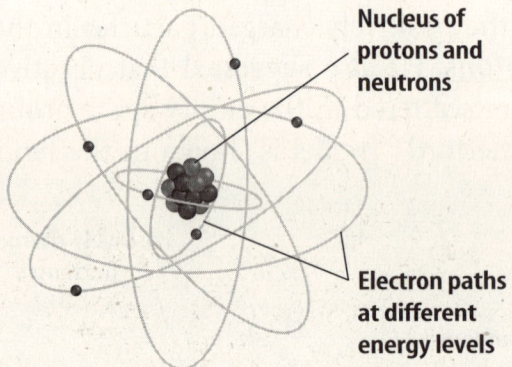

Nucleus of protons and neutrons

Electron paths at different energy levels

Picture This

10. **Compare and Contrast** How is Bohr's atomic model different from the modern atomic model?

Bohr's model:

_____

_____

Modern model:

_____

_____

What is the modern atomic model?

Today, scientists realize that electrons have characteristics similar to both waves and particles. So, electrons do not orbit the nucleus of an atom in paths. Instead, electrons move in a cloud around the nucleus, as shown in the figure. The dark area shows where the electron is most likely to be in the electron cloud.

Modern Model of an Atom

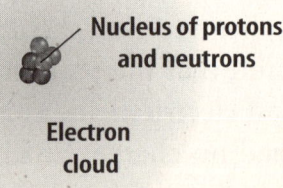

Nucleus of protons and neutrons

Electron cloud

● After You Read

Mini Glossary

atom: a small particle that makes up most types of matter

electron: an invisible particle with a negative charge around the nucleus of an atom

law of conservation of matter: matter is not created or destroyed—matter only changes form

matter: anything that has mass and takes up space

neutron: (NEW trahn) a neutral particle in the nucleus of an atom

nucleus: (NEW klee us) the positively charged, central part of an atom

proton: positively charged particle in the nucleus of an atom

1. Review the terms and their definitions in the Mini Glossary. Write a sentence to explain the law of conservation of matter.

_____

_____

2. Fill in each blank in the concept map.

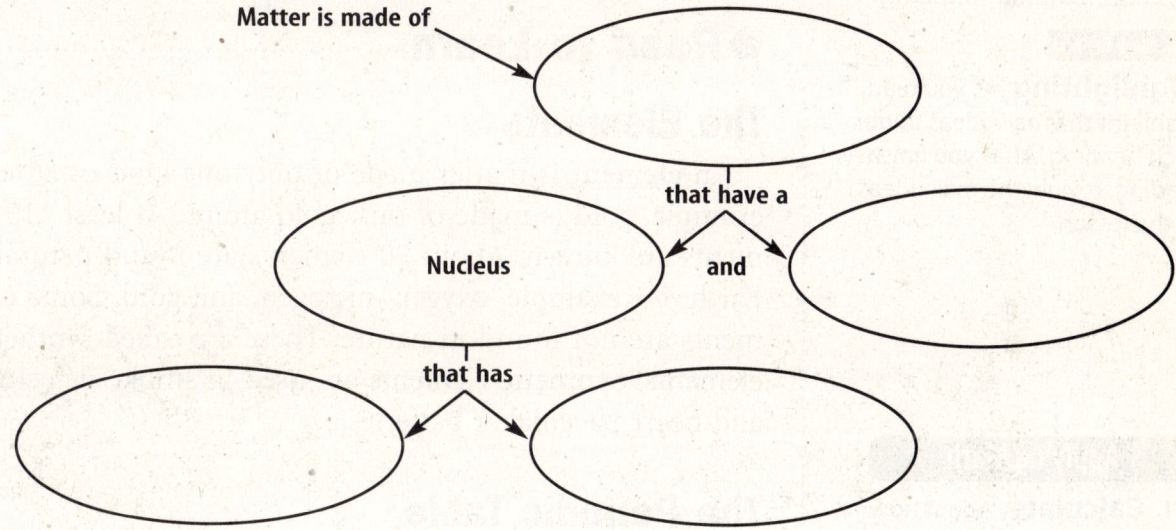

Matter is made of

that have a

Nucleus   **and**

that has

3. How could you explain the modern atomic model to another student?

_____

_____

_____

Science Online Visit **ips.msscience.com** to access your textbook, interactive games, and projects to help you learn more about the structure of matter.

End of Section

Atoms, Elements, and the Periodic Table

section ② The Simplest Matter

What You'll Learn

- about elements in the periodic table
- what atomic mass and atomic number are
- what an isotope is
- about metals, metalloids, and nonmetals

Mark the Text

Highlighting As you read, highlight the main ideas under each heading. After you finish reading, review the main ideas of the lesson.

Applying Math

1. **Calculate** About how many elements are synthetic elements?

● Before You Read

Have you ever taken something apart to see what it was made of? Describe a time when you did this.

_____

_____

● Read to Learn

The Elements

An **element** is matter made of only one kind of atom. For example, gold is made of only gold atoms. At least 110 elements are known. About 90 elements are found naturally on Earth, for example, oxygen, nitrogen, and gold. Some elements are not found in nature. These are called synthetic elements. Synthetic elements are used in smoke detectors and heart pacemaker batteries.

The Periodic Table

How would you find a certain book in a library? If you look at the books on the shelves as you walk past, you probably won't find the book you want. Libraries organize the books to help you quickly find the ones you want.

Scientists organize information about the elements, too. They created a chart called the periodic table of the elements. Each element in the chart has a chemical symbol. The symbols have one to three letters. The symbol for oxygen is O. The symbol for aluminum is Al. Scientists all over the world use these chemical symbols.

How are elements listed in the periodic table?

The periodic table has rows and columns that show how the elements relate to one another. The elements are grouped by their properties. The rows go from left to right and are called periods. The elements in a period have the same number of energy levels. The columns go up and down and are called groups. The elements in a group have similar properties related to their structures. They also tend to form similar bonds. The figure may help you remember the difference between periods and groups in the periodic table.

G R O U P

PERIOD

Picture This

2. Locate To help you remember the location of periods and groups in a periodic table, draw an arrow pointing left and right through PERIOD. Then draw an arrow pointing up and down through GROUP.

Identifying Characteristics

Each element is different and has unique properties. These differences can be described by looking at the relationships between the atomic particles, or parts of the atoms, in each element. Numbers in the periodic table describe these relationships.

What is the atomic number of an element?

The figure shows the periodic table block for chlorine. The symbol for chlorine is Cl. The number at the top, 17, is the atomic number for chlorine. The **atomic number** is the number of protons in the nucleus of each atom of an element. So, every chlorine atom has 17 protons in its nucleus. Each element in the periodic table has a different atomic number. This means that each element has a different number of protons in its nucleus. ☑

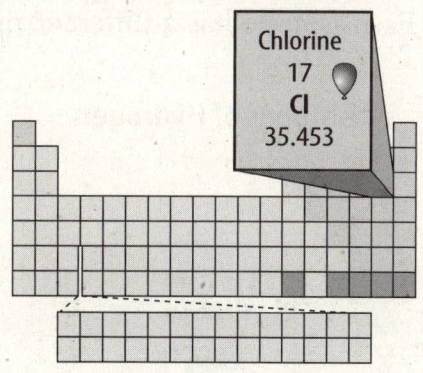

Chlorine
17
Cl
35.453

Picture This

3. Locate Circle the atomic number for chlorine in the periodic table block.

What is an isotope?

The atomic number, or number of protons in the nucleus, is always the same for an element. But the number of neutrons in the nucleus is not always the same. For example, some chlorine atoms have 18 neutrons and some have 20 neutrons. Chlorine-35 is a chlorine atom that has 18 neutrons. Chlorine-37 has 20 neutrons. These two chlorine atoms are isotopes. **Isotopes** (I suh tohps) are atoms of the same element that have different numbers of neutrons.

☑ Reading Check

4. Define What does the atomic number of an atom represent?

_____

_____

_____

Picture This

5. **Read a Table** How many neutrons does chlorine-37 have?

_____

Picture This

6. **Understanding Figures** What is the mass number of the hydrogen isotope deuterium?

What is a mass number?

You can refer to a certain isotope by using its mass number. An atom's **mass number** is the number of protons plus the number of neutrons in its nucleus. Look at the table below. Chlorine-35 has a mass number of 35 because the number of protons (17) plus the number of neutrons (18) equals 35. The isotope is named chlorine-35 because its mass number is 35.

Isotope	Number of protons		Number of neutrons	Mass number
Chlorine-35:	17	+	18	35
Chlorine-37:	17	+	20	37

Every particle in the nucleus adds to the mass of an atom. So, if an atom has more neutrons, its mass is greater. If it has fewer neutrons, its mass is less. The mass of chlorine-37 is greater than the mass of chlorine-35.

Hydrogen is the first element in the periodic table. Hydrogen has three isotopes with mass numbers of 1, 2, and 3, shown below. Every hydrogen atom always has one proton. Each isotope has a different number of neutrons.

Isotopes of Hydrogen

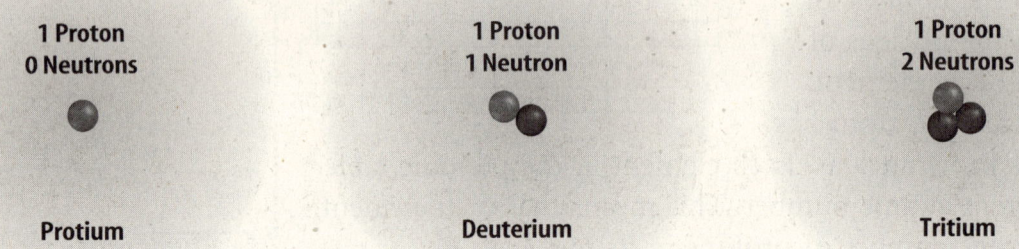

1 Proton
0 Neutrons

Protium

1 Proton
1 Neutron

Deuterium

1 Proton
2 Neutrons

Tritium

What is atomic mass?

The number below an element's chemical symbol is the atomic mass. **Atomic mass** is the average mass of all the isotopes of an element. The atomic mass takes into account how often the isotopes are found. For chlorine, the atomic mass is 35.45 u. The letter *u* stands for "atomic mass unit," the unit of measure for atomic mass.

Classification of Elements

The elements are divided into three classes or categories—metals, metalloids (ME tuh loydz), and nonmetals. The elements in each category have similar properties.

What are some properties of metals?

Metals are elements that have a shiny or metallic appearance and are good conductors of heat and electricity. All metals, except mercury, are solids at room temperature. Metals also are malleable (MAL yuh bul). Malleable means they can be bent and pounded into shapes. Metals are ductile. Ductile means they can be stretched into wires without breaking. Gold, silver, iron, copper, and lead are examples of metals. Most of the elements in the periodic table are metals. ☑

What are some properties of nonmetals?

Nonmetals are elements that usually look dull and are poor conductors of heat and electricity. They are brittle, which means they cannot change shape easily without breaking. You cannot stretch or bend brittle materials. Many nonmetals are gases at room temperature.

Nonmetals are important to life. Look at the figure. More than 97 percent of your body is made up of different nonmetals. Examples of nonmetals include chlorine, oxygen, hydrogen, nitrogen, and carbon. Most of the elements on the right side of the periodic table are nonmetals.

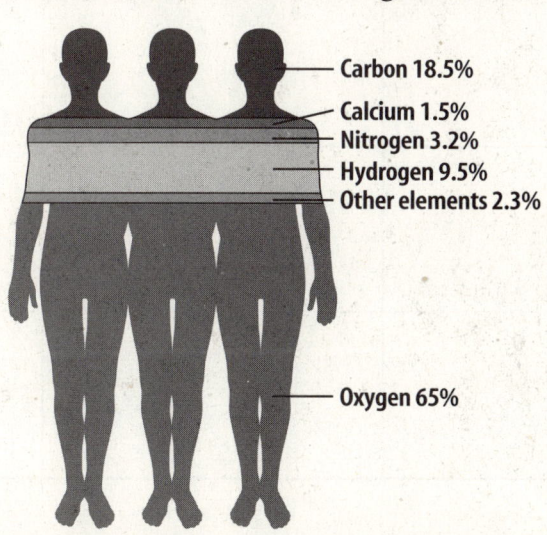

Carbon 18.5%
Calcium 1.5%
Nitrogen 3.2%
Hydrogen 9.5%
Other elements 2.3%

Oxygen 65%

What are some properties of metalloids?

Metalloids are elements that have properties of both metals and nonmetals. All metalloids are solids at room temperature. Some metalloids are shiny. Many metalloids can conduct heat and electricity, but not as well as metals can. On the periodic table, metalloids are found between metals and nonmetals. Silicon is an example of a metalloid. It is used in electronic circuits in computers and televisions.

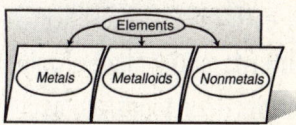

Compare and Contrast
Make the Foldable below to help you understand how metals, metalloids, and nonmetals are alike and different.

> Elements
> Metals | Metalloids | Nonmetals

Reading Check

7. **Conclude** Most of the elements in the periodic table are metals. Is this sentence true or false? Circle your answer.

   True   **False**

Picture This

8. **Compare** Which nonmetal is found the most in your body?

_____

● After You Read

Mini Glossary

atomic mass: the average mass of all the isotopes of an element

atomic number: the number of protons in the nucleus of each atom of an element

element: matter made of only one kind of atom

isotope: (I suh tohps) atoms of the same element that have a different number of neutrons

mass number: the number of protons plus the number of neutrons in the nucleus of each atom of an element

metalloids: elements that have properties of both metals and nonmetals

metals: elements that have a shiny, or metallic, appearance and are good conductors of heat and electricity

nonmetals: elements that usually looks dull and are poor conductors of heat and electricity

1. Review the terms and their definitions in the Mini Glossary. Which two terms describe numbers that appear in an element's square on the periodic table?

_____

2. Complete the table to identify properties of metals, metalloids, and nonmetals.

Properties	Metals	Metalloids	Nonmetals
Appearance— how they look			
Ability to conduct heat and electricity			
Ability to bend and stretch			

3. You were asked to highlight the main ideas under each heading. How did you decide what the main ideas were?

_____

_____

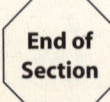
End of Section

Science Online Visit **ips.msscience.com** to access your textbook, interactive games, and projects to help you learn more about elements and the periodic table.

chapter 3

Atoms, Elements, and the Periodic Table

section 3 Compounds and Mixtures

Before You Read

If you mix together salt and water, what happens to the salt? What happens to the water?

____

____

____

What You'll Learn

- what a compound is
- how different types of mixtures are similar and different

Read to Learn

Substances

Scientists classify matter depending on what it is made of and how it behaves. Matter that has the same composition and properties throughout is a **substance.** Elements such as gold and aluminum are substances. Substances also can be two or more elements combined, like brass. Brass is made of copper and zinc.

Study Coach

Create a Quiz As you read this section, write a quiz question for each paragraph. After you finish reading the section, answer your quiz questions.

What is a compound?

A **compound** is a substance whose smallest unit is made up of atoms of more than one element bonded together. Water is a compound. It is made up of hydrogen and oxygen. Hydrogen and oxygen are both colorless gases. But, when they are combined, they make water. Water is sometimes written as H_2O. H stands for hydrogen and O stands for oxygen. Many compounds have properties that are different from those of its elements. For example, water is different from the gases hydrogen and oxygen. It also is different from hydrogen peroxide (H_2O_2), another compound made from hydrogen and oxygen.

FOLDABLES™

C Identify Make the following Foldables from quarter-sheets of notebook paper to help you identify the differences between H_2O, C_2O, and CO.

What is a chemical formula?

Compounds have chemical formulas. A chemical formula shows the elements that make up a compound. It also shows how many atoms of each element are in the compound. For example, H_2O is the chemical formula of water. The small number to the right of an element tells how many atoms are in one unit, or molecule, of that compound. When no number is written, the molecule has one atom of that element. So, a molecule of water is made up of two atoms of hydrogen and one atom of oxygen.

H_2O_2 is the chemical formula for hydrogen peroxide. It has two hydrogen atoms and two oxygen atoms. So, the elements hydrogen and oxygen form two compounds—water and hydrogen peroxide. Look at the figures below to see the differences in their structure. The properties of water are very different from the properties of hydrogen peroxide.

Picture This

1. **Circle** the chemical formula in each figure.

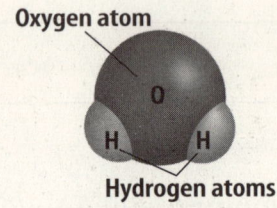

Oxygen atom

Hydrogen atoms

Hydrogen atoms

Oxygen atoms

H_2O Water H_2O_2 Hydrogen Peroxide

Think it Over

2. **Identify** What is the chemical formula for carbon dioxide?

Are compounds always the same?

A given compound always is made of the same elements and in the same proportion, or ratio. For example, one unit of water is always made of two hydrogen atoms and one oxygen atom. You can write the number of molecules of water that you have by putting a number in front of the formula. So, 6 H_2O means you have six molecules of water.

Mixtures

A mixture is made when two or more substances come together but do not combine to make a new substance. The substances can be elements, compounds, or elements and compounds. The proportions of the substances in a mixture can be changed without changing the identity of the mixture, unlike in a compound. Sand and water form a mixture. If you add more sand to the mixture, you still have a mixture of sand and water.

Air Another example of a mixture is air. Air is made of nitrogen, oxygen, and many other gases. There can be different amounts of these gases in the mixture. But you still have a mixture of air.

You see mixtures every day. A salad of lettuce, tomatoes, and cucumbers is a mixture. The mixture may have more tomatoes than cucumbers, but it is still a salad.

How can mixtures be separated?

You can separate many mixtures. A mixture of solids can be separated by using different screens or filters. For example, you could separate a mixture of pebbles and sand by pouring the mixture through a screen. The screen can catch the pebbles, but let the sand go through.

You also can use a liquid to separate some mixtures of solids. If you add water to a mixture of sugar and sand, only the sugar will dissolve in the water. Then, you can pour the mixture through a filter that catches the sand. Next, you can separate the sugar from the water by heating it. As shown in the figure, even your blood is a mixture that can be separated.

What are homogeneous and heterogeneous mixtures?

Homogeneous Mixtures Homogeneous means "the same throughout." So, homogeneous mixtures are those that look the same throughout. You cannot see the different parts of the mixture. Since you can't see the different parts, you might not know it is a mixture. Homogeneous mixtures can be solids, liquids, or gases. Brass, sugar water, and air are mixtures.

Heterogeneous Mixtures Heterogeneous means "completely different." Heterogeneous mixtures have larger parts that are different from each other. You can see the different parts of a heterogeneous mixture. Vegetable soup is a heterogeneous mixture. ☑

FOLDABLES

D Contrast Make the following 2-tab Foldable to help you learn the differences between homogeneous mixtures and heterogeneous mixtures.

Homogeneous Mixtures | Heterogeneous Mixtures

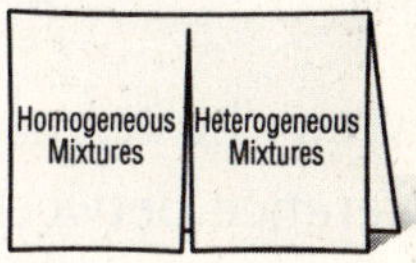

Reading Check

3. **Explain** How is a heterogeneous mixture different from a homogeneous mixture?

_____

_____

_____

● After You Read

Mini Glossary

compound: a substance whose smallest unit is made up of atoms of more than one element bonded together

mixture: made when two or more substances come together but do not combine to make a new substance

substance: matter that has the same composition and properties throughout

1. Review the terms and definitions in the Mini Glossary. In your own words, describe the difference between a compound and a mixture.

_____

_____

_____

2. Complete the chart below to compare the substances discussed in this section.

Substance	Definition	Examples
Element		
Compound		
Mixture		

End of Section

 Visit **ips.msscience.com** to access your textbook, interactive games, and projects to help you learn more about compounds and mixtures.

States of Matter

section ❶ Matter

● Before You Read

Think about your classroom. On the lines below, describe some of the things in your classroom that take up space.

_____

_____

What You'll Learn
■ that matter is made of particles that are always moving
■ how the particles are arranged in the three states of matter

● Read to Learn

What is matter?

Look around you. Maybe you see a glass of water. Maybe you see books on a shelf. These are examples of matter. **Matter** is anything that takes up space and has mass. You cannot always see matter. For example, air is matter.

What determines a material's state of matter?

All matter is made up of tiny particles such as atoms, molecules, or ions. Each particle attracts other particles. These particles are always moving. A material's state of matter is determined by the movement of the particles and the strength of attraction between them.

There are four different states, or forms, of matter. They are solid, liquid, gas, and plasma. Plasma only happens at very high temperatures. It is found in stars, lightning, and neon lights and is not common on Earth. This chapter will focus on the three main states of matter—solid, liquid, and gas.

Solids

A **solid** is matter with a definite shape and volume. What happens to a rock when you put it in a bucket? It does not change shape or size. A solid does not change to take the shape of the container it is in. This is because the particles of a solid are packed close together.

Mark the Text

Identify States of Matter As you read the section, draw a circle around the name of each state of matter. Then underline the definition of each state.

FOLDABLES

Ⓐ **Find Main Ideas** Make the following Foldable to record the main ideas about solids, liquids, and gases. Be sure to include examples.

Do the particles in a solid move?

The particles in all types of matter are always moving. A solid's particles are vibrating in place, however, they do not have enough energy to move out of their fixed positions.

What are crystalline solids?

Some solids have particles arranged in a three-dimensional pattern. This repeating pattern is called a crystal. Solids with this pattern are crystalline solids. Sodium chloride, or table salt, is an example. You can see the arrangement of the

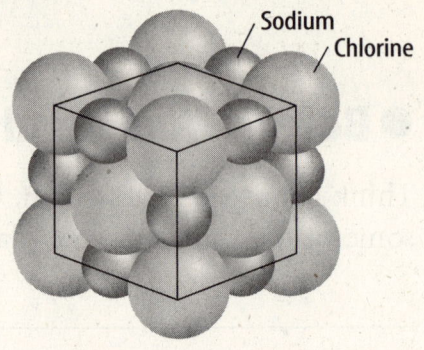

Sodium
Chlorine

particles in sodium chloride in the figure. They are in the shape of a cube. Sugar, sand, and snow are crystalline solids.

What are amorphous solids?

Some solids come together without forming crystals. They are called amorphous (uh MOR fuhs) solids. Their large particles are arranged randomly (in no certain order). Rubber, plastic, and glass are amorphous solids.

Liquids

You use liquids every day. Water is a liquid and so is orange juice. A **liquid** is matter that has a definite volume but no definite shape. A liquid takes the shape of its container but keeps the same volume. What happens if you pour 50 ml of orange juice from a bottle into a glass? You still have the same volume of orange juice, 50 ml, but the shape of the juice changes. ☑

How are particles arranged in a liquid?

The particles in a liquid move more freely than those in a solid. So a liquid can have different shapes. The particles in a liquid have enough energy to move past one another. But they do not have enough energy to move far apart. The figure shows the arrangement of the particles in a liquid.

Liquid

Picture This

1. **Identify** Write the name of an item in the classroom that is close to the same shape as this crystal of sodium chloride.

_____

☑ Reading Check

2. **Describe** Circle the sentence(s) that are true about a liquid.

It can change shape.
It has a definite shape.
Its volume stays the same.

Picture This

3. **Contrast** How are particles in a liquid different from particles in a solid?

_____

_____

_____

Do all liquids flow like water?

You know that honey flows slower than water. Other liquids do too. Some liquids flow more easily than others. **Viscosity** is how much a liquid resists flowing. The slower a liquid flows, the higher its viscosity. Honey has a high viscosity. It does not flow easily. Water has a low viscosity. It flows very easily. Viscosity describes the attraction between the particles of a liquid. For many liquids, viscosity increases as the liquid becomes colder.

What is surface tension?

Did you know that a needle will float on the surface of water? It floats because the particles on the surface of a liquid pull themselves together and resist being pushed apart. This happens because of the attractive forces between the particles. Particles below the surface of a liquid are pulled in all directions. But particles at the surface of a liquid are pulled toward the center of the liquid and sideways along the surface. There are no particles above to pull on them. **Surface tension** is the uneven forces acting on the particles on the surface of a liquid. Surface tension makes it seem like there is a thin film stretched across the surface of a liquid. ☑

Gases

Gas is matter that does not have a definite shape or volume. The particles in gas are far apart, as shown in the figure. Gas particles move quickly in all directions. They spread out evenly as far apart as possible. A gas will fill the container it is in. A gas can expand or be compressed.

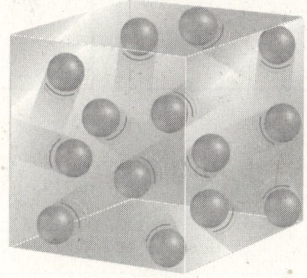

Gas

Think of a balloon filled with air. What happens if you squeeze the air into a smaller part of the balloon? The gas particles get closer together. This happens because you decreased the volume of the container the gas was in. Most gases are invisible. The air you breathe is a mixture of gases.

What is vapor?

Water is a liquid at room temperature. But water can also be a gas. The gas state of water is called water vapor. Vapor is matter that is in the gas state but is usually found as a liquid or solid at room temperature.

Copyright © Glencoe/McGraw-Hill, a division of The McGraw-Hill Companies, Inc.

💡 Think it Over

4. Compare Which has the higher viscosity, mayonnaise or honey?

_____

✔ Reading Check

5. Determine Is the following sentence true or false? Particles below the surface of a liquid are pulled in all directions.

_____

Picture This

6. Describe the arrangement and movement of the particles in a gas.

_____

_____

● After You Read

Mini Glossary

gas: matter that does not have a definite shape or volume

liquid: matter that has a definite volume but no definite shape that can flow

matter: anything that takes up space and has mass

solid: matter with a definite shape and volume

surface tension: the uneven forces acting on the particles on the surface of a liquid

viscosity: how much a liquid resists flowing

1. Read the key terms and definitions in the Mini Glossary above. On the lines below, tell how a solid and a liquid are similar.

   _____

   _____

2. Complete the chart below. Identify the three main states of matter and give two examples of each.

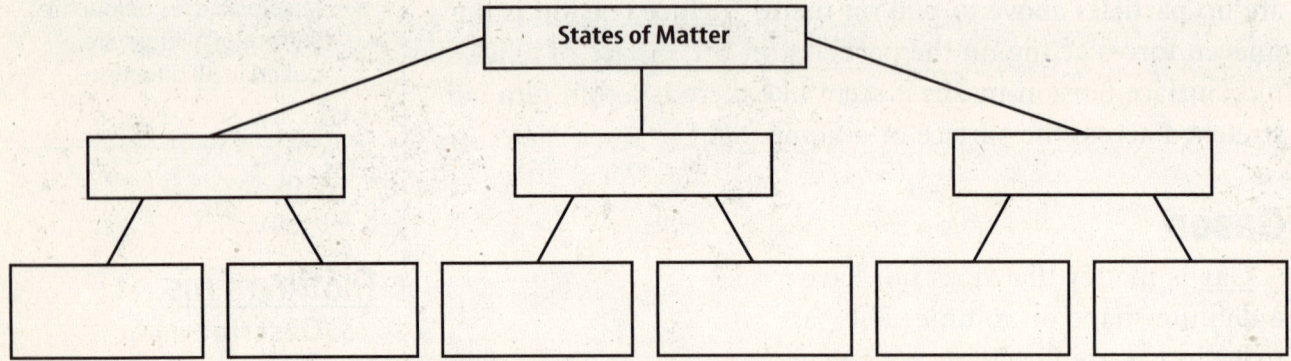

States of Matter

3. Think of a way of organizing the traits of solids, liquids, and gases to help you remember their characteristics.

   _____

   _____

   _____

End of Section

Science Online Visit **ips.msscience.com** to access your textbook, interactive games, and projects to help you learn more about matter.

States of Matter

section ② Changes of State

● Before You Read

How could you turn an ice cube into water? How could you turn water into an ice cube?

● Read to Learn

Thermal Energy and Heat

Imagine a swan ice sculpture. As time passes, drops of water begin to fall from the sculpture. Drip by drip, the swan becomes a puddle of liquid water. What makes matter change from one state to another? To answer this question, you need to think about the particles that make up matter.

How does energy affect particles?

Energy is the ability to do work or cause change. The energy of motion is called kinetic energy. Particles in matter are always moving. How much they move depends on how much kinetic energy they have. Particles with more kinetic energy move faster and farther apart. Particles with less kinetic energy move slower and stay closer together.

What is thermal energy?

<u>Thermal energy</u> is the total kinetic energy of all the particles in a sample of matter. Thermal energy depends on the number of particles in a substance and the amount of energy each particle has. The thermal energy of a substance changes if the number of particles changes. It also changes if the amount of energy each particle has changes. Suppose you have one cup of warm water and one cup of hot water. The hot water has more thermal energy. When you have the same size sample, the warmer substance has more thermal energy.

FOLDABLES

Ⓑ Compare and Contrast Make the following Foldable to compare and contrast thermal energy and temperature.

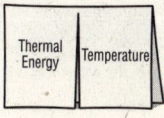

1. Calculate What is the average of the following five numbers: 6, 5, 3, 5, 6? Show your work.

_____

_____

_____

Picture This

2. Infer Which tea has more thermal energy?

_____

Think it Over

3. Predict On a bright, sunshiny day at the beach, which will heat up more quickly, the water or the sand?

_____

What is temperature?

Not all of the particles in a sample have the same amount of energy. Some have more energy than others. <u>Temperature</u> is the average kinetic energy of all the particles of a substance. You find an average by adding a group of numbers and dividing the total by the number of items in the group. For example, the average of the numbers 2, 4, 8, and 10 is $(2 + 4 + 8 + 10) \div 4 = 6$. Temperature is different from thermal energy, because thermal energy is a total and temperature is an average.

The iced tea in the figure is colder than the hot tea. In other words, the temperature of the iced tea is lower than the temperature of the hot tea. So the average kinetic energy of the particles in the iced tea is less than the average kinetic energy of the particles in the hot tea.

Particles in Motion

What is heat?

What happens when you stand close to a fire? You get warm. When a warm object is close to a cooler object, thermal energy moves from the warm object to the cooler one. <u>Heat</u> is the movement of thermal energy from a substance at a higher temperature to a substance at a lower temperature. When a substance is heated, it gains thermal energy. This means its particles move faster. The temperature of the substance rises. A substance loses thermal energy when it is cooled. Its particles move more slowly and the temperature of the substance drops.

Specific Heat

The specific heat of a substance is the amount of heat needed to raise the temperature of 1 g of the substance 1°C. Substances that have a low specific heat cool down and heat up quickly. They need only small amounts of heat to make their temperatures rise. A substance with a high specific heat cools down and heats up slowly. A larger amount of heat is needed to make its temperature rise or fall. Water has a high specific heat. Most metals and sand have a low specific heat.

Changes Between the Solid and Liquid States

Matter can change from one state to another when thermal energy is absorbed or released. This change is known as change of state.

What is melting?

As ice is heated, it absorbs thermal energy. The temperature of the ice rises. At some point, the temperature stops rising. The ice begins to change into liquid water. **Melting** is the change from the solid state to the liquid state. The temperature at which a substance changes from a solid to a liquid is called the melting point. The melting point of water is 0°C.

Amorphous solids melt differently than crystalline solids. Amorphous solids do not have crystal structures to break down. They do not melt into liquids. They simply get softer and softer. For example, glassblowers can shape glass into beautiful vases while it is hot because glass is an amorphous solid.

What is freezing?

A liquid can be changed back into a solid by cooling it. **Freezing** is the change from the liquid state to the solid state. As the liquid cools, it loses thermal energy. Its particles slow down and come closer together. Attractive forces begin to trap particles and crystals form. Freezing and melting are opposite processes.

As you can see in the graph, the temperature at which a substance changes from the liquid state to the solid state is called the freezing point. The freezing point of the liquid state of a substance is the same temperature as the melting point of the solid state. For example, solid water melts at 0°C and liquid water freezes at 0°C.

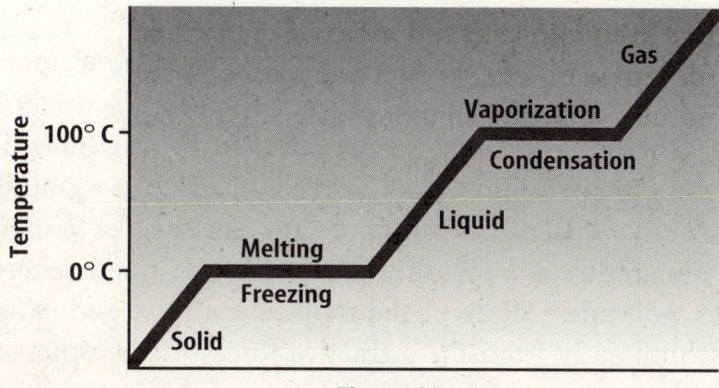

Think it Over

4. **Infer** Is an ice cube that is melting gaining or losing thermal energy?

_____

_____

Picture This

5. **Reading a Graph** Look at the graph. What two processes, besides melting and freezing, happen at the same temperature?

_____

_____

When does the temperature change again?

The temperature of a substance stays the same while it is freezing. Energy is released during freezing because particles in a liquid have more energy than particles in a solid. This energy is released into the surroundings. The temperature of the substance begins to decrease again after all of the liquid has become solid.

Changes Between the Liquid and Gas States

It rained overnight. You and your friends have fun jumping in puddles the next morning. But by afternoon, the puddles are gone. The liquid water in the puddles changed into a gas. Matter changes between the liquid and gas states through vaporization and condensation.

How does a liquid change to a gas?

When liquid water is heated, its temperature rises until it is 100°C. At this point, liquid water changes into water vapor. **Vaporization** is the change from a liquid to a gas. The temperature of a substance does not change during vaporization. But, the substance absorbs thermal energy. This energy makes the particles move faster until they have enough energy to escape the liquid as gas particles.

Vaporization

There are two forms of vaporization. Vaporization below the surface of a liquid is called boiling. When a liquid boils, bubbles within the liquid rise to the surface, as shown in the figure. The temperature at which a liquid boils is called the boiling point.

What is evaporation?

Vaporization at the surface of a liquid is called evaporation. Evaporation happens at temperatures below the boiling point. Evaporation explains how puddles dry up. Imagine that you could see individual water molecules in a puddle. You would see that they move at different speeds. Remember temperature is a measure of the average kinetic energy of the molecules. Some of the molecules that are moving fastest pull away from the attractive forces of the other molecules and escape from the surface of the water. ☑

<u>Picture This</u>

6. Label Draw arrows to show the direction of the movement of bubbles in the boiling water.

7. Name the two forms of vaporization.

_____

_____

During evaporation, the fastest molecules also must be close to the surface of the liquid. They also have to be moving in the right direction and they have to keep from hitting other molecules as they leave. The particles that are still in the liquid are the slower, cooler ones. Evaporation cools the liquid and anything near the liquid. Evaporation cools you when you sweat. Perspiration evaporates from your skin.

What is condensation?

What happens to a glass of cold lemonade on a hot day? The outside of the glass becomes covered with drops of water. What happened? As a gas cools, its particles slow down. The particles slow down enough for their attractions to bring them together. When the particles come together, they form droplets of liquid. This process is called condensation. **Condensation** is the change from a gas to a liquid. It is the opposite of vaporization. ☑

As a gas condenses to a liquid, it releases the thermal energy that it absorbed when it became a gas. The temperature of the substance does not change during condensation. The decrease in energy changes the arrangement of the particles. After the change of state is complete, the temperature continues to drop.

Condensation formed the water droplets on your glass of lemonade. Condensation also is how rain forms. Water vapor in the atmosphere condenses to make water droplets in clouds. When the droplets are large enough, they fall to the ground as rain.

Changes Between the Solid and Gas States

Some substances can change from the solid state to the gas state without ever becoming a liquid. This process is called sublimation. During sublimation, the particles on the surface of the solid gain enough energy to become a gas. ☑

One example of a substance that goes through sublimation is dry ice. Dry ice is the solid form of carbon dioxide. At room temperature and pressure, carbon dioxide is not a liquid. It is a gas. Therefore, as dry ice absorbs thermal energy from the objects around it, it changes directly into a gas. When dry ice becomes a gas, it absorbs thermal energy from water vapor in the air. The loss of thermal energy causes the water vapor to cool and condense into liquid water droplets. This causes fog to form.

☑ Reading Check

8. **Explain** What is the opposite of vaporization?

_____

_____

☑ Reading Check

9. **Determine** Why are some substances able to go directly from a solid state to a gas state?

_____

_____

_____

● After You Read
Mini Glossary

condensation: the change from a gas to a liquid
freezing: the change from the liquid state to the solid state
heat: movement of thermal energy from a substance at a higher temperature to a substance at a lower temperature
melting: the change from the solid state to the liquid state

temperature: the average kinetic energy of all the particles of a substance
thermal energy: the total kinetic energy of all the particles in a sample of matter
vaporization: the change from a liquid to a gas

1. Review the terms and their definitions in the Mini Glossary. How is freezing related to melting?

_____

_____

2. Above each arrow, write the name of the process needed to make the change in states of matter.

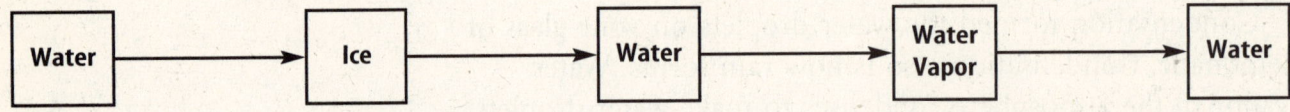

| Water | → | Ice | → | Water | → | Water Vapor | → | Water |

3. You were asked to highlight each way that matter can change states. How did highlighting help you to learn the ways?

_____

_____

_____

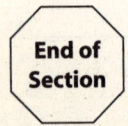
End of Section

Science Online Visit **ips.msscience.com** to access your textbook, interactive games, and projects to help you learn more about changes of state.

62 States of Matter

chapter 4 States of Matter

section 3 Behavior of Fluids

Before You Read

What happens to a balloon if you keep blowing air into it? On the lines below, describe what happens and why.

____

____

____

What You'll Learn
- why some things float and others sink
- how pressure is moved through fluids

Read to Learn

Mark the Text

Locating Information Underline every heading in the reading that asks a question. Then, highlight the answers to those questions as you find them.

Pressure

Suppose you and your friends want to play volleyball, but the ball is flat. You pump air into the ball until it is firm. The ball is firm because of the movement of the air particles inside the ball. The air particles inside the ball bump into each other and against the walls of the ball. When the particles bump into the walls of the ball, they push with a force on the walls. The force pushes the surface of the ball outward. The forces of all the individual particles add together to make up the pressure of the air inside the ball.

What is pressure?

Pressure is equal to the force put on a surface divided by the total area over which the force is applied.

$$\text{pressure} = \frac{\text{force}}{\text{area}}$$

When force is measured in newtons (N) and area is measured in square meters (m^2), pressure is measured in newtons per square meter (N/m^2). This unit of pressure is called a pascal (Pa). A more useful unit when talking about atmospheric pressure is the kilopascal (kPa), which is 1,000 pascals.

Applying Math

1. **Compute** A person standing on one foot is applying a force of 500 N. If the foot covers 100 cm^2, what is the pressure? Show your work.

____

____

How are force and area related to pressure?

The formula for pressure tells you that pressure depends on the amount of force and the area over which the force is applied. This means that as the force increases over a given area, pressure increases. If the force decreases in that same area, pressure decreases.

The opposite is true if the force stays the same, but the area over which it is applied changes. If a force is applied to a smaller area, pressure increases. If the same amount of force is spread out over a larger area, pressure decreases. ☑

The figures below show this. The force of the dancer's weight remains the same. However, the area where the force is applied changes. Her pointed toes have less area than her flat feet. So, the pressure of the dancer's weight on pointed toes is greater than the pressure on her flat feet.

Reading Check

2. **Explain** What happens to pressure when the area decreases but the force stays the same?

Picture This

3. **Identify** In the figure, circle the greater area amount. Put a box around the greater pressure amount.

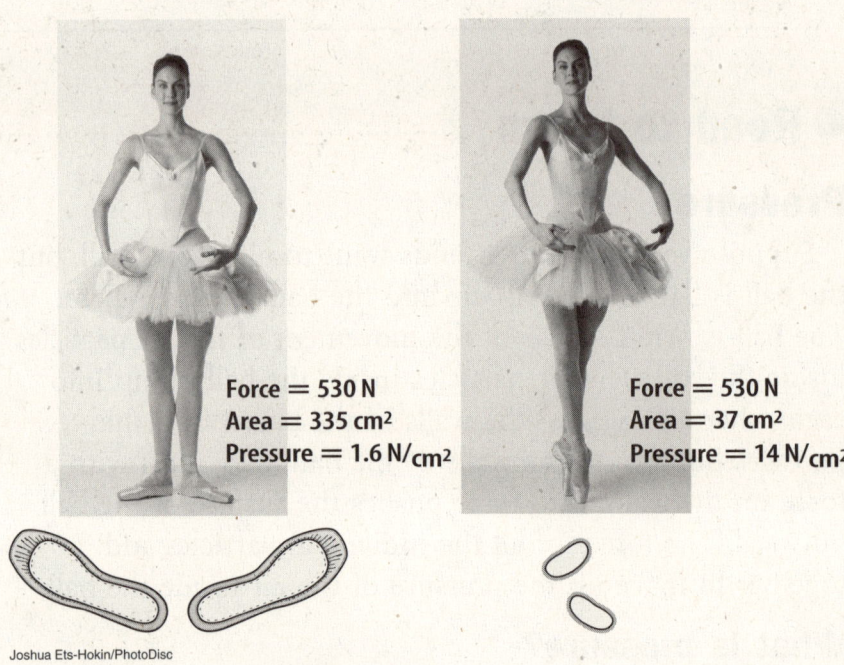

Force = 530 N
Area = 335 cm²
Pressure = 1.6 N/cm²

Force = 530 N
Area = 37 cm²
Pressure = 14 N/cm²

Joshua Ets-Hokin/PhotoDisc

What is atmospheric pressure?

You cannot see it and you usually cannot feel it, but the air around you presses on you with great force. Air pressure on objects is known as atmospheric pressure because air makes up the atmosphere around Earth. Atmospheric pressure is 101.3 kPa at sea level. So, air puts a force of about 101,000 N on every square meter it touches. This is about the weight of a large truck.

How does air pressure help you?

Air pressure helps you. Air pressure allows you to drink from a straw. Look at the figure. When you first suck on a straw, you remove the air from it. Air pressure pushes down on the liquid in your glass and forces it up into the straw. If you tried to drink through a straw in a sealed, airtight

Richard Hutchings

container, it would not work. The air would not be able to push down on the surface of the drink.

Why don't you feel the force of air?

You don't feel the force of air because pressure from the fluids in your body balances it. The fluids in your body put an outward pressure on your body. This pressure balances the pressure from the air on the surface of your body.

How does atmospheric pressure change?

Atmospheric pressure changes with altitude. Altitude is the height above sea level. As altitude increases, atmospheric pressure decreases, because there are fewer air particles in a given volume. Since there are fewer particles, they bump into each other less often, and therefore there is less pressure.

A French physician named Blaise Pascal was the first one to test this idea. He partially filled a balloon with air. The balloon was carried to the top of a mountain. The figure shows what happened. The balloon expanded while being carried up the mountain. The amount of air inside the balloon stayed the same. But, the air pressure

pushing in on it from the outside decreased. This allowed the particles of air inside the balloon to spread out further.

Picture This

4. **Label** By the arrow in the figure, write what is pushing down on the liquid in the glass.

Picture This

5. **Infer** What will happen as the hiker brings the balloon back down the mountain?

_____

_____

How does air pressure affect travelers?

Have you ever been in an airplane? Have you driven up a mountain? If so, you have probably felt a popping sensation in your ears. As the air pressure drops, the air pressure in your ears increases. Soon the air pressure in your ears is greater than the air pressure outside your body. When some air is released from your ears, you hear a pop. This release of air makes the pressure inside and outside your ears the same. The pressure in an airplane is controlled so the pressure does not change greatly during a flight. ☑

Changes in Gas Pressure

The pressure of a gas in closed containers can change just like atmospheric pressure can change. The pressure of a gas in a closed container changes with volume and temperature.

What happens if the volume of a gas in a closed container decreases?

If you squeeze part of a filled balloon, the rest of the balloon gets firmer. When you squeeze a balloon, you decrease its volume. The same number of particles is now in a smaller space. The particles bump into each other and the walls of the container more often. This increases the pressure. Any time you decrease the volume of a space without changing its temperature, pressure increases.

What happens if the volume of a gas increases?

Look at the figures below. They show a piston moving and changing the pressure of the gas particles. If you make a container larger and do not change its temperature, the particles will bump into each other less often. Therefore, the pressure will be less. So, as volume increases, pressure decreases.

☑ **Reading Check**

6. **Explain** What causes your ears to pop when you are flying?

_____

_____

_____

Picture This

7. **Interpret** Circle the piston that has the least pressure.

As volume increases, pressure decreases.

How does temperature affect pressure?

Recall that temperature rises as the kinetic energy of the particles in a substance increases. The greater the kinetic energy, the faster the particles move. The faster the particles move, the more they bump into each other. This makes the pressure greater, even though the volume of the gas stays the same. If the temperature of a gas in a closed container increases, the pressure of the gas also will increase.

Float or Sink

Water pressure pushes on you in all directions when you are under water. Water pressure increases as you go deeper in the water. The pressure pushing up on the bottom of an object becomes greater than the pressure pushing down on it. This is because the bottom of the object is deeper in the water.

What makes an object float in water?

Suppose you throw a small log in a lake. As shown in the figure, the water pressure under the log is greater than the water pressure above the log. This pushes the log up and makes it float. **Buoyant force** is the force that pushes up on an object in a fluid.

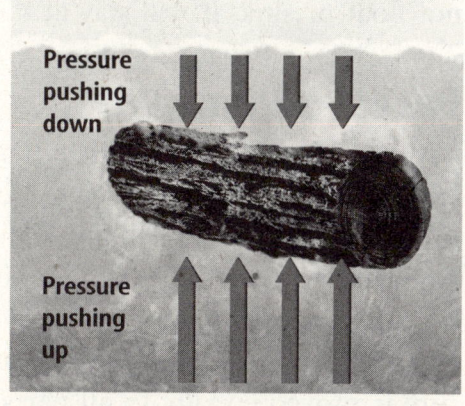

Pressure pushing down

Pressure pushing up

Buoyant force cannot make everything float. If the buoyant force is equal to the weight of an object, the object will float like the person shown below. But if the buoyant force is less than the weight of an object, the object will sink.

Weight

Buoyant force

???????????

Picture This

8. **Determine** In the first figure, why are the arrows under the log longer than the arrows above the log?

_____

_____

_____

Picture This

9. **Describe** What would happen to the person in the second figure if the weight arrow were longer than the buoyant force arrow?

_____

_____

What is Archimedes' Principle?

What determines the buoyant force? **Archimedes'** (ar kuh MEE deez) **principle** states that the buoyant force of an object is equal to the weight of the fluid removed by the object. Think about a beaker that is filled to the top with water. If you put an object in the beaker, some water will spill out. If you weigh the spilled water, you will find the buoyant force on the object.

What is density?

Understanding density can help you decide if an object will float. **Density** is mass divided by volume.

$$density = \frac{mass}{volume}$$

If an object is less dense than the fluid it is in, it will float. If an object is more dense than the fluid it is in, it will sink. What if an object has the same density as the fluid? It will not float or sink. It will stay at the same level in the fluid.

Pascal's Principle

What happens if you squeeze a plastic container filled with water? If the container is closed, the water has nowhere to go. The pressure in the water increases by the same amount everywhere in the container—not just where you squeeze. **Pascal's principle** states that when a force is applied to a fluid in a closed container, the increase in pressure is moved equally to all parts of the fluid.

How do hydraulic systems work?

Have you ever wondered how a car is raised and lowered at the mechanic's shop? A device called a hydraulic (hy DRAW lihk) system is used. It uses Pascal's principle to increase force. Look at the figure of the hydraulic lift on the next page. There is a downward force on the piston on the left. This increases the pressure in the fluid in the tube. The increased pressure is moved to the piston on the right. Why is the piston on the right able to lift the car? Recall that pressure is equal to force divided by area. ☑

$$pressure = \frac{force}{area} \text{ or force} = pressure \times area$$

Applying Math

10. Calculate You are given a sample of a solid that has a mass of 12.0 g. Its volume is 4.0 cm³. What is its density in g/cm³? Will it float in water? (The density of water is 1.0 g/cm³.) Show your work.

_____

_____

_____

FOLDABLES™

C Organize Information Write down information about Archimedes' and Pascal's principles on two quarter sheets of paper.

✔ Reading Check

11. Identify What does a hydraulic system increase using Pascal's principle?

_____

_____

Greater Area If the two pistons on the tube have the same area, the force on both pistons will be the same. But the piston on the right has a greater surface area. If you multiply the same pressure by a larger area, the force is greater. So the force on the right will be greater.

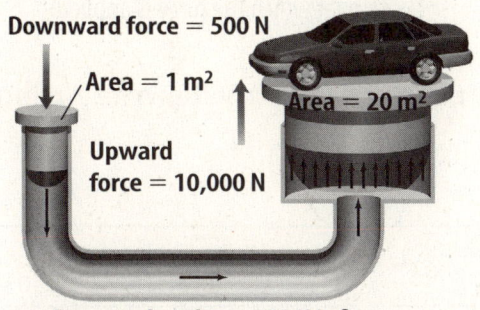

Hydraulic Lift

Downward force = 500 N

Area = 1 m²

Area = 20 m²

Upward force = 10,000 N

Pressure in tube = 500 N/m²

What are force pumps?

If you punch a hole in the top of a closed milk carton, the milk is pushed out the hole when you squeeze the carton. This is known as a force pump. A force pump makes it possible to squeeze toothpaste out of a tube.

A heart has two force pumps. One force pump pushes the blood from the heart to the lungs. The blood picks up oxygen in the lungs and returns to the heart. Another force pump pushes the blood with oxygen in it to the rest of the body. The two force pumps in a heart are shown below.

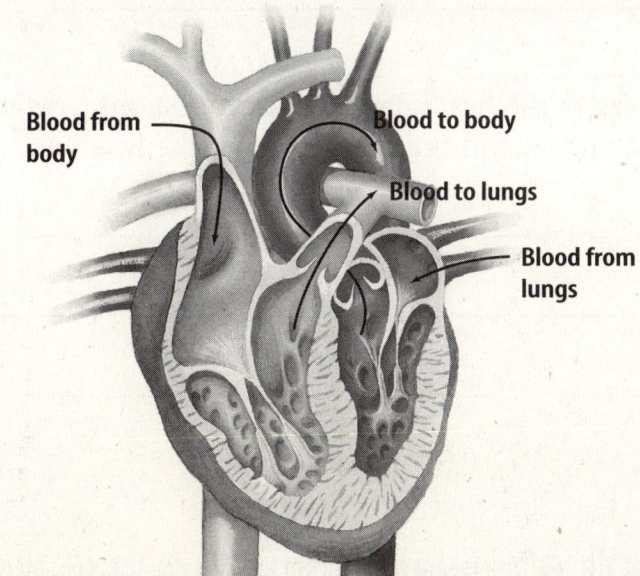

Blood from body

Blood to body

Blood to lungs

Blood from lungs

Picture This

12. **Calculate** How many times greater is the upward force on the car than the downward force on the piston? (Hint: Divide the upward force by the downward force.)

    a. 5
    b. 10
    c. 20
    d. 25

Picture This

13. **Interpreting Diagrams** Use a highlighter to trace the path of the blood as it enters and leaves the heart. Start where the blood comes in from the body.

● After You Read

Mini Glossary

Archimedes' principle: the buoyant force of an object is equal to the weight of the fluid removed by the object

buoyant force: the force that pushes up on an object that is in a fluid

density: mass divided by volume

Pascal's principle: when a force is applied to a fluid in a closed container, the increase in pressure is moved equally to all parts of the fluid

pressure: the force put on a surface divided by the total area over which the force is applied

1. Review the terms and their definitions in the Mini Glossary. Rewrite Archimedes' principle in your own words.

_____

_____

2. Complete the table by circling whether pressure increases or decreases as a result of the event.

Event	Pressure
Force increases	(increases) / decreases
Force decreases	increases / decreases
Area over which force is applied increases	increases / decreases
Volume decreases	increases / decreases
Temperature increases	increases / decreases

3. You were asked to highlight the answers to the headings that were questions as you read. How did this help you make sure you understood the main ideas of the heading?

_____

_____

_____

End of Section

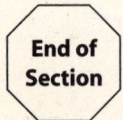 Science Online Visit **ips.msscience.com** to access your textbook, interactive games, and projects to help you learn more about behavior of fluids.

Matter—Properties and Changes

section ❶ Physical Properties

● Before You Read

Choose an object you can see. Describe it on the lines below. What is its color, shape, and size?

_____

_____

What You'll Learn

■ the common physical properties of matter
■ how to find the density of a substance
■ how acids and bases are alike and different

● Read to Learn

Physical Properties

Suppose you saw a dinosaur skeleton in a museum. How would you describe it to a friend? You could talk about its color, shape, size, and if it was rough or smooth. These are physical properties, or characteristics, of the skeleton. You can use your senses to describe physical properties. A **physical property** is something you can observe about matter without changing what the matter is. All matter has physical properties.

What are some physical properties?

You probably know about some physical properties like color, shape, smell, and taste. There are other physical properties like mass, volume, and density. Mass is the amount of matter in an object. *m* is the symbol for mass. A bowling ball has more mass than a balloon. Volume is the amount of space an object occupies. *V* is the symbol for volume. A swimming pool holds a larger volume of water than a paper cup does. **Density** is the amount of mass in a certain volume. *D* is the symbol for density. A bowling ball is more dense than a balloon. To find the density of an object, divide its mass by its volume. Use the formula below to calculate density.

$$\text{Density} = \text{mass/volume}$$

$$D = \frac{m}{V}$$

Mark the Text

Identify Main Ideas As you read this section, highlight each sentence that describes a physical property.

FOLDABLES™

Ⓐ Build Vocabulary Make a Foldable to help you learn the vocabulary terms in this chapter.

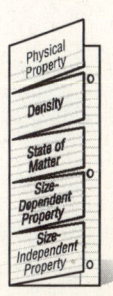

Think it Over

1. **Evaluate** Which has the higher density, a golf ball or a table-tennis ball?

_____

_____

☑ **Reading Check**

2. **Classify** Which is *not* a state of matter? (Circle your answer.)
   a. gas
   b. liquid
   c. water
   d. solid

How does mass affect density?

A bowling ball and a balloon have about the same volume. To decide which one has the higher density, you compare their masses. Because they are about the same volume, the one with more mass has the higher density. The bowling ball has a higher density. Suppose you want to compare two bowling balls. They are the same size, shape, color, and volume. But, one bowling ball has more mass than the other. Although the volumes are the same, the densities of the bowling balls are different because their masses are different.

How can density be used?

Density is a physical property of an object. It can be used to identify an unknown substance or element. For example, the density of silver is 10.5 g/cm^3 at 20°C. Recall that density is a unit of mass divided by a unit of volume. Also, scientists usually report the density of matter at a certain temperature.

Suppose you wanted to know if a ring is made of pure silver. You can find the density of the ring by dividing its mass by its volume. If the ring's density is any value other than 10.5 g/cm^3, the ring is not pure silver.

What are the states of matter?

State of matter is another physical property. The **state of matter** tells you whether matter is a solid, a liquid, or a gas. The state of matter depends on the temperature and pressure of the matter. ☑

Water can be a solid, a liquid, or a gas. Ice is water in the solid state. The water you drink is in the liquid state. You cannot see water as a gas. It is vapor in the air. A water molecule is always the same, whether the water is a solid, a liquid, or a gas. A water molecule always has two hydrogen atoms and one oxygen atom no matter what state water is in.

What are some other physical properties of matter?

A **size-dependent property** is a physical property that changes when the size of an object changes. One wooden block has a volume of 30 cm^3 and a mass of 20 g. Another wooden block has a volume of 60 cm^3 and a mass of 40 g. The volume and mass of the block change when the size changes. Volume and mass are size-dependent properties.

What is a size-independent property?

If you calculate the density of both wooden blocks, you'll find each one has a density of 0.67 g/cm^3. The density did not change with the size of the block. Density is an example of a size-independent property. A **size-independent property** is a physical property that does not change when an object changes size. The table below shows examples of other size-dependent and size-independent properties.

Physical Properties	
Type of Property	**Property**
Size-dependent properties	length, width, height, volume, mass
Size-independent properties	density, color, state

Physical Properties of Acids and Bases

One way to describe matter is to classify it as either an acid or a base. Acids and bases can be strong or weak. The strength of an acid or base can be determined by finding its pH.

What is the pH scale?

The pH scale is used to measure the strength of an acid or a base. The pH scale has a range of 0 to 14. The pH of acids ranges from 0 to just below 7. The pH of bases ranges from just above 7 to 14. Matter with a pH of exactly 7 is neutral. It is neither an acid nor a base. Pure water is a neutral substance. Its pH is exactly 7.

What are acids?

What do you think of when you hear the word *acid*? Do you think of a dangerous chemical that can burn your skin, make holes in your clothes, and even destroy metal? Some acids, such as concentrated hydrochloric acid, can harm you. Other acids are not harmful.

Some acids can be eaten. When you drink a soft drink, you drink carbonic and phosphoric (fahs FOR ihk) acids. When you eat a citrus fruit, such as an orange or a grapefruit, you eat citric and ascorbic (uh SOR bihk) acids.

Picture This

3. Use Tables Name a size-independent property other than density.

_____

_____

Think it Over

4. Classify A substance has a pH of 3. Is this substance an acid or a base?

_____

_____

FOLDABLES™

B Classify Make the following Foldable to help classify materials as acids or bases.

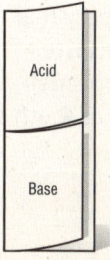

What are some physical properties of acids?

Most acids have a distinctive taste and smell. Imagine cutting into a lemon. You will notice a sharp smell. The smell of a lemon comes from the citric acid in the fruit. If you take a bite of a lemon, you would immediately notice a very sour taste. Most citrus fruits taste sour because they contain citric acid. The figure below shows examples of some citrus fruits.

If you bit into a citrus fruit, then rubbed your teeth back and forth, they would squeak. This is a physical property of acids. The sharp smell and the sour taste also are physical properties of acids.

Picture This

5. Identify Which citrus fruit in the figure do you think tastes the most sour?

_____

Acids and Aging Vitamin C and alpha-hydroxy acids are also found in fruit. These acids are sometimes used in skin creams. It is believed these acids slow down the aging process.

What are some physical properties of bases?

Bases are different substances than acids. The physical properties of bases are different from the physical properties of acids. You may have a common base in your house—ammonia (uh MOH nyuh). Ammonia is sometimes used for household cleaning. If you got a cleaner that contained ammonia on your fingers and rubbed them together, your fingers would feel slippery.

Another common base is soap. When you rub wet soap on your hands, they also feel slippery. This slippery feeling is a physical property of bases. You shouldn't taste soap, but if you did, you'd notice a bitter taste. A bitter taste is another physical property of bases. ☑

Remember that you should never taste, touch, or smell anything in lab unless your teacher tells you to.

✔ Reading Check

6. Determine Circle a physical property of bases.
a. slippery feel
b. sour taste
c. sharp smell
d. makes teeth squeak

After You Read

Mini Glossary

density: the amount of mass in a given volume

physical property: any characteristic of matter that can be observed without changing what the matter is

size-dependent property: a physical property that changes when the size of an object changes

size-independent property: a physical property that does not change when an object changes size

state of matter: a physical property that describes whether a sample of matter is a solid, a liquid, or a gas

1. Review the terms and their definitions in the Mini Glossary. Write a sentence using one of the terms to describe a property of an object.

_____

_____

_____

2. Use the outline below to help you review what you have read. Fill in the blanks where information is missing.

Physical Properties

I. Common Physical Properties

   A. Color

   B. _____

   C. _____

   D. _____

   E. Mass

   F. _____

   G. _____

   H. State

     1. Solid

     2. _____

     3. _____

II. Size-Dependent Properties

   A. _____

   B. _____

   C. _____

   D. _____

   E. _____

III. Size-Independent Properties

   A. _____

   B. _____

   C. _____

Science Online Visit **ips.msscience.com** to access your textbook, interactive games, and projects to help you learn more about physical properties.

End of Section

Matter—Properties and Changes

section ❷ Chemical Properties

What You'll Learn

- some chemical properties of matter
- the chemical properties of acids and bases
- how a salt is formed

Study Coach

Create a Quiz After you read this section, write a quiz question for each paragraph. After you write the quiz, answer the questions.

💡 **Think it Over**

1. **Identify** Give an example of a chemical property.

_____

_____

● **Before You Read**

Chemical properties can tell you whether one substance will react with another. Describe a situation where it would be useful to know the chemical properties of a substance.

_____

_____

● **Read to Learn**

A Complete Description

In the last section, you learned that an object can be described by its physical properties. You also learned about states of matter. Water can be a solid, liquid, or gas. You learned that acids have a sour taste. You can give a more complete description of something by also telling how it behaves.

What are some examples of chemical properties?

If you strike a match, it will probably burn. When a match burns, phosphorus (FAHS frus) and wood in the match combine with oxygen in the air to make new compounds. Phosphorus and wood can burn. The ability to burn is a chemical property. A **chemical property** is any characteristic of matter that allows it to change to a different type of matter.

If you leave a slice of apple on a table, the apple will turn brown. Compounds in the apple react with oxygen to form new compounds. The ability to react with oxygen is another chemical property.

Knowing the chemical properties of a material is important. Gasoline has the ability to burn easily. Gas pumps have warning labels that tell customers not to get near them with anything that might start the gasoline burning.

How are chemical properties used?

Look at the figure below. You probably would want to wear a bracelet made of gold rather than one made of iron. Why? An iron bracelet would not be as nice to look at or as valuable as a bracelet made of gold.

Also, iron has a chemical property that makes it a poor choice for jewelry. Think about what happens to an iron object when it is left outside in the rain. It rusts. Iron rusts easily because of its high reactivity (ree ak TIH vuh tee) with oxygen and water. Gold has a low reactivity, so it is a good choice for jewelry. The <u>reactivity</u> of a substance is how easily that substance reacts with another substance.

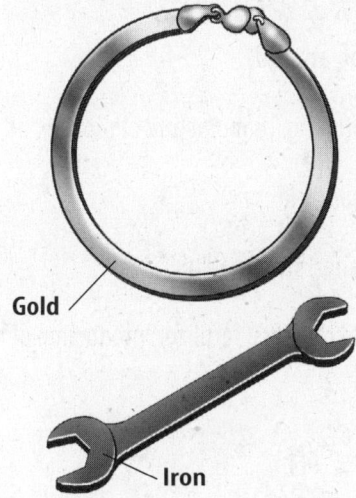

Gold

Iron

Picture This

2. Infer If a material rusts easily, does it have a high reactivity or a low reactivity?

_____

_____

How can chemical properties keep swimming pools clean?

Chlorine has a high reactivity. When chlorine combines with water, it forms a compound called hypochlorous acid. Hypochlorous acid kills bacteria, insects, algae, and plants. Many people add this chlorine compound to the water in swimming pools to help keep it clean. Hypochlorous acid in water can be harmful to humans. It forms compounds that can make swimmers' eyes and skin burn.

Do acids and bases have chemical properties?

You learned that one physical property of acids is that they taste sour. One physical property of bases is that they feel slippery. Acids and bases also have chemical properties. The chemical properties of acids and bases are what make them both useful but sometimes harmful.

Think it Over

3. Infer What could happen if too much chlorine is added to a swimming pool?

_____

_____

_____

_____

Common Acids and Bases		
Name of Acid	Formula	Where It's Found
Acetic acid	CH_3COOH	Vinegar
Acetylsalicylic acid	$C_9H_8O_4$	Aspirin
Ascorbic acid (vitamin C)	$C_6H_8O_6$	Citrus fruits, tomatoes
Carbonic acid	H_2CO_3	Carbonated drinks
Hydrochloric acid	HCl	Gastric juice in stomach
Name of Base	Formula	Where It's Found
Aluminum hydroxide	$Al(OH)_3$	Deodorant, antacid
Calcium hydroxide	$Ca(OH)_2$	Leather tanning, manufacture of mortar and plaster
Magnesium hydroxide	$Mg(OH)_2$	Laxative, antacid
Sodium hydroxide	NaOH	Drain cleaner, soap making
Ammonia	NH_3	Household cleaners, fertilizer, production of rayon and nylon

Picture This

4. Use Tables Where can hydrochloric acid be found?

_____

_____

Acids and Bases The table above shows some common acids and bases, their chemical formulas, and where they can be found. Knowing the chemical properties of acids and bases can help you use them safely.

What are some chemical properties of acids?

Many acids react with, or corrode, certain metals. Have you ever used aluminum foil to cover leftover tomato sauce? The next day, you may have seen small holes in the foil where the tomato sauce touched it. The acids in the sauce reacted with the aluminum and made holes in it. Acids in tomatoes, oranges, and other foods are not harmful to you. But, some acids are harmful.

Many acids can damage plant and animal tissue. When rain contains acid, it is called acid rain. Small amounts of nitric (NI trihk) acid and sulfuric (sul FYOOR ihk) acid are found in acid rain. Both of these acids are strong acids. Plants and animals can be harmed in areas where acid rain falls. Sulfuric acid causes burns when it touches skin.

What are some chemical properties of bases?

You may think that acids are more dangerous than bases. That is not true. A strong base is just as dangerous as a strong acid. Sodium hydroxide (hi DRAHK side) is a strong base. Sodium hydroxide can damage living tissue.

Remember that ammonia is a base. It's possible to get a bloody nose if you smell strong ammonia. If you touch a strong base, it will burn your skin. Ammonia feels slippery because it reacts with the proteins in the tissues in your skin. This reaction can damage your skin.

What are salts?

Acids and bases react with each other when they combine. When an acid reacts with a base, they form water and a compound called a salt. A **salt** is a compound made of a metal and a nonmetal that forms when an acid and a base react.

The figure below shows uses for some salts. The salt you shake on your food is the most common salt. Table salt is called sodium chloride. Sodium chloride is formed when the base sodium hydroxide reacts with hydrochloric acid. Chalk, called calcium carbonate, is a useful salt. Another salt, ammonium chloride, is used in some batteries.

Aaron Haupt

FOLDABLES

C Construct a Venn Diagram Make the following Foldable to compare and contrast acids, bases, and salts.

Acid

H2O + Salt

Base

Picture This

5. **Identify** Circle the sodium chloride in the figure.

● After You Read

Mini Glossary

chemical property: any characteristic of matter that allows it to change to a different type of matter

reactivity: how easily a substance reacts with another substance

salt: a compound made of a metal and a nonmetal that forms when an acid and a base react

1. Review the terms and their definitions in the Mini Glossary. Write a sentence that describes a chemical property of acids.

   _____

   _____

   _____

2. Write the letter of the material in Column 2 that has the chemical property described in Column 1.

Column 1	Column 2
_____ 1. ability to react with oxygen and water	**a.** acid
_____ 2. ability to form a salt when combined with an acid	**b.** chlorine
_____ 3. ability to burn	**c.** iron
_____ 4. ability to kill bacteria	**d.** base
_____ 5. ability to react with metals	**e.** wood

3. You were asked to write a quiz question for each paragraph in this section. How did this help you learn the content of the section?

   _____

   _____

   _____

End of Section

Science○**nline** Visit **ips.msscience.com** to access your textbook, interactive games, and projects to help you learn more about chemical properties.

chapter 5

Matter—Properties and Changes

section ❸ Physical and Chemical Changes

◉ Before You Read

Write what you think is an example of a physical change. After you read this section, check and see if you were correct.

_____

_____

What You'll Learn

■ how to identify physical and chemical changes
■ how physical and chemical changes affect the world you live in

◉ Read to Learn

Physical Change

Sometimes a building must be destroyed to make space for a new one. Experts know exactly how to blow up buildings so that no one gets hurt and no other buildings are damaged. The second before a building is blown up, you see the building. In a few seconds, you see a pile of steel and concrete. The appearance of the building has changed.

What is a physical change?

The building that was destroyed underwent a physical change. A **physical change** is any change in size, shape, form, or state where the matter itself does not change into something else. The matter stays the same. In a physical change, only the physical properties change. When a building is blown up, the materials in the building stay the same. They just look different. ☑

How can you recognize a physical change?

Just look to see if the matter has changed size, shape, form, or state. If any of these things have changed, there was a physical change. If you cut a watermelon into pieces, the watermelon changes size and shape, but it is still a watermelon. That is a physical change. Put one of the pieces into your mouth and bite it. You change the watermelon's size and shape again, but you don't change the watermelon into something else.

Mark the Text

Identify Details As you read this section, highlight facts about physical changes in one color. Highlight facts about chemical changes in a different color.

☑ Reading Check

1. **Determine** Which is *not* a physical change? (Circle your answer.)
   a. change in size
   b. change in color
   c. change in shape
   d. change in form

What is a change of state?

Another example of a physical change is matter changing state. During a change of state, matter changes from one state to another.

Suppose you and your friends make snow cones. A snow cone is a mixture of frozen water, sugar, food coloring, and flavoring. When you first make the snow cone, the water is solid ice. If you take your snow cone outside in the hot sunshine, the ice begins to melt. The solid water changes state and becomes liquid water.

A water molecule that is frozen in the solid state has two hydrogen atoms and one oxygen atom. A water molecule that is in the liquid state also has two hydrogen atoms and one oxygen atom. When water changes state, it does not change chemically. Only its form changes.

The solid water in the snow cone turns into liquid water and drips onto the hot sidewalk. As the liquid water warms even more, it changes state again. It evaporates. The liquid water becomes a gas.

As you can see, matter can change from a solid to a liquid and from a liquid to a gas. Matter also can change from a gas to a liquid. Dew forms when water vapor in the air changes to liquid water. You see dew in the mornings as drops of water on grass and plants. Matter can change from a liquid to a solid. Liquid water freezes and turns to ice, or solid water. Liquid metal cools to become solid metal. These are all examples of changes of state.

Chemical Changes

Look at the figure below. An apple cut in half and left out turns brown. Shiny copper pennies turn dull and dark over time. What do these changes have in common? Each of these changes is a chemical change. A **chemical change** occurs when one type of matter changes into a different type of matter with different properties.

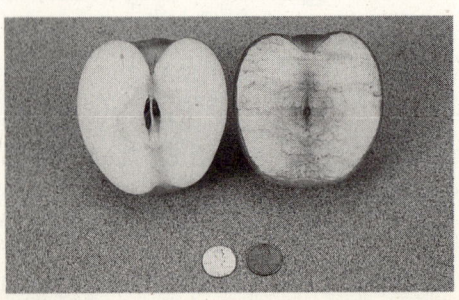

Matt Meadows

Copyright © Glencoe/McGraw-Hill, a division of The McGraw-Hill Companies, Inc.

Picture This

3. **Recognize Cause and Effect** Circle the apple half and the penny in the figure that have undergone a chemical change.

What are some examples of chemical change?

Chemical changes are happening all the time. Chemical changes take place around you and inside you. A chemical change occurs when plants use photosynthesis to make food. When you eat fruits and vegetables produced by photosynthesis, chemical changes occur inside your body. Your body chemically changes these foods so they can be used by your cells.

Some chemical changes occur slowly. Iron rusting is a chemical change. You can't see the process of iron rusting because it happens too slowly. It may take a few years for an object made of iron to rust. ☑

Some chemical changes happen quickly. When you light a gas grill, the gas immediately burns. When matter burns, it changes chemically. Reactions happen at different rates and produce different products but, they are all chemical changes.

What forms during a chemical change?

Signs of a physical change are easy to see. Ice melts and paper is cut. Something changes shape, size, form, or state. Signs of a chemical change are not always so easy to see. If a new type of matter forms that is chemically different from the original matter, then a chemical change has happened. New matter must form, or the change is not a chemical one. ☑

Once matter undergoes a chemical change, it is difficult to change it back. When wood combines with oxygen and burns, the wood and oxygen change into ash and gases. You can't put the ash and gases back together to make wood.

When you bake a cake, changes happen that make the liquid batter become solid. This change is more than a physical change. A chemical change happens when the baking powder mixes with water. This mixture forms bubbles that make the cake rise. The raw egg in the batter undergoes chemical changes that makes the egg solid. These changes cannot be reversed.

What are signs of a chemical change?

When wood burns and a cake bakes, you can tell a chemical change occured. You see the new substances. It is not always this easy to tell when a chemical change has occured and new substances have formed. What signs should you look for?

☑ **Reading Check**

4. **Classify** Circle an example of a chemical change.
   a. ice melting
   b. iron rusting
   c. water freezing
   d. a building being destroyed

☑ **Reading Check**

5. **Describe** one thing that must happen for a chemical change to take place.

_____

_____

_____

6. Identify What forms of energy can be given off during a chemical change?

_____

_____

_____

Energy When a chemical change occurs, energy is either given off or taken in. The energy can be in the form of light, heat, or sound. It's easy to tell that energy is given off when something burns. You see the energy as light and feel the energy as heat. ☑

Sometimes the energy change is hard to notice. The change may be quite small or happen very slowly. Remember rust is a chemical change. There is an energy change when an object rusts, but it happens so slowly you don't notice it.

Gases and Solids During some chemical changes a gas or solid forms. In a chemical change, the new gas or solid is a new substance. It is not the result of a change of state.

Chemical and Physical Changes in Nature

A change in color can be a sign of a chemical change. Leaves change color in the fall. Leaves have red, yellow, and orange pigments in them year-round. During the spring and summer, leaves also contain large amounts of green chlorophyll. This chlorophyll hides the colors of the pigments. You see the leaves as green.

In the fall, changes in temperature and rainfall amounts cause trees to stop producing chlorophyll. When chlorophyll is no longer made, the yellow, red, and orange pigments can be seen.

What is physical weathering?

Some physical changes happen quickly. Others take place over a long time. Physical weathering is a physical change that often happens very slowly. Physical weathering is responsible for much of the shape of Earth's surface.

You can see examples of physical weathering in your own school yard. Soil is produced by physical weathering. Wind blows against rocks. Water from rain falls on rocks. Over a long period of time, water and wind break the rocks into smaller and smaller pieces. Finally, the pieces become soil. Waves in the ocean pound on rocks at the shore and break them up. This weathering makes sand.

Water also breaks up rocks in another way. Water fills cracks in rocks. When the temperature is cold enough, the water freezes. When the temperature warms, the water thaws. When this happens over and over, the rock splits into smaller pieces.

💡 **Think it Over**

7. Explain how physical weathering changes the shape of Earth's surface.

_____

_____

_____

_____

No matter how small the pieces of rock are, they are still made of the same material as the larger rocks. The rocks have undergone a physical change. Gravity, plants, animals, and the movement of land during earthquakes also cause physical changes on Earth.

What is chemical weathering?

The figure below shows formations inside a cave. These formations are chemical weathering. Water moves slowly through layers of rock above the cave. Minerals dissolve in the water as the water passes through the rock layers. This water finally reach the cave. When the water evaporates, the minerals are left behind. These minerals build up slowly. They form the icicle shapes, called stalactites, hanging from the cave ceiling. This type of chemical weathering does not happen quickly. It took many, many years for these stalactites to form.

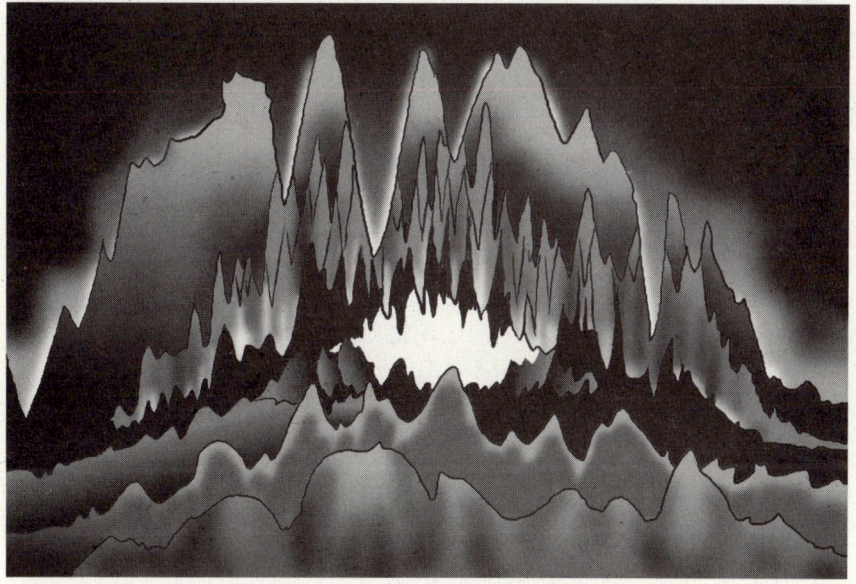

Picture This

8. Observe Circle an example of a stalactite in the figure.

Cave formations happen naturally. Some chemical weathering is not a natural process. The acid in rain is produced by cars, factories, and other industries. Acid rain can chemically weather marble buildings, statues, and other outdoor objects.

Think it Over

9. Identify Give one example of chemical weathering.

_____

_____

● After You Read

Mini Glossary

chemical change: occurs when one type of matter changes into a different type of matter with different properties

physical change: any change in size, shape, form, or state in which the matter itself does not change into a different substance

1. Review the terms and their definitions in the Mini Glossary. Write a sentence that describes a chemical change you have seen.

_____

_____

_____

2. Complete the diagram below to show how matter in one state can change into a different state.

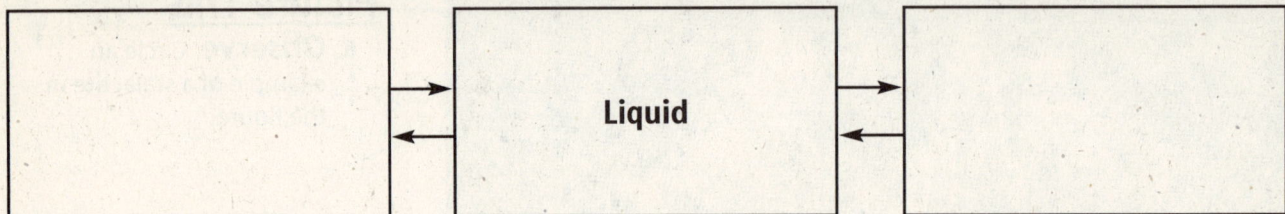

| | | Liquid | | |

3. You were asked to highlight facts about physical changes in one color and to highlight facts about chemical changes in a different color. How would this help you study for a test?

_____

_____

_____

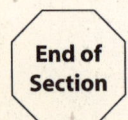

End of Section

Science nline Visit **ips.msscience.com** to access your textbook, interactive games, and projects to help you learn more about physical and chemical changes.

Atomic Structure and Chemical Bonds

section ❶ Why do atoms combine?

● Before You Read

Think of a rock. How would you describe it? Now think of a balloon. How is a balloon different from a rock?

_____

_____

What You'll Learn

■ how electrons are arranged in an atom
■ the amount of energy electrons have
■ how the periodic table is organized

● Read to Learn

Atomic Structure

You might be surprised to learn that all matter contains mostly empty space. Even solids like rocks and metals have empty space. How can this be? Although there might be little or no space between atoms, there is a lot of empty space within each atom.

At the center of an atom is the nucleus. It contains protons and neutrons. The rest of the atom is empty except for electrons. Electrons are extremely small compared to the nucleus. The **electron cloud** is the space around the nucleus where the electrons travel. However, the exact location of any one electron cannot be determined and their paths are not well-defined, as shown in the figure.

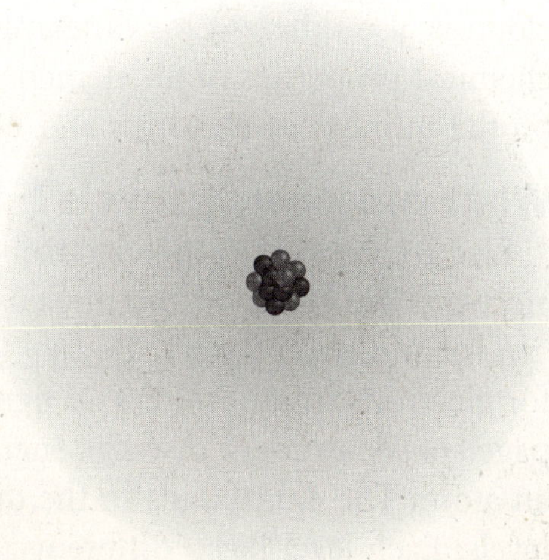

Imagine that the nucleus of an atom is the size of a penny. Electrons would be smaller than grains of dust. The electron cloud would go out as far as 20 football fields.

Mark the Text

Highlight As you read highlight important sentences and terms. When you are finished reading, review what you have highlighted.

FOLDABLES

Ⓐ Organize Information
Make the following layered book Foldable using four sheets of notebook paper to help you organize information about atoms.

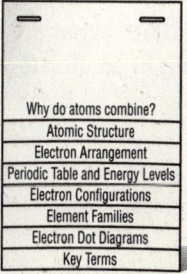

| Why do atoms combine? |
| Atomic Structure |
| Electron Arrangement |
| Periodic Table and Energy Levels |
| Electron Configurations |
| Element Families |
| Electron Dot Diagrams |
| Key Terms |

Copyright © Glencoe/McGraw-Hill, a division of The McGraw-Hill Companies, Inc.

Where in the electron cloud are electrons?

You might think that electrons are like the planets that circle the Sun, but they are not. Planets travel in predictable orbits. You can tell exactly where a planet will be at any time. This is not true for electrons. Although electrons do travel in predictable areas, it is impossible to tell exactly where any one electron will be at any time. So, scientists use a mathematical model to predict where an electron might be.

What makes the atoms of elements different?

The atoms of every element are different. Each element has a certain number of protons, neutrons, and electrons in its atoms. The number of protons is always the same as the number of electrons in neutral atoms. The figure shows a two-dimensional model of the electron structure of a lithium atom. A neutral lithium atom has three protons and four neutrons in its nucleus. Three electrons move around its nucleus.

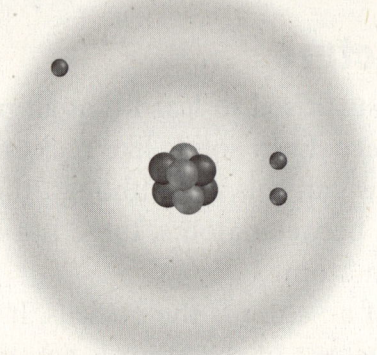

Electron Arrangement

You know that different elements have different numbers of electrons. The way electrons are arranged in the electron cloud is also different for different elements. The physical and chemical properties of an element can be different, depending on the number of electrons and how they are arranged.

What are energy levels?

Electrons move in the electron cloud in an atom. Some electrons are closer to the nucleus than others. So, electrons can be in different areas. <u>Energy levels</u> are the different areas for electrons in an atom. The figure at the top of the next page shows a model of what energy levels might look like in an atom. The dark bands in the diagram represent the energy levels. Each level has a different amount of energy. ☑

Picture This

1. Identify Circle the electrons in the figure.

 **Reading Check**

2. Define What are energy levels in an atom?

_____

_____

_____

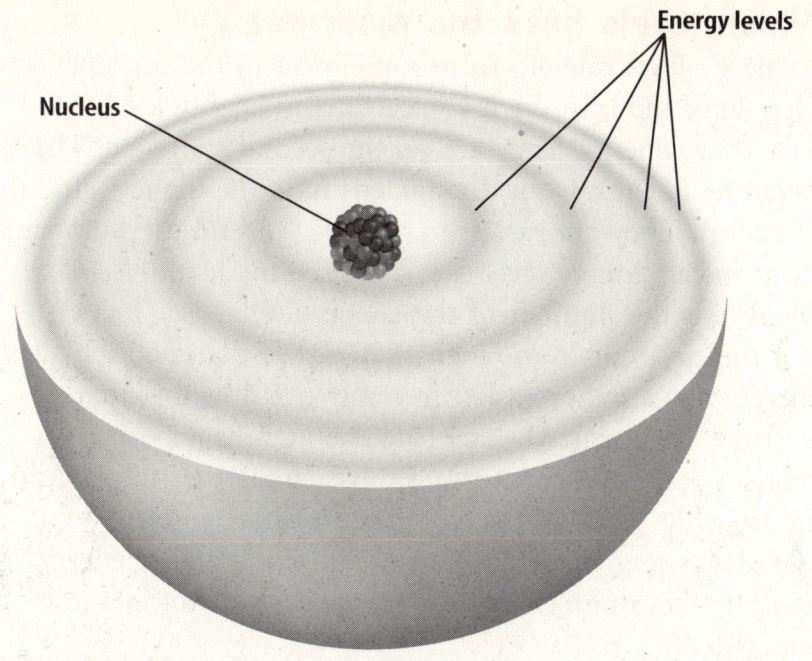

Nucleus

Energy levels

Picture This

3. **Label** energy level one on the model of the atom.

How many electrons are in an energy level?

Each energy level can hold only a certain number of electrons. Energy levels farther away from the nucleus can hold more electrons. Energy levels close to the nucleus hold fewer electrons. Energy level one, the closest to the nucleus, can hold one or two electrons. Energy level two can hold up to eight electrons. Energy level three can hold up to 18 electrons. Energy level four can hold up to 32 electrons.

How can energy levels be represented?

Look at the stairway in the figure below. This stairway shows the maximum number of electrons each energy level can hold. The electrons in each energy level have different amounts of energy.

Picture This

4. **Apply** Chlorine has 17 electrons. What is the highest energy level in which you would find electrons in a chlorine atom?

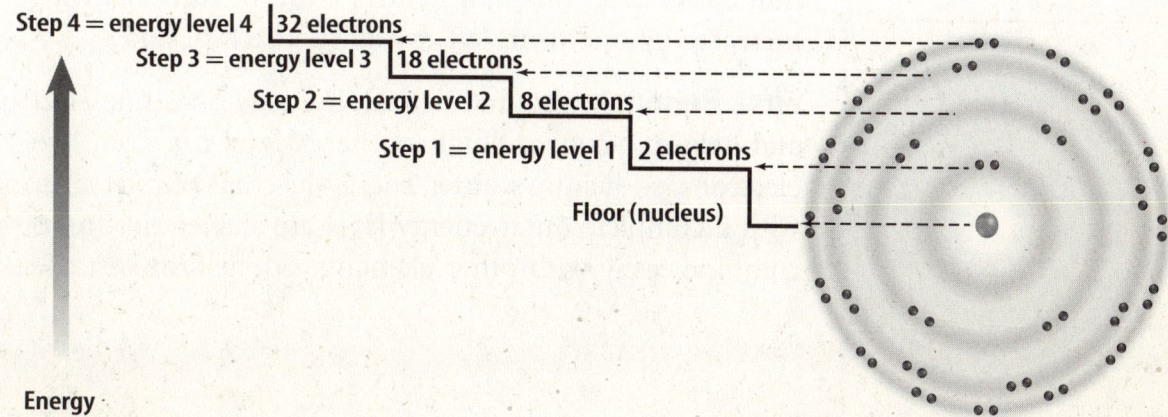

Step 4 = energy level 4 | 32 electrons
Step 3 = energy level 3 | 18 electrons
Step 2 = energy level 2 | 8 electrons
Step 1 = energy level 1 | 2 electrons
Floor (nucleus)

Energy

_____

_____

_____

Which levels have the most energy?

Energy level one electrons are closest to the nucleus. They have the least energy. Electrons in energy level two have more energy. They are on the second stair step. The electrons farthest from the nucleus have the highest amount of energy. They are easiest to remove. To find the number of electrons an energy level can hold, use the formula $2n^2$, where n is the number of the energy level.

If the electrons in the highest energy level have the highest energy, why are they easiest to remove? Remember that electrons are negatively charged. The nucleus is positively charged because it contains protons. The farther the electrons are from the nucleus, the less they are attracted to it. So, it takes less energy to remove an electron in a higher energy level. It takes more energy to remove an electron in a lower energy level.

Periodic Table and Energy Levels

The periodic table has a lot of data about the elements. You can use it to understand energy levels, too. Remember that the atomic number of an element is the number of protons in an atom of the element. The number of protons is equal to the number of electrons in a neutral atom. So, you can look at the atomic number of an element and find how many protons and electrons it has. For example, oxygen is atomic number 8. This means that oxygen has eight protons in its nucleus and eight electrons. ✓

_____

_____

_____

Electron Configurations

In the periodic table, the elements are arranged in a certain order. Part of the periodic table is shown on the next page. Look at the horizontal rows, or periods. The number of electrons in a neutral atom of each element increases by one from left to right in each period.

First Period In the first period, hydrogen has one electron and helium has two. The first energy level can hold two electrons, so helium's outer energy level is complete. Atoms with a complete outer energy level are stable and do not combine easily with other elements. So, helium is stable.

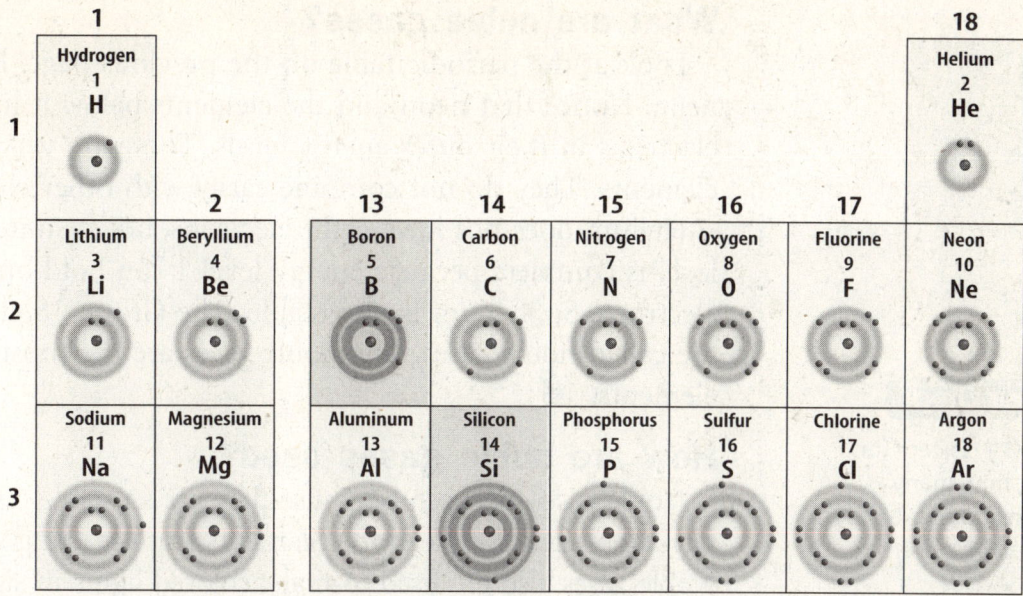

Second Period In the second period, or row, lithium has three electrons. Two electrons fill energy level one. This leaves only one electron for energy level two. Energy level two can hold up to eight electrons. Look at each element to the right of lithium. The electrons begin to fill energy level two. On the right side of the periodic table, neon has a total of 10 electrons. Two are in energy level one. Eight are in energy level two. Because the outer energy level of neon is complete, neon is a stable element.

Third Period In the third period, the electrons begin filling energy level three. On the right side, argon has eight electrons in energy level three. Is it full? No, energy level three can hold 18 electrons. An atom with exactly eight electrons in its outer energy level is stable. Argon is a stable element. Each period in the periodic table ends with a stable element.

Element Families

Each column in the periodic table is an element group, or family. The first element family begins with lithium. Hydrogen is separate from the first family. The elements in each family have the same number of electrons in their outer energy level. So the members of each element family have similar chemical properties. ☑

Dmitri Mendeleev, a Russian chemist, noticed this repeating pattern of properties. In 1869, he created his first periodic table of elements using this pattern. Mendeleev's table is much like the one used today.

Picture This

7. **Identify** Circle all of the elements that have three electrons in their outer energy level.

✔ **Reading Check**

8. **Determine** What is true about elements in a family?

_____

_____

_____

What are noble gases?

Look at the periodic table on the previous page. Find neon. Notice that neon and the elements below it have eight electrons in their outer energy levels. These are very stable elements. They do not combine easily with other elements.

Helium does not have eight electrons, but its outer energy level is complete because energy level 1 can hold only two electrons. So, helium is also stable. The Group 18 elements are called noble gases. The noble gases are the most stable elements. ☑

How are noble gases used?

Noble gases are useful because they are so stable. Lightbulbs have noble gases to keep the filaments from reacting with air. Noble gases also are used to make colored lights in signs. If an electric current passes through the noble gases, they glow with different colors. Neon makes orange-red, argon makes lavender, and helium makes a yellow-white light.

What are halogens?

The elements in Group 17 are called halogens. Look at the elements in Group 17 in the periodic table on the previous page. Notice that their outer energy levels have seven electrons. If they gain one electron, they become stable. The halogens are very reactive.

Fluorine is the most reactive halogen because its outer energy level is closest to the nucleus. The other halogens get less reactive as their outer energy levels get farther away from the nucleus. So, chlorine in period 3 is less reactive than fluorine in period 2.

What are alkali metals?

Look at the elements lithium and sodium in the periodic table on the previous page. Except for hydrogen, the elements in this group are called alkali metals. Alkali metals have one electron in their outer energy levels. When these elements react, this one electron in the outer energy level is removed.

Remember that it is easier to remove electrons that are farther away from the nucleus because less energy is needed. Elements at the bottom of the group give up the one electron in their outer energy levels more easily than those at the top. The elements toward the bottom of this group are more reactive. For example, sodium in period 3 is more reactive than lithium in period 2. This is the opposite of the halogens.

9. Identify Except for helium, how many electrons do noble gases have in their outer energy levels?

_____

_____

Think it Over

10. Infer Think about the electrons in the outer energy levels of halogens and alkali metals. Why do you think that halogens and alkali metals would react very well with each other?

_____

_____

_____

_____

Electron Dot Diagrams

You have read that the properties of an element depend on the number of electrons in its outer energy level. That's why chemists often make models of atoms showing only the electrons in the outer energy level. These models are called electron dot diagrams. An **electron dot diagram** is the chemical symbol for an element surrounded by as many dots as there are electrons in its outer energy level.

Why not draw dots for all of the electrons in an atom? You could do that. But for some elements, you would have to draw a lot of dots. For example, iodine has 53 electrons. So, you would have to draw 53 dots. What really matters is the number of electrons in the outer energy level. These are the electrons that determine how an element can react. Iodine has seven electrons in the outer energy level. Drawing seven dots is easier than drawing 53 dots. ☑

How do you write electron dot diagrams?

How many dots do you draw? Where do you put them? First, you need to know how many electrons are in the outer energy level. You can use the periodic table to find out. To make the diagram, write the symbol for the element. For boron, you write the letter B. Boron has three electrons in its outer energy level. So, you need to draw three dots. Start by making one dot above the letter B. Next, go clockwise and draw a dot to the right of the B. Then, draw a dot below the B. If there are more electrons, keep drawing dots clockwise around the symbol. If there are more than four electrons, draw dots in pairs. The diagram below shows electron dot diagrams for boron, carbon, nitrogen, and oxygen.

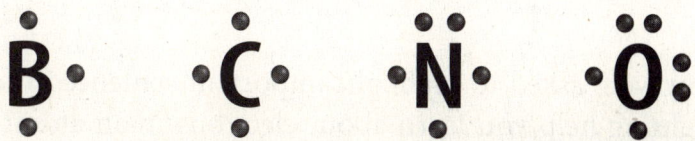

How do you use electron dot diagrams?

You can use electron dot diagrams to show how atoms bond with each other. A **chemical bond** is the force that holds two atoms together. Chemical bonds hold atoms together like glue holds things together. Atoms bond with other atoms in ways that make each atom more stable. That means the outer energy levels of bonded atoms becomes like those of the noble gases.

☑ **Reading Check**

11. **Explain** What does the number of electrons in the outer energy level of an atom show?

_____

_____

Picture This

12. **Identify** How many electrons does oxygen have in its outer energy level?

_____

● After You Read

Mini Glossary

chemical bond: the force that holds two atoms together

electron cloud: the space around the nucleus where the electrons travel

electron dot diagram: the chemical symbol for an element surrounded by as many dots as there are electrons in its outer energy level

energy levels: the different areas for electrons in an atom

1. Review the terms and their definitions in the Mini Glossary. Where in an atom are the energy levels located? Answer in a complete sentence.

   _____

   _____

   _____

2. Below is the electron dot diagram for nitrogen and the symbol P for the element phosphorus. Both nitrogen and phosphorus are in Group 15 on the periodic table. Complete the electron dot diagram for phosphorus.

   :
   •N•
   •

   P

3. Look at the dot diagrams above. What do nitrogen and phosphorus have in common?

   _____

   _____

   _____

4. At the beginning of the section, you were asked to highlight important sentences and terms as you read. How did highlighting help you learn about electrons in an atom?

   _____

   _____

   _____

End of Section

Science Online Visit **ips.msscience.com** to access your textbook, interactive games, and projects to help you learn more about how atoms combine.

Atomic Structure and Chemical Bonds

section ② How Elements Bond

● Before You Read

What does it mean when two things are bonded? What are some things you might use to bond two items?

_____

_____

● Read to Learn

Ionic Bonds—Loss and Gain

Elements that join by chemical bonds do not fall apart easily. Atoms form bonds by using the electrons in their outer energy levels. Elements can bond in four different ways—they can lose electrons, gain electrons, pool electrons, or share electrons with another element.

Lose Electrons Sodium chloride forms when sodium and chlorine atoms bond as shown below. Sodium is a soft, silvery metal. It is a member of the alkali metal family. It reacts violently when added to water or chlorine. Sodium is so reactive because it has only one electron in its outer energy level. Removing this one electron empties the outer energy level. What is left is the completed energy level below it. Sodium is now stable, like neon. Remember, neon is a stable noble gas.

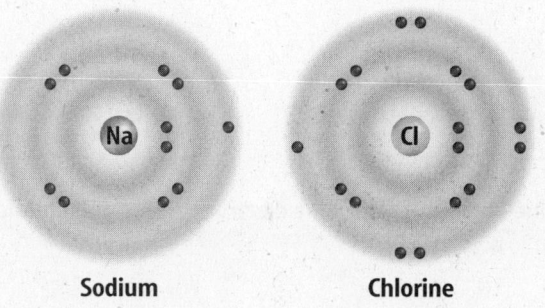

Sodium   Chlorine

What You'll Learn
- about ionic and covalent bonds
- about compounds and molecules
- polar and nonpolar covalent bonds
- chemical shorthand

Study Coach

Create a Quiz As you read the section, write quiz questions about the main ideas and vocabulary terms. When you finish reading, answer your quiz questions.

FOLDABLES

ⓑ Find Main Ideas Make the following Foldable out of notebook paper. Make quarter-sheets and a half-sheet to list the main ideas on covalent bonds, chemical shorthand, and how atoms become stable.

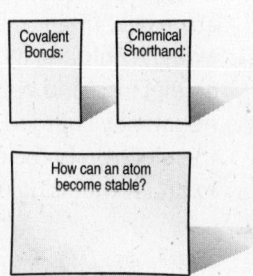

Gain Electrons Chlorine bonds in the opposite way. It gains an electron. Chlorine has seven electrons in its outer energy level. It gains an electron to have a complete outer energy level. When this happens, chlorine becomes stable. It then has the same number of electrons as the noble gas argon.

What are ions?

When a sodium atom loses an electron, it becomes more stable. But, the atom has one fewer electron than protons. This changes the balance of electric charge. The atom becomes a positively charged ion. When a chlorine atom gains an electron, it has one more electron than protons. This makes the chlorine atom a negatively charged ion. An **ion** (I ahn) is an atom that is no longer neutral because it has gained or lost an electron. The figure below shows how sodium ions (Na^+) and chloride ions (Cl^-) are formed. ✔

✔ **Reading Check**

1. **Describe** Which of the following does not describe an ion: positive, negative, or neutral?

_____

How Ions Form

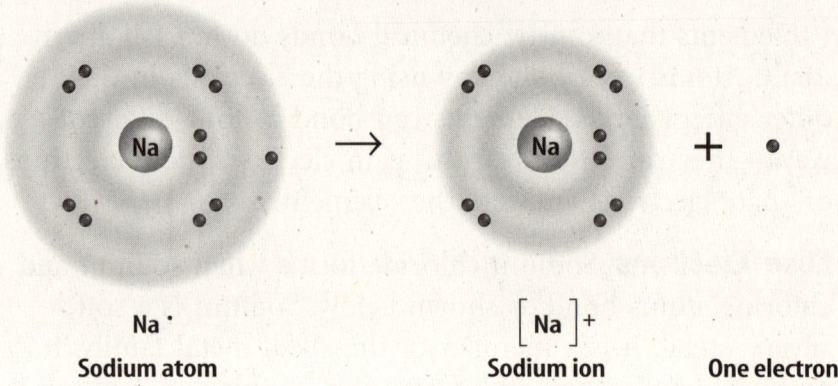

Na ⋅
Sodium atom

$\left[Na \right]^+$
Sodium ion

One electron

Picture This

2. **Draw and Label** A sodium atom loses an electron and becomes positively charged. Circle the electron in the sodium atom (Na) that is taken away. A chlorine atom gains an electron and becomes negatively charged. Circle the electron in the chloride ion (Cl–) that is gained.

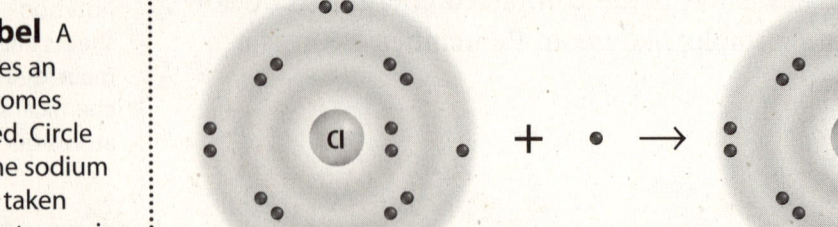

:Cl⋅
Chlorine atom

One electron

$\left[:Cl:\right]^-$
Chloride ion

How do ions form bonds?

Positive sodium ions and negative chloride ions are strongly attracted to each other. An **ionic bond** is the attraction that holds negative ions and positive ions close together.

The figure below shows how sodium and chloride ions form an ionic bond. The compound sodium chloride, or table salt, is formed. A **compound** is a pure substance that has two or more elements that are chemically bonded.

$$Na^\bullet \quad + \quad \cdot \ddot{\underset{\cdot\cdot}{Cl}} : \quad \rightarrow \quad [Na]^+ \left[: \ddot{\underset{\cdot\cdot}{Cl}} : \right]^-$$

Sodium ion   Chloride ion   Sodium Chloride

Can elements gain or lose more than one electron?

The element magnesium (Mg) in Group 2 has two electrons in its outer energy level. Magnesium can lose these two electrons to have a completed outer energy level. The symbol for a magnesium ion is Mg^{2+}. The 2+ shows that the ion has lost two electrons.

One Ionic Bond When magnesium loses its two electrons, the electrons could be gained by an oxygen atom. Oxygen needs to gain two electrons to become stable. So, a magnesium ion, Mg^{2+}, and an oxide ion, O^{2-}, can form an ionic bond to make magnesium oxide (MgO). The 2+ charge of the magnesium ion and the 2− charge of the oxide ion balance each other.

Two Ionic Bonds A single magnesium ion (Mg^{2+}) also can bond with two chlorine ions (Cl^-). The 2+ charge of the magnesium ion is balanced by the two negative charges of the two chlorine ions. Each chlorine ion gains one electron. This ionic bond between magnesium and chlorine forms the compound magnesium chloride ($MgCl_2$). ☑

Metallic Bonding—Pooling

In the examples above, a metal formed ionic bonds with nonmetals. Metals can form bonds with other metal atoms, but in a different way.

Picture This

3. **Use a Diagram** In the first dot diagram, notice the dot next to the symbol for sodium, Na. Draw an arrow from that dot to the place in the dot diagram for chlorine, Cl, where it will be located in sodium chloride.

✔ Reading Check

4. **Explain** In magnesium chloride, what balances the 2+ charge of the magnesium ion?

_____

_____

How do metals bond?

In metal atoms, the electrons are not held tightly to the outer energy levels of the atoms. Instead, they move freely among all the metal ions. These moving electrons form a pool of shared electrons. The figure below is an example of a pool of electrons in the metal silver. **Metallic bonds** form when metal atoms share their pooled electrons.

Picture This

5. **Describe** How would you describe the outer electrons of the silver atoms—attached to certain atoms or free?

_____

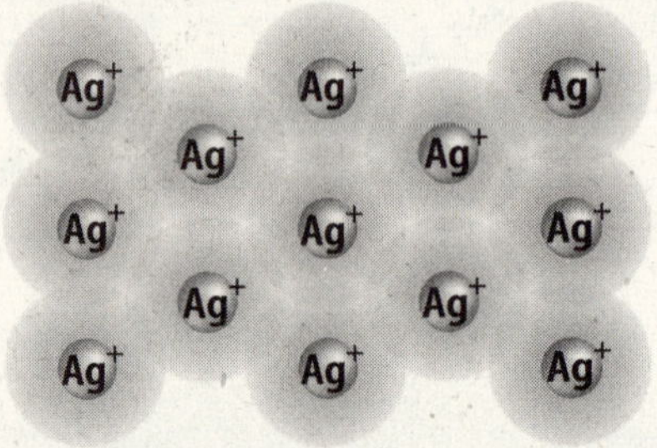

_____

What are some properties of metals?

Metallic bonds cause metals to have special properties. The pooled electrons let the atoms slide past each other to stretch and not break. So, metals can be hammered into sheets without breaking. They can also be drawn into wires without breaking. Metallic bonds also let metals conduct electricity well. An electric current in solids is a flow of electrons. The outer electrons in metal atoms move easily from one atom to the next to conduct electric current.

Covalent Bonds—Sharing

Some atoms don't gain or lose electrons very easily. For example, carbon has six total electrons. Four of these electrons are in the outer energy level. To be more stable, carbon would either have to gain four electrons or lose four. Losing or gaining four electrons would take a lot of energy. But, carbon can form a bond by sharing electrons.

What is a covalent bond?

Atoms of many elements become more stable by sharing electrons. A **covalent** (koh VAY luhnt) **bond** is a chemical bond that forms between nonmetal atoms when they share electrons. Shared electrons are attracted to the nuclei of both atoms. They move back and forth between the outer energy levels of each atom in the covalent bond. So, each atom is stable some of the time. Compounds held together with covalent bonds are called molecular compounds. ✔

✔ **Reading Check**

6. **Explain** What type of atoms share electrons on a covalent bond?

_____

_____

Neutral Particles The atoms in a covalent bond form a neutral particle. The particle is neutral because it has the same number of positive and negative charges. A **molecule** (MAH lih kyewl) is the neutral particle formed when atoms share electrons. A molecule is the basic unit of a molecular compound. The figure below shows how molecules form by sharing electrons. Notice in the figure that none of the atoms are ions. That's because no electrons are gained or lost when a molecule forms. Solids that are crystals, such as sodium chloride, are not called molecules, because their basic units are ions, not molecules.

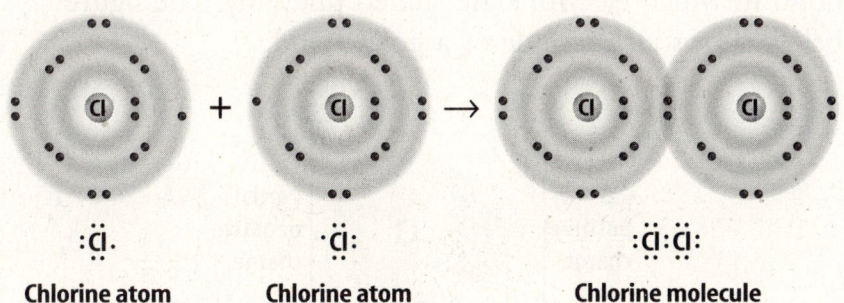

| Chlorine atom | Chlorine atom | Chlorine molecule |

Picture This

7. **Evaluate** Look at the figure. How many electrons are shared between the two chlorine atoms in a chlorine molecule?

_____

What are double and triple bonds?

Sometimes an atom shares more than one electron with another atom. Look at the molecule of carbon dioxide shown below. Each oxygen atom shares two electrons with the carbon atom. The carbon atom shares two of its electrons with each oxygen atom. When two pairs of electrons form a covalent bond, it is called a double bond. A triple bond happens when three pairs of electrons are shared in a covalent bond. The nitrogen molecule in the figure below is an example of a triple bond.

| Carbon atom | Oxygen atoms | Carbon dioxide molecule |

| Nitrogen atoms | Nitrogen molecule |

Picture This

8. **Interpret Scientific Illustrations** How many pairs of electrons are shared between the nitrogen atoms in the triple bond of the nitrogen molecule?

_____

Polar and Nonpolar Molecules

Atoms in a covalent bond don't always share electrons equally. Some atoms have a greater attraction for electrons than others do. For example, hydrogen and chlorine can form a covalent bond. But, chlorine attracts electrons more strongly than hydrogen does. So, the shared electron pair spends more time around the chlorine atom than the hydrogen atom.

Since the electron pair spends more time around the chlorine atom, this end of the molecule has a slight negative charge. The hydrogen end of the molecule has a slight positive charge. This happens because the hydrogen atom is without its electron most of the time. A **polar bond** is a bond in which electrons are shared unevenly. The figure below shows an example of a polar bond.

Partial positive charge — H Cl — Partial negative charge

Picture This

9. **Use Models** How would you describe the sharing of the electrons in the hydrogen chloride molecule—even or uneven?

_____

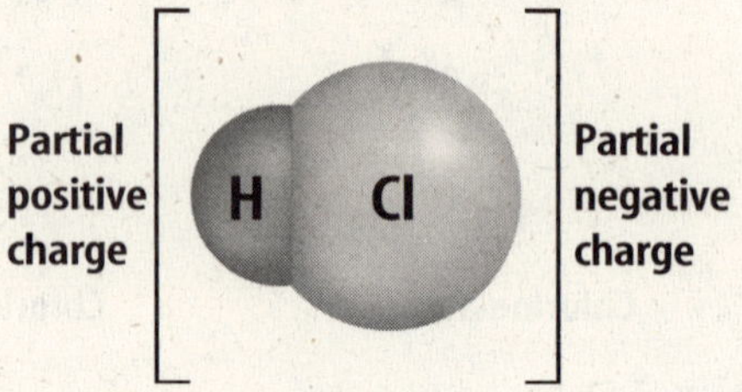

What makes water molecules polar?

Water molecules form when hydrogen and oxygen share electrons. Water molecules are polar because the electrons are shared unevenly. The oxygen atom has a greater share of the electrons than the hydrogen atom. Look at the figure below. Water molecules can be attracted to positive and negative charges because they are polar. Many of the physical properties of water are due to the fact that the molecule is polar. Molecules that do not have uneven charges are called nonpolar molecules. An example of a nonpolar bond is the triple bond in a nitrogen molecule.

Think it Over

10. **Explain** Water molecules attract each other because they are polar. Many water molecules are attracted to many other water molecules. Explain which charges are attracted to each other.

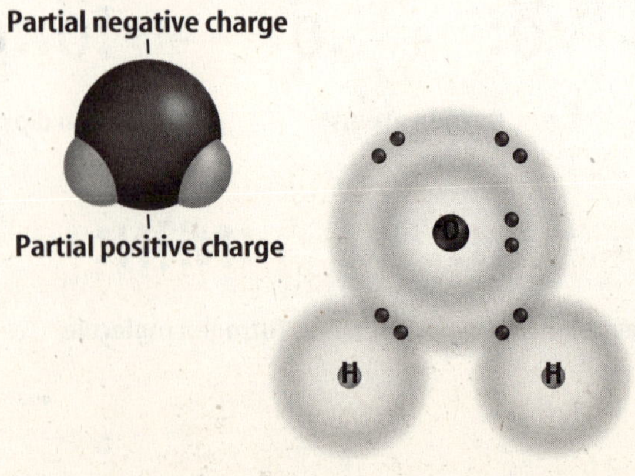

Partial negative charge

Partial positive charge

Chemical Shorthand

In medieval times, alchemists (AL kuh mists) were the first to study chemistry. The alchemists learned about the properties of some elements. They used symbols to represent elements and chemical processes. Scientists today use symbols, too. The table below shows both ancient and modern symbols for some elements. The periodic table includes the symbol for each element. The symbols are usually one or two letters. Often, the symbol is the first letter of the name for the element. For example, the symbol for oxygen is O. Sometimes, the name for the element in another language is used. For example, the symbol for potassium is K. The Latin word for potassium is kalium.

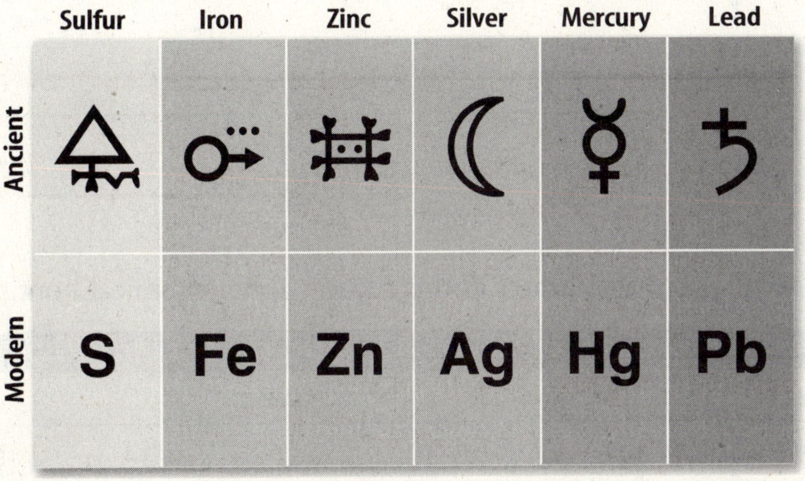

	Sulfur	Iron	Zinc	Silver	Mercury	Lead
Ancient						
Modern	S	Fe	Zn	Ag	Hg	Pb

Picture This

11. **Determine** Which is probably most easily understood by people today, the ancient symbols or the modern symbols?

_____

What are chemical formulas?

Symbols and numbers are used to show the elements in compounds. A **chemical formula** is a set of chemical symbols and numbers that shows which elements are in a compound and how many atoms of each element are in it. For example, two hydrogen atoms in a covalent bond are represented by the chemical formula H_2. The H stands for hydrogen. The subscript 2 tells you that there are two hydrogen atoms. A subscript is a number that is written a little below a line of text. Another example of a chemical formula is H_2O, or water. The formula tells you there are two hydrogen atoms and one oxygen atom in a water molecule. Notice that when symbols don't have a subscript, like the O in H_2O, there is only one atom. ☑

☑ **Reading Check**

12. **Infer** What does a chemical formula tell you about a compound?

_____

_____

_____

_____

● After You Read

Mini Glossary

chemical formula: a set of chemical symbols and numbers that shows which elements are in a compound and how many atoms of each element are in it

compound: a pure substance that has two or more elements that are chemically bonded

covalent (koh VAY luhnt) bond: a chemical bond that forms between atoms when they share electrons

ion (I ahn): an atom that is no longer neutral because it has gained or lost an electron

ionic bond: attraction that holds negative ions and positive ions close together

metallic bond: a chemical bond that forms when metal atoms share their pooled electrons

molecule (MAH lih kyewl): a neutral particle formed when atoms share electrons

polar bond: a bond in which electrons are shared unevenly

1. Review the terms and their definitions in the Mini Glossary. Choose two terms that describe different kinds of chemical bonds and write a sentence or two that tells how they are different.

_____

_____

2. Fill in the table below to summarize what you learned in this section about chemical bonds.

Chemical Bonds		
Type of Bond	Reaction	Example
Ionic bond	A negative ion and a positive ion come together.	
Metallic bond		
Covalent bond		hydrogen, water
Polar bond		

End of Section

Science Online Visit **ips.msscience.com** to access your textbook, interactive games, and projects to help you learn more about how elements bond.

chapter 7 Chemical Reactions

section ❶ Chemical Formulas and Equations

● Before You Read

What happens to wood when it burns? Describe the changes on the lines below.

_____

_____

● Read to Learn

Physical or Chemical Change?

Have you ever seen smoke from a campfire? Smoke is a clue that a chemical reaction is taking place. There are always clues when a chemical reaction is happening.

Matter can change in two ways. It can have a physical change or a chemical change. Physical changes only affect physical properties. For example, the newspaper in the first figure is folded. It is a different shape, but it is still a newspaper. This is a physical change.

Chemical changes produce new substances. The newspaper in the second figure is burning. Burning is a chemical change because new substances are produced. The properties of the new substances are different from the properties of the original substances. A **chemical reaction** is a process that produces chemical change.

Physical Change

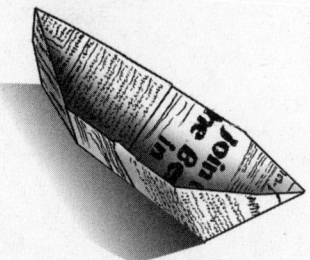

Chemical Change

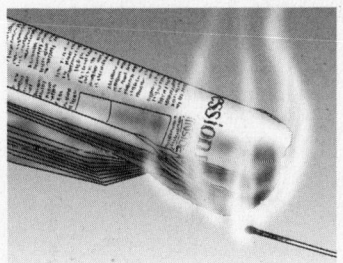

What You'll Learn

- identify a chemical reaction
- how to read a balanced chemical equation
- how reactions release or absorb energy
- the law of conservation of mass

▶ **Mark the Text**

Highlight Chemical Equations Highlight each chemical equation in the section. Read the chemical equation two or three times to make sure you understand what is happening in the chemical reaction.

FOLDABLES™

Ⓐ Organize Information Make the following Foldable to help you organize information in this section. Write information about each topic in the Foldable.

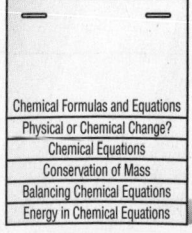

| Chemical Formulas and Equations |
| Physical or Chemical Change? |
| Chemical Equations |
| Conservation of Mass |
| Balancing Chemical Equations |
| Energy in Chemical Equations |

What are clues that a chemical reaction is happening?

You can use your senses to help you know if a chemical reaction is happening. When you watch a firefly glow, you are watching a chemical reaction. Two chemicals combine in the firefly's body and give off light. Your senses of touch and smell help detect chemical reactions in a fire. You smell the smoke and feel the heat. Have you ever tasted sour milk? If so, you have tasted the results of a chemical reaction. You can also hear a chemical reaction happening. The hissing sound of burning firewood is from a chemical reaction.

Chemical Equations

How can you describe a chemical reaction? First, you need to know which substances are reacting. You also need to know which substances are formed in the reaction. The substances that react are called reactants (ree AK tunts). **Reactants** are the substances that exist before the reaction starts. **Products** are the substances that are formed in the reaction.

Look at the figure below. A chemical reaction happens when you mix baking soda and vinegar. It bubbles and foams. The bubbles tell you that a chemical reaction happened.

Baking soda and vinegar are the common names for the reactants in this reaction. They also have chemical names. Baking soda is sodium hydrogen carbonate (often called sodium bicarbonate). Vinegar is a solution of acetic (uh SEE tihk) acid in water. What are the products of the reaction? You can see that bubbles form. What else is happening?

Chemical reaction with baking soda and vinegar

Think it Over

1. **Summarize** What senses can you use to find clues that a chemical reaction is happening?

_____

_____

Picture This

2. **Describe** what you can see in the chemical reaction between vinegar and baking soda.

_____

_____

What is a chemical equation?

The bubbles from the baking soda and vinegar tell you a gas was produced. But, they do not tell you what kind of gas. Are bubbles of gas the only product? More happens in a chemical reaction than you can see with your eyes. Chemists try to find out what reactants are used and what products are formed in a chemical reaction. Then, they write a chemical equation. A **chemical equation** tells chemists the reactants, products, and amounts of each substance in the reaction. Some equations tell the physical state of a substance. ☑

How do words describe a chemical reaction?

Words can be used in an equation to name the reactants and products in a chemical reaction. The reactants in the equation are listed on the left side of an arrow. The reactants have plus signs between them. The products are on the right side of the arrow and also have plus signs between them. The arrow stands for the changes that happen during the chemical reaction. The arrow means *produces*.

You can begin to think of changes as chemical reactions even if you do not know the names of all the substances in the reaction. The table below shows word equations for chemical reactions you might see around your home.

Reactions Around the Home		
Reactants		**Products**
Baking soda + Vinegar	→	Gas + White solid
Charcoal + Oxygen	→	Ash + Gas + Heat
Iron + Oxygen + Water	→	Rust
Silver + Hydrogen sulfide	→	Black tarnish + Gas
Gas (kitchen range) + Oxygen	→	Gas + Heat
Sliced apple + Oxygen	→	Apple turns brown

When are chemical names used?

Chemical names are usually used in word equations instead of common names like baking soda and vinegar. In the baking soda and vinegar reaction, you already know that the chemical names are sodium hydrogen carbonate and acetic acid. The chemical names of the products are sodium acetate, water, and carbon dioxide. The word equation for the reaction is:

Acetic acid + Sodium hydrogen carbonate →
    Sodium acetate + Water + Carbon dioxide

3. Explain What do chemists learn from chemical equations?

_____

_____

_____

Picture This

4. Apply Use words to write a chemical equation for what happens when you peel a banana but do not eat it right away.

_____

_____

Why do chemists use chemical formulas?

The word equation for the reaction of baking soda and vinegar is long. So, chemists replace the chemical names with chemical formulas in the equation. The chemical equation for the reaction between baking soda and vinegar is:

$$CH_3COOH + NaHCO_3 \rightarrow CH_3COONa + H_2O + CO_2$$

| Acetic acid (vinegar) | Sodium hydrogen carbonate (baking soda) | Sodium acetate | Water | Carbon dioxide |

What are subscripts?

Look at the small numbers in the formula above. These numbers are called subscripts. They tell you the number of atoms of each element in that compound. For example, the subscript 2 in CO_2 means each molecule of carbon dioxide has two oxygen atoms. If an atom has no subscript, then there is only one atom of that element in the compound. There is only one atom of carbon in carbon dioxide.

Conservation of Mass

What happens to the atoms in the reactants when they are changed into products? The law of conservation of mass says that the mass of the products has to be the same as the mass of the reactants. French chemist Antoine Lavoisier proved that nothing is lost or created in chemical reactions. Chemical equations are like math equations. In math equations, the right and left sides of the equation are equal. In chemical equations, the number and kind of atoms are equal on both sides. The figure shows that every atom that is on the reactant side of the equation is also on the product side.

Think it Over

5. Identify How many hydrogen atoms are there in sodium acetate? How many sodium atoms (Na) are there?

Picture This

6. Identify How many hydrogen atoms are in the reactants in the figure? How many hydrogen atoms are in the products?

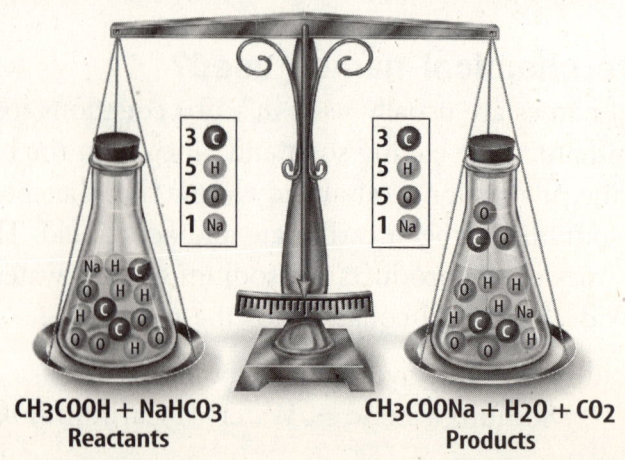

CH3COOH + NaHCO3
Reactants

CH3COONa + H2O + CO2
Products

Balancing Chemical Equations

You need to follow the law of conservation of mass when you write a chemical equation. Look back at the figure on the previous page. Count the number of each type of atom on each side of the scale. There are equal numbers of each kind of atom on each side. This means the equation is balanced. So, the law of conservation of mass is followed. ☑

Not all chemical equations are balanced so easily. The following unbalanced equation shows what happens when silver tarnishes by reacting with sulfur compounds in air.

$$Ag + H_2S \rightarrow Ag_2S + H_2$$

Silver Hydrogen   Silver   Hydrogen
        sulfide   sulfide

How do you balance an equation?

Count the number of each type of atom in the reactants and products above. The reactants and products have the same numbers of hydrogen and sulfur atoms. But, there is one silver atom on the reactant side and two silver atoms are on the product side. This cannot be true. A chemical reaction cannot create a silver atom. This equation does not represent the reaction correctly. Place a 2 in front of the reactant Ag. Now see if the equation is balanced. Count the number of atoms of each type again.

$$2Ag + H_2S \rightarrow Ag_2S + H_2$$

The equation is now balanced. There are equal numbers of silver atoms in the reactants and the products. To balance chemical equations, numbers are placed before the formulas. These numbers, called coefficients, show how many molecules of a compound there are. Never change the subscripts in a formula. This changes the identity of the compound.

Practice balancing equations with the following:

$$CH_4 + O_2 \rightarrow CO_2 + H_2O$$

Count the number of carbon, hydrogen and oxygen atoms on each side. There are 2 more hydrogen atoms in the reactants. Multiply H_2O by 2 to give 4 hydrogen atoms.

$$CH_4 + O_2 \rightarrow CO_2 + 2H_2O$$

Now there are 2 oxygen atoms in the reactants and 4 in the products. Multiply O_2 by 2 to give 4 oxygen atoms. The balanced equation is:

$$CH_4 + 2O_2 \rightarrow CO_2 + 2H_2O$$

✔ Reading Check

7. Explain What law is followed when an equation is balanced?

_____

_____

_____

Applying Math

8. Apply Balance this equation:
$$HCl + Cu \rightarrow CuCl_2 + H_2$$

_____

_____

Energy in Chemical Reactions

Often, energy is released or absorbed in a chemical reaction. A welding torch burns hydrogen and oxygen to produce high temperatures. The energy for the torch is released when oxygen and hydrogen combine to form water.

$$2H_2 + O_2 \rightarrow 2H_2O + \text{energy}$$

Where does released energy come from?

Think about the chemical bonds that break and form when atoms gain, lose, or share electrons. When a chemical reaction happens, bonds break in the reactants. New bonds form in the products. In reactions that release energy, the products are more stable. Their bonds have less energy than the bonds of the reactants. The extra energy is released. It can be released in forms like light, sound, and heat.

How is energy absorbed in a reaction?

In reactions where energy is absorbed, the reactants are more stable. Their bonds have less energy than the bonds in the products.

$$\underset{\text{water}}{2H_2O} + \text{energy} \rightarrow \underset{\text{Hydrogen}}{2H_2} + \underset{\text{Oxygen}}{O_2}$$

In this reaction, electricity supplies the extra energy needed to break water into hydrogen and oxygen, as shown in the figure.

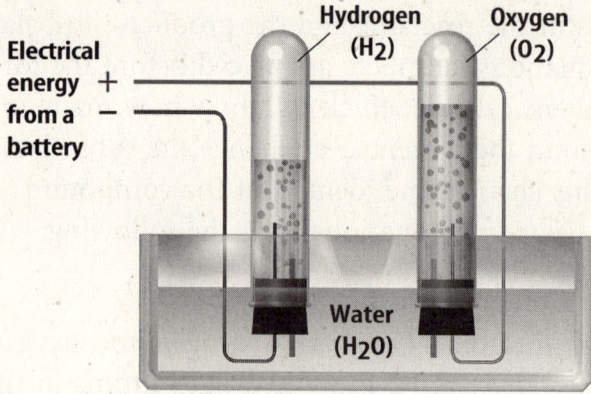

Reactions can release or absorb many forms of energy, including electricity, light, thermal energy, and sound. Special terms are used when thermal energy is gained or lost in reactions. **Endothermic** (en doh THUR mihk) **reactions** absorb it. **Exothermic** (ek soh THUR mihk) **reactions** release it. *Therm* means heat, as in thermometers.

Think it Over

9. **Describe** a reaction where energy is released as thermal energy.

_____

_____

Think it Over

10. **Identify** Which is more stable in this reaction—water or hydrogen?

_____

Picture This

11. **Determine** Where does the electrical energy come from to break water into hydrogen and oxygen atoms?

_____

What is an example of an exothermic reaction?

You probably already know of reactions that release thermal energy. Burning is an exothermic reaction. A substance combines with oxygen to produce thermal energy. Light, carbon dioxide, and water are also produced.

Sometimes thermal energy is released quickly. For example, charcoal lighter fluid combines with oxygen and produces enough of it to start a charcoal fire within a few minutes. Other materials also combine with oxygen, but they release thermal energy so slowly that you cannot see or feel it happen. This is what happens when iron combines with oxygen in the air to form rust.

What is an example of an endothermic reaction?

Sometimes heat energy must be added for a reaction to take place. The way a cold pack works is an example of an endothermic process. A cold pack is made of a thick plastic outer pouch filled with water. The pouch with water surrounds a thin plastic inner pouch filled with ammonium nitrate. When you squeeze the cold pack, the inner pouch breaks. The ammonium nitrate mixes with the water and dissolves. As the ammonium nitrate dissolves, it absorbs thermal energy from its surroundings. So, the cold pack absorbs thermal energy from your skin and your skin feels cold.

How is energy written in an equation?

The word *energy* in equations can be either a reactant or a product. When it is written as a reactant, it is something needed for the reaction to happen. For example, electricity is needed to break water into hydrogen and oxygen. It is important to know that energy must be added for this to happen.

In the equation for an exothermic reaction, energy often is written with the products. This tells you that energy is released. You include energy when you write the reaction that occurs between oxygen and methane in natural gas when you cook on a gas range. This thermal energy is what cooks your food.

$$CH_4 \ + \ 2O_2 \ \rightarrow \ CO_2 \ + \ 2H_2O \ + \ \text{energy}$$
Methane   Oxygen   Carbon   Water
                  dioxide

You don't have to write the word *energy* in an equation. But, if you do, it helps you remember that energy is an important part of the equation.

12. Identify What is released during an exothermic reaction?

_____

_____

Think it Over

13. Describe In what type of equation is energy written with the reactants?

_____

_____

● After You Read

Mini Glossary

chemical equation: a written form that tells the reactants, products, physical state, and amounts of each substance in the reaction

chemical reaction: a process that produces chemical change

endothermic (en doh THUR mihk) reaction: a reaction that absorbs heat energy

exothermic (ek soh THUR mihk) reaction: a reaction that releases heat energy

product: a substance that is formed in the reaction

reactant (ree AK tunt): a substance that is there before the reaction starts

1. Read the key terms and definitions in the Mini Glossary above. In your own words, describe how a reactant and a product are related.

_____

_____

_____

2. Balance each equation in the table.

Equation	Balanced Equation
$H_2 + O_2 \rightarrow H_2O$	
$H_2 + Cl_2 \rightarrow HCl$	
$Al + CuCl_2 \rightarrow AlCl_3 + Cu$	

3. How did highlighting the chemical equations in the section help you to understand chemical equations?

_____

_____

_____

End of Section

Science Online Visit **ips.msscience.com** to access your textbook, interactive games, and projects to help you learn more about chemical formulas and equations.

Chemical Reactions

section ❷ Rates of Chemical Reactions

● Before You Read

Chemical reactions can happen quickly or slowly. Describe a
chemical reaction that happens quickly.

_____

_____

What You'll Learn

- how to measure the speed of a chemical reaction
- how chemical reactions can be speeded up or slowed down

● Read to Learn

How Fast?

Fireworks explode one after another quickly. Old copper
pennies slowly darken. How long you fry an egg makes a
difference in what the yolk is like. These are common
chemical reactions in your life. Time has something to do
with each example. Not all chemical reactions happen at
the same speed.

Some reactions, like fireworks, need help to get going.
Other chemical reactions seem to start by themselves. In
this section, you also will learn what makes reactions speed
up or slow down once they are going.

Activation Energy—Starting a Reaction

Before a reaction can start, molecules of the reactants have
to bump into each other, or collide. The reactants must smash
into each other with a certain amount of energy. If they do
not have enough energy, the reaction will not happen. Why is
this true?

Old bonds must break in the reactants so new bonds can
form. This takes energy. To start any chemical reaction, at
least some energy is needed. **Activation energy** is the energy
needed to start a reaction. Even reactions that release energy
need activation energy to start.

Study Coach

Make Flash Cards For each
paragraph, think of a question
that might be on a quiz. Write
the question on one side of a
flash card. Write the answer on
the other side. Quiz yourself
until you know all the answers.

FOLDABLES

Ⓑ Organize Information
Use two half sheets of paper to
organize information about
activation energy and enzymes
as catalysts.

Enzymes as Catalysts

Reactions Have an
Activation Energy

Gasoline One example of a reaction that needs energy to start is the burning of gasoline. Because gasoline needs energy to start burning, there are signs at filling stations warning you not to smoke. Other signs tell you to turn off your car, not to use mobile phones, and not to get in your car until you are finished fueling.

Reaction Rate

Many physical processes can be measured by rate. A rate tells you how much something changes over a given period of time. For example, you ride your bike at a rate, or speed. Rate is the distance you ride divided by the time it takes you to ride that distance. You may ride at a rate of 20 km/h.

Chemical reactions have rates too. The **rate of reaction** tells how fast a reaction happens after it has started. You can measure the rate of reaction two ways. You can measure how fast one of the reactants is used up. You also can measure how fast one of the products is made. Both measurements tell how the amount of a substance changes per unit of time.

Reaction rate is important to companies that make products. The faster the product can be made, the less it usually costs. But sometimes fast rates of reaction are not good. A fast reaction rate can cause food to spoil quickly. A slower reaction rate will help food stay fresh longer.

How does temperature affect rate?

Increasing Temperature Most chemical reactions speed up when temperature increases. The high temperature inside an oven speeds up the chemical reactions that turn liquid batter into a cake. Atoms and molecules move faster at higher temperatures, as shown in the figure below. Faster molecules crash into each other more often than slower molecules. They also crash with more energy. These crashes often have enough energy to break the old bonds.

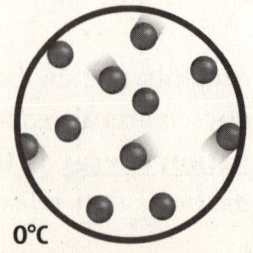
0°C

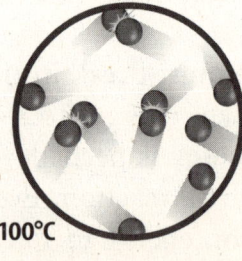
100°C

💡 **Think it Over**

1. **Apply** Why do you think that the faster a product can be made, the less it usually costs?

_____

_____

_____

Picture This
2. **Analyze** At which temperature are the molecules moving slower?

_____

Decreasing Temperature Lowering the temperature slows down most reactions. Food spoiling is a chemical reaction. Putting food in a refrigerator or freezer lowers its temperature. This slows the rate of reaction.

How does concentration affect rate?

When reactant atoms and molecules are closer together, there is a greater chance they will collide and react faster. Think about a crowd of people leaving a baseball game. When you try to walk through the crowd, you will probably bump into people. If it were not so crowded, you would be less likely to bump into people.

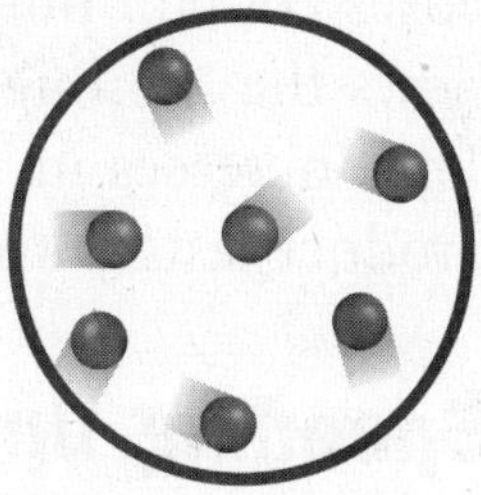

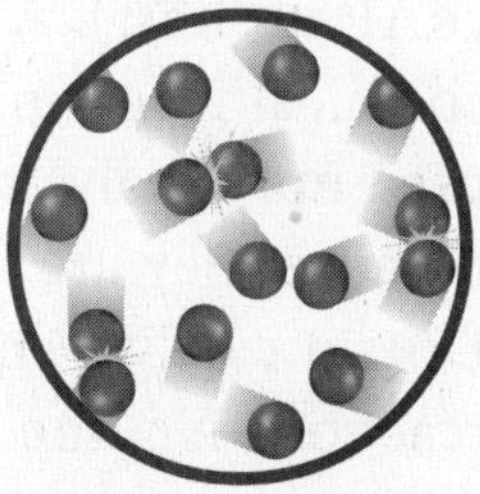

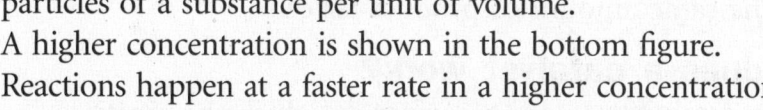

Concentration is the amount of substance in a certain volume. The figure on top shows molecules at a low concentration. If you increase the concentration, you increase the number of particles of a substance per unit of volume. A higher concentration is shown in the bottom figure. Reactions happen at a faster rate in a higher concentration.

How does surface area affect rate?

Surface area also affects how fast a reaction happens. You can quickly start a campfire with small twigs, but starting a fire with only large logs would probably not work.

The figure on the left below shows the iron atoms in a steel beam. Most of the iron atoms are stuck inside the beam. They cannot react. Only the atoms in the outer layer of the reactant material can react. If more molecules are out in the open, the reaction speeds up. This happens with the iron atoms in steel wool, a mass of woven steel fibers, in the figure on the right below. Because more of its iron atoms are open to oxygen in the air, it will form rust faster.

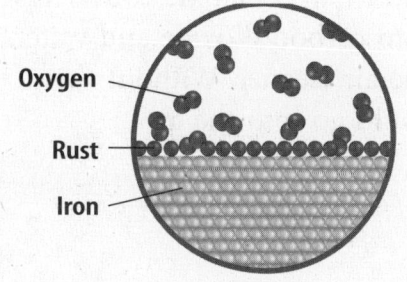
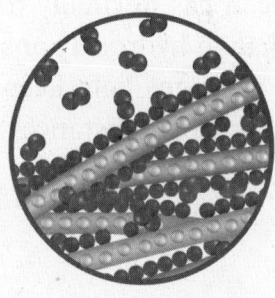

Oxygen

Rust

Iron

3. **Explain** Why does lowering temperature slow down a chemical reaction?

_____

_____

_____

Picture This

4. **Describe** In which circle are molecules more likely to bump into each other—the top circle or the bottom circle?

_____

Picture This

5. **Compare** How is the steel wool different from the steel beam?

_____

_____

_____

💡 **Think it Over**

6. **Analyze** Why would you want to speed up the reactions that change harmful substances to less harmful ones in car exhaust?

_____

_____

Slowing Down Reactions

An **inhibitor** is a substance that slows down a chemical reaction. When an inhibitor is added to a reaction, it can take longer to form a certain amount of product. Some inhibitors completely stop reactions. Many cereals and cereal boxes contain butylated hydroxytoluene, or BHT. The BHT slows the spoiling of the cereal and the packaging material. This increases its shelf life. The shelf life is how long a product lasts without spoiling.

Speeding Up Reactions

You can speed up a chemical reaction by adding a catalyst (KAT uh lihst). A **catalyst** is a substance that speeds up a chemical reaction without changing permanently or being used up. Catalysts are not shown in chemical equations because they are not changed or used up in the reaction. Does a catalyst change how much product is made by a reaction? No, the same amount of product is made as would be made without the catalyst. But, with the catalyst, it will make the same amount of product faster.

How does a catalyst work?

Many catalysts provide a surface on which the reaction occurs. Sometimes the catalyst holds the reacting molecules in a specific position that makes it better for the reaction to occur. Other catalysts reduce the activation energy needed to start a reaction.

How do catalytic converters work?

Catalysts are used in the exhaust systems of cars and trucks. They help fuel combustion, or burning. The exhaust goes through the catalyst. Beads coated with metals like platinum or rhodium are often the catalysts. Catalysts speed up the reactions that change the harmful substances in exhaust to less harmful substances. For example, they change carbon monoxide, a dangerous gas, into carbon dioxide, a gas normally found in air. Catalytic converters also change hydrocarbons into carbon dioxide and water. These reactions help keep the air cleaner. Without them, more harmful substances would go into the air.

What are enzymes?

Some of the best catalysts are at work in the reactions that take place in your body. These catalysts are called enzymes. **Enzymes** are large protein molecules that speed up reactions needed for your cells to work correctly. They help your body change food to fuel. They also help your body build bone and muscle tissue and change extra energy to fat. They even make other enzymes. ☑

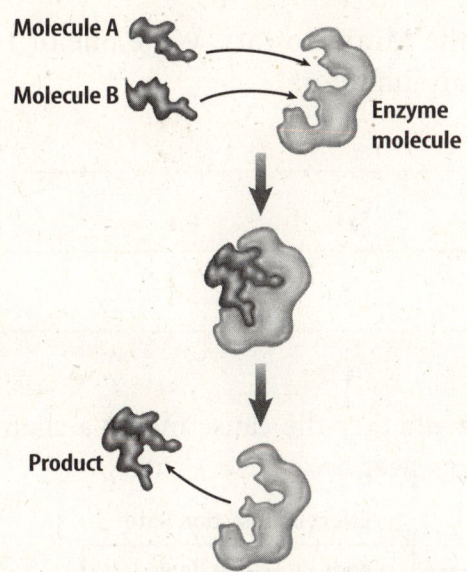

Molecule A

Molecule B

Enzyme molecule

Product

Without enzymes, reactions would happen too slowly to be useful or they would not happen at all. Enzymes work like other catalysts. They make a chemical reaction go faster by bringing certain molecules together. The figure shows how molecules that have the right shape can fit on the surface of an enzyme. Now the molecules can react and form a new product. Enzymes are chemical specialists. They exist to carry out each type of reaction in your body.

What are other uses for enzymes?

Enzymes work in other substances, too. One group of enzymes, called proteases (PROH tee ay ses), specializes in protein reactions. They work within cells to break down proteins. Proteins are large, complicated molecules. The meat tenderizer used on meat has proteases. The proteases break down the proteins in meat, making it more tender. Contact lens cleaning solutions also contain proteases. They break down proteins from your eyes that can collect on your contact lenses and cloud your view.

✔ **Reading Check**

7. **Identify** What is one example of how enzymes help your body?

_____

_____

Picture This

8. **Describe** what enzymes do to molecules.

_____

_____

● After You Read

Mini Glossary

activation energy: the energy needed to start a reaction

catalyst (KAT uh lihst): a substance that speeds up a chemical reaction without changing permanently or being used up

concentration: the amount of substance in a certain volume

enzyme: a large protein molecule that speeds up a reaction needed for cells to work correctly

inhibitor: a substance that slows down a chemical reaction

rate of reaction: tells how fast a reaction happens after it has started

1. Review the terms and their definitions in the Mini Glossary. Write one or two sentences comparing and contrasting a catalyst and an inhibitor.

   _____

   _____

   _____

2. In the cause and effect chart below, decide whether the cause makes a chemical reaction speed up or slow down. Circle the correct answer.

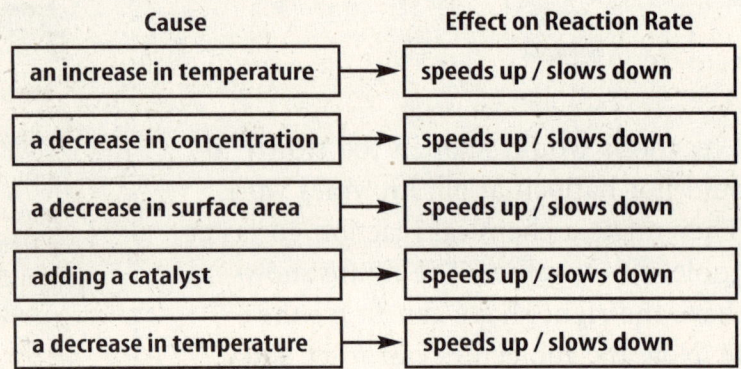

Cause	Effect on Reaction Rate
an increase in temperature	speeds up / slows down
a decrease in concentration	speeds up / slows down
a decrease in surface area	speeds up / slows down
adding a catalyst	speeds up / slows down
a decrease in temperature	speeds up / slows down

3. You were asked to write questions and answers on flash cards. How did making the flash cards help you learn about rates of chemical reactions?

   _____

   _____

   _____

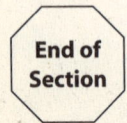

End of Section

Science Online Visit **ips.msscience.com** to access your textbook, interactive games, and projects to help you learn more about rates of chemical reactions.

Substances, Mixtures, and Solubility

section ❶ What is a solution?

● Before You Read

Do you add sugar to your tea? How do you know that the white substance will make your drink sweeter?

_____

_____

● Read to Learn

Substances

Water, salt water, and pulpy orange juice are different liquids. Their differences can be explained by chemistry. Think about pure water. If you freeze it, melt it, or boil it, it is still water. But, if you boil salt water, the water turns to gas and leaves the salt behind. If you strain pulpy orange juice, it loses its pulp. How does chemistry explain these differences? The answer has to do with the chemical makeup of these materials.

What are atoms, substances, and elements?

Atoms Recall that atoms are the basic building block of matter. Each atom has its own chemical and physical properties. These properties are determined by the number of protons the atom has.

Substances A <u>substance</u> is matter that has the same fixed makeup and properties throughout. A substance cannot be broken down into simpler parts by a physical process. For example, you can freeze, boil, stir, and filter water, but it is still water. The only way to change a substance is by a chemical process. The table on the next page shows some examples of physical processes and chemical processes.

What You'll Learn

- the differences between substances and mixtures
- two types of mixtures
- how solutions form
- different types of solutions

▶ **Mark the Text**

Underline As you read, underline words and sentences that you think are important to remember. After you read, review what you have underlined.

FOLDABLES™

Ⓐ **Classify** Use quarter-sheets of paper to help you organize definitions and examples of substances and mixtures.

Substance	Mixture
Heterogeneous Mixture	Homogeneous Mixture

Picture This

1. **Explain** How do physical processes differ from chemical processes?

_____

_____

_____

_____

Physical Processes (do not change substances)	Chemical Processes (do change substances)
Boiling	Burning
Changing pressure	Reacting with other chemicals
Cooling	Reacting with light
Sorting	

Elements An element is an example of a pure substance. An element cannot be broken down into simpler substances. The number of protons in an element cannot change unless the element changes.

What are compounds?

Water is a compound. A compound is a substance made of two or more elements that are chemically combined. The makeup of a compound is always the same. For example, a water molecule always has two hydrogen atoms combined with one oxygen atom. All water, whether frozen, liquid, or vapor, has the same ratio of hydrogen atoms to oxygen atoms.

Think it Over

2. **Classify** Why is water a compound and not an element?

_____

_____

Mixtures

Imagine drinking a glass of salt water. You would know right away that it is not pure water. Salt water is not a pure substance. It is a mixture of salt and water. Mixtures are made when two or more substances come together but do not chemically bond together to make a new substance. The substances can be separated by physical processes. For example, you can boil salt water to separate the salt from the water.

Mixtures do not contain an exact amount of each substance like a compound. Lemonade can be weak tasting or strong tasting. It depends on how much lemon juice is added to the water. It also can be sweet or sour, depending on how much sugar is added. No matter how strong, weak, sweet, or sour, it is still lemonade.

What are heterogeneous mixtures?

Some mixtures are easy to see. A watermelon is a mixture of fruit and seeds. But, the fruit and seeds aren't mixed evenly. A **heterogeneous** (he tuh ruh JEE nee us) **mixture** is a mixture where the substances are not mixed evenly. The substances in heterogeneous mixtures are usually easy to tell apart. A bowl of cereal with milk is another example of a heterogeneous mixture.

What are homogeneous mixtures?

When you mix sugar and water together you don't see the sugar particles floating in the water. Sugar water is a homogeneous (ho muh JEE nee us) mixture. A **homogeneous mixture** has two or more substances in which the molecules mix evenly but do not bond together. Another name for a homogeneous mixture is a **solution**. The figure shows the mixture of sugar and water molecules in a solution of sugar water.

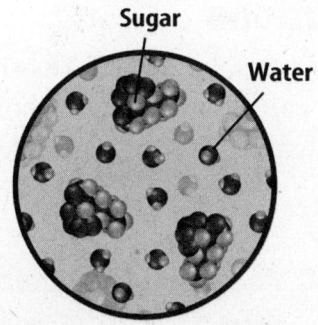

Sugar

Water

Picture This

3. Describe Look at the figure. How would you describe the sugar and water molecules in a solution of sugar water? Circle the answer.

   a. not mixed evenly
   b. combined
   c. mixed evenly
   d. compounded

How Solutions Form

When you mix sugar and water together, you can't see the sugar particles in the water. The sugar doesn't actually disappear. The sugar molecules spread out until they are evenly spaced throughout the water molecules, forming a solution. This is called dissolving. The substance in a solution that dissolves, or seems to disappear, is called the **solute**. The substance that dissolves the solute in a solution is the **solvent**. In the sugar water solution, the sugar is the solute and water is the solvent.

How can solids form from solutions?

Sometimes, a solute can come back out of a solution and form a solid. This process is called crystallization.

Crystallization Crystallization happens because of a physical change. For example, crystallization can happen when a solution is cooled. Crystallization also can happen when some of the solvent evaporates. A stalactite, or hanging rock, in a cave is an example of crystallization. Minerals dissolve in water as it flows through rocks. When the solution drips from the ceiling of the cave, some of the water evaporates. The minerals in the solution crystallize to form the stalactite.

Think it Over

4. Apply Minerals dissolve in water as it flows through rocks at the top of the cave. In this solution, what is the solute and what is the solvent?

_____

_____

_____

Precipitate Formation When some solutions are mixed, a chemical change happens and a solid forms. A solid that forms when solutions are mixed and a chemical change happens is a **precipitate** (prih SIH puh tut). Precipitate formation is different from crystallization because a chemical change takes place. A precipitate can form in a shower. Minerals that are dissolved in water can react chemically with soap. This chemical reaction forms a precipitate called soap scum.

Types of Solutions

Not all solutions are solid solutes dissolved in liquid solvents. Solutions can be made up of combinations of solids, liquids, and gases. See the examples in the table.

Picture This

5. **Interpret Data** Name two solutions that have carbon dioxide as one of the solutes.

____

____

____

Examples of Common Solutions

Solution	Solvent/ State	Solute/ State	State of Solution
Earth's atmosphere	nitrogen/gas	oxygen/gas carbon dioxide/gas argon/gas	gas
Carbonated beverage	water/liquid	carbon dioxide/gas	liquid
Brass	copper/solid	zinc/solid	solid

Liquid Solutions

Sugar water and salt water are examples of solutions with liquid solvents and solid solutes. The solute in a solution can be a solid, another liquid, or even a gas. The state of the solution will usually be the same as the state of the solvent. For example, sugar is a solid and water is a liquid. When sugar and water are mixed together to form a solution, the solution is a liquid, not a solid. ☑

Reading Check

6. **Explain** What does the state of a solution usually depend on?

____

____

What are liquid-gas and liquid-liquid solutions?

Liquid-Gas Carbonated drinks are examples of solutions with liquid solvents and gas solutes. The gas solute is carbon dioxide. Water is the liquid solvent. Carbon dioxide gives the drinks their fizz.

Liquid-Liquid Vinegar is an example of a liquid-liquid solution. Water is the liquid solvent and acetic acid is the liquid solute. In vinegar, only 5 percent of the solution is acetic acid. Water makes up 95 percent of the solution.

Gaseous and Solid Solutions

Gas Solutions Sometimes, a small amount of one gas is dissolved in a larger amount of another gas. This is a gaseous solution, also called a gas-gas solution. The air you breathe is a gaseous solution. About 78 percent of air is nitrogen, which is the solvent. About 20 percent of air is oxygen, which is one of the solutes. Other solutes in air are carbon dioxide, argon, and some other gases in small amounts.

Solid Solutions There are also solid solutions. In a solid solution, the solvent is solid. The solute can be a solid, liquid, or gas. The most common solid solutions are solid-solid solutions. Both the solvent and solute are solids. Steel is a solution of carbon dissolved in iron. A solid-solid solution made from two or more metals is called an alloy. Brass is an alloy of zinc dissolved in copper. The figure shows what microscopic views of steel and brass might look like. ☑

Reading Check

7. Identify What states can the solutes be in a solid solution?

_____

_____

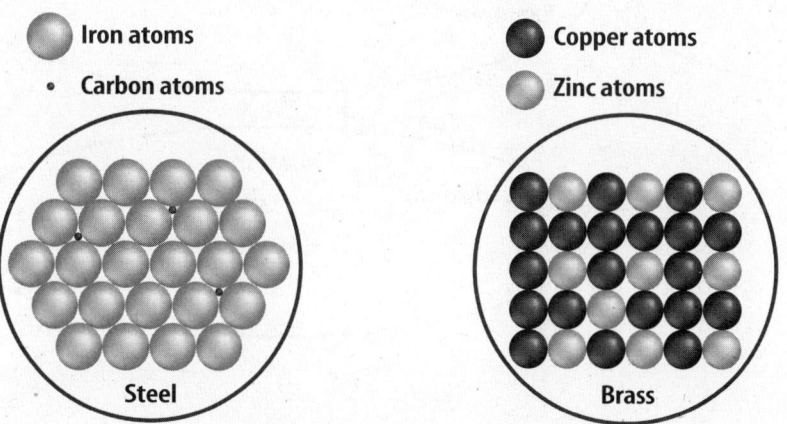

Iron atoms

Carbon atoms

Copper atoms

Zinc atoms

Steel

Brass

Applying Math

8. Interpret a Scientific Illustration Look at the sample of brass in the figure. What is the ratio of copper atoms to the total number of atoms?

_____

● After You Read

Mini Glossary

heterogeneous mixture: a mixture where the substances are not mixed evenly

homogeneous mixture: a mixture that has two or more substances in which the molecules mix evenly, but do not bond together

precipitate: a solid that forms when solutions are mixed and a chemical change happens

solute: the substance in a solution that dissolves, or seems to disappear

solution: another name for a homogeneous mixture

solvent: the substance that dissolves the solute in a solution

substance: matter that has the same fixed makeup and properties throughout

1. Review the terms and their definitions in the Mini Glossary. Write a sentence using at least one glossary term to describe the mixture of vegetables in a salad.

_____

_____

2. Fill in the graphic organizer with important facts about mixtures and solutions

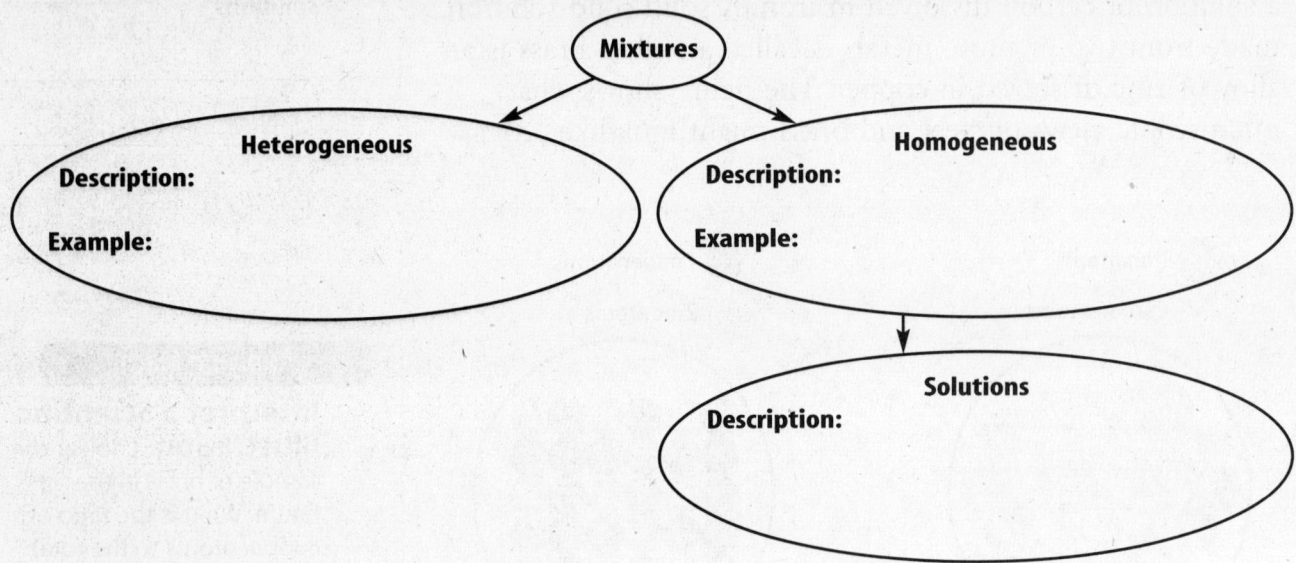

3. As you read this section, you underlined words and sentences that you thought were important. How did you decide what to underline?

_____

_____

 Visit **ips.msscience.com** to access your textbook, interactive games, and projects to help you learn more about weathering.

chapter 8

Substances, Mixtures, and Solubility

section ❷ Solubility

● Before You Read

What happens when you put one teaspoon of sugar in a glass of water? What would happen if you put one cup of sugar in a glass of water?

_____

_____

● Read to Learn

Water—The Universal Solvent

Water is a solvent for many solutions, including fruit juice and vinegar. A solution in which water is the solvent is called an **aqueous** (A kwee us) solution. Water dissolves things so well it is often called the universal solvent.

What are molecular compounds?

You know that atoms can join with other atoms to form compounds. When certain atoms form compounds, they share electrons. Sharing electrons is called covalent (co VAY lent) bonding. Compounds that have covalent bonds are called molecular compounds, or molecules.

Nonpolar Molecules In some molecules, the atoms share their electrons equally. When the electrons are shared evenly, the molecule is called nonpolar. Look at the hydrogen molecule. A hydrogen molecule is nonpolar.

Hydrogen Molecule

What You'll Learn

- why water is a good solvent
- how much of a solute will dissolve in a solvent
- how temperature affects chemical reactions

Study Coach

Make Flash Cards As you read, write important questions on note cards. Write the answer to each question on the back of the card. After you read, see if you can answer your questions without looking at the answers.

FOLDABLES

Ⓑ **Compare and Contrast**
Make the following Foldable to show how molecular compounds and ionic compounds are alike and different.

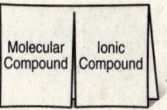

Molecular Compound | Ionic Compound

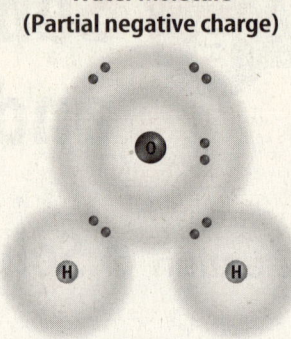

(Partial positive charge)

Picture This

1. **Circle** the shared electrons in the figure.

Polar Molecules The atoms in some molecules do not share their electrons equally. Look at the water molecule. Two hydrogen atoms share electrons with one oxygen atom. But, the electrons spend more time around the oxygen atom than they spend around the hydrogen atom. This makes the oxygen part of the water molecule have a slightly negative charge. The hydrogen parts have a slightly positive charge. The total charge of the water molecule is neutral. Molecules that have slightly positive and slightly negative charges are called polar molecules. The bonds between its atoms are called polar covalent bonds.

What are ionic bonds?

Some atoms do not share electrons when they join to form compounds. Instead, these atoms lose or gain electrons making the number of protons and electrons in the atom no longer equal. The atom becomes either positively charged or negatively charged. Atoms with a charge are called ions. Bonds between ions are called ionic bonds. The compound that is formed is called an ionic compound. Table salt is an ionic compound. It is made of sodium ions and chloride ions. Each sodium atom loses an electron to a chlorine atom and becomes positively charged. The chlorine atoms gain the electrons from the sodium atoms and become negatively charged. ☑

☑ **Reading Check**

2. **Summarize** Write the correct words to complete the sentence on the lines below: Atoms in an ionic compound ___a.___ or ___b.___ electrons.

a. _____

b. _____

How does water dissolve ionic compounds?

Remember that a water molecule is polar. It attracts positive and negative ions. Look at the figure. When sodium chloride, or table salt, is added to water, the sodium (Na) ions and chloride (Cl) ions are attracted by the water molecules. The slightly negative end of a water molecule attracts positive sodium ions (Na^+). The slightly positive end of a water molecule attracts negative chloride ions (Cl^-). When an ionic compound is mixed with water, the ions are pulled apart, or dissolved, by the water molecules.

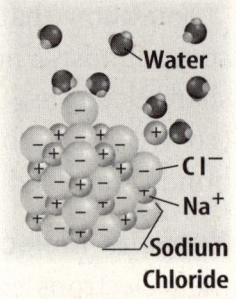

Water
Cl^-
Na^+
Sodium
Chloride

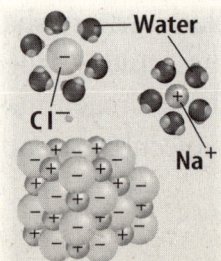

Water
Cl^-
Na^+

Picture This

3. **Highlight** Use a highlighter to circle the part of the figure that shows the dissolved table salt.

How does water dissolve molecular compounds?

Water does dissolve molecular compounds, like sugar, that are not made of ions. But, water does not break sugar molecules apart as it does in ionic compounds. Sugar molecules are polar, like water molecules. Polar water molecules are attracted to the positive and negative ends of the sugar molecules. The water molecules move in between sugar molecules and spread them apart. When this happens, the sugar dissolves.

What will dissolve?

When you put a teaspoon of sugar in iced tea and stir it, what happens? The sugar dissolves. Why doesn't the spoon dissolve? A substance that dissolves in another substance is soluble, or able to be dissolved in that substance. Sugar is soluble in water. You would say the metal of the spoon is insoluble in water because it does not dissolve.

What does "like dissolves like" mean?

Chemists use the phrase "like dissolves like" to remember which solvents can dissolve which solutes. "Like dissolves like" means polar solvents dissolve polar solutes. Also, nonpolar solvents dissolve nonpolar solutes. Sugar and water are both polar, so water dissolves sugar. Why does water dissolve sodium chloride? Remember, the sodium and chloride ions have charges like polar molecules. So, water reacts with these ions in the same way it reacts with other polar molecules. ☑

Have you ever mixed oil and water in a glass? They do not form a solution. Instead, the two liquids separate and form layers in the glass. Oil molecules are nonpolar. Polar water molecules are not attracted to them. So, oil will not dissolve in water.

How much will dissolve?

__Solubility__ (sahl yuh BIH luh tee) is the measurement that describes how much solute dissolves in a given amount of solvent. Usually, solubility is the amount of solute that can dissolve in 100 g of solvent at a certain temperature. Some solutes are highly soluble. This means that a large amount of solute can be dissolved in 100 g of solvent. For example, 63 g of potassium chromate will dissolve in 100 g of water at 25°C. But, some solutes are not very soluble. Only 0.00025 g of barium sulfate will dissolve in 100 g of water at 25°C. This solubility is so low that it is called insoluble.

✔ **Reading Check**

4. **Recall** What does "like dissolves like" mean?

_____

_____

_____

_____

FOLDABLES

C **Organize Information**
Use two quarter-sheets of paper to write information about solutions and how they dissolve.

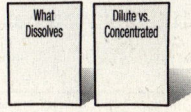

How does temperature affect solubility?

The solubility of many solutes changes with the temperature of the solvent. Sugar dissolves faster in hot tea than it does in iced tea. Also, more sugar can dissolve in hot tea than in iced tea. The solubility and the rate that sugar dissolves in water increases as temperature increases.

This is not true of all solutes. The graph shows the solubility at different temperatures of several solutes in water. An increase in temperature decreases the solubility of a gas in a liquid-gas solution. That is why cold sodas fizz less than warm sodas when you open them.

Applying Math

5. Interpret Data How many grams of sugar will dissolve in 100 g of water at 60°C?

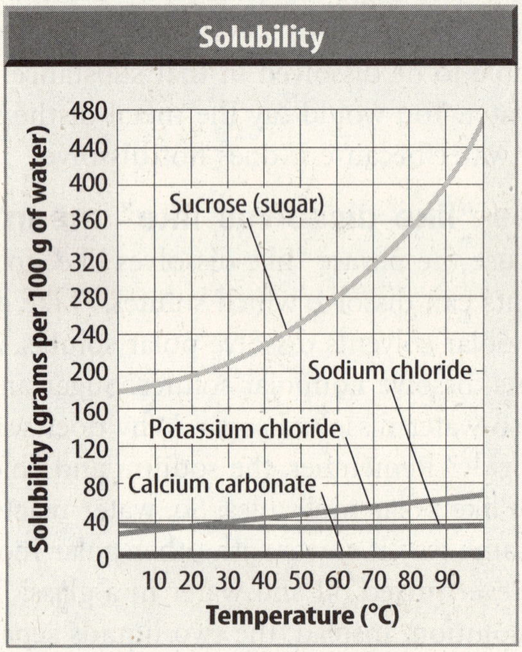

Solubility

Solubility (grams per 100 g of water)

480
440
400
360
320
280
240
200
160
120
80
40
0

Sucrose (sugar)

Sodium chloride

Potassium chloride

Calcium carbonate

10 20 30 40 50 60 70 80 90

Temperature (°C)

What are saturated solutions?

If you add potassium chromate to 100 g of water at 25°C, only 63 g of it will dissolve. A solution that contains all of the solute that it can hold under the given conditions is **saturated**. When a solution has less solute than what is needed to become saturated, it is called an unsaturated solution. Look at the graph above. A saturated sugar water solution would contain about 204 g of sugar in 100 g of water at 25°C. If less sugar is used, the solution is unsaturated.

A hot solvent usually can hold more solute that a cold solvent. When a saturated solution cools, some of the solute usually falls out of solution. But, if a saturated solution is cooled slowly, sometimes the extra solute can stay dissolved. This is called a supersaturated solution.

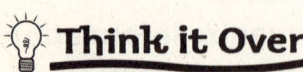

Think it Over

6. Apply Tell whether this solution is saturated or unsaturated: 50 g of sugar in 100 g of water at 25°C.

Rate of Dissolving

A solute dissolves faster when the solution is shaken or stirred or heated. These actions make the surfaces of the solute come into contact with the solvent more quickly. You can do the same thing by breaking up or grinding the solute into smaller pieces. For example, granules of sugar dissolve more quickly than sugar cubes. Grinding increases the surface area of the solute that is exposed to the solvent.

Chemical reactions happen when molecules bump into each other. At colder temperatures, chemical reactions happen more slowly. Refrigerators slow the chemical reactions that cause food to spoil. So, food stays fresh longer in a refrigerator than at room temperature.

Concentration

The **concentration** of a solution tells you how much solute is in a solution compared to the amount of solvent. A concentrated solution has a lot of solute for a given amount of solvent. A dilute solution has little solute for a given amount of solvent.

How do you measure concentration?

Doctors need to give the exact concentration of medicines so patients will be treated correctly. One way to give an exact concentration is to give the percentage of the volume of the solution that is made of solute. The label of a fruit juice container might say, "contains 10 percent juice." This means that 10 percent of the container is the solute: juice. Ninety percent is the solvent: water and other substances like sugar.

How do solute particles affect solvents?

Solutes often change the freezing and boiling points of solvents. When liquids freeze, their molecules arrange themselves in certain ways. Solute particles can change the way the molecules arrange themselves and lower the freezing point.

When liquids boil, their molecules gain enough energy to move from the liquid state to the gaseous state. Solute particles can interfere with the change from liquid to gaseous state. More energy is needed to make the solvent particles escape from the liquid. This increases the boiling point. ☑

💡 Think it Over

7. Infer A metal worker needs an acid to dissolve metal. Does she probably need an acid that is concentrated or dilute?

_____

☑ **Reading Check**

8. Determine What do solutes often change in solvents?

_____

_____

● After You Read

Mini Glossary

aqueous: a solution in which water is the solvent

concentration: how much solute is in a solution compared to the amount of solvent

saturated: a solution that contains of all the solute that it can hold under given conditions

solubility: the measurement that describes how much solute dissolves in a given amount of solvent

1. Review the terms and their definitions in the Mini Glossary. Use two of the terms in one or two sentences to describe a bottle of orange juice.

_____

_____

_____

2. In the graphic organizer, explain how water dissolves the types of compounds shown.

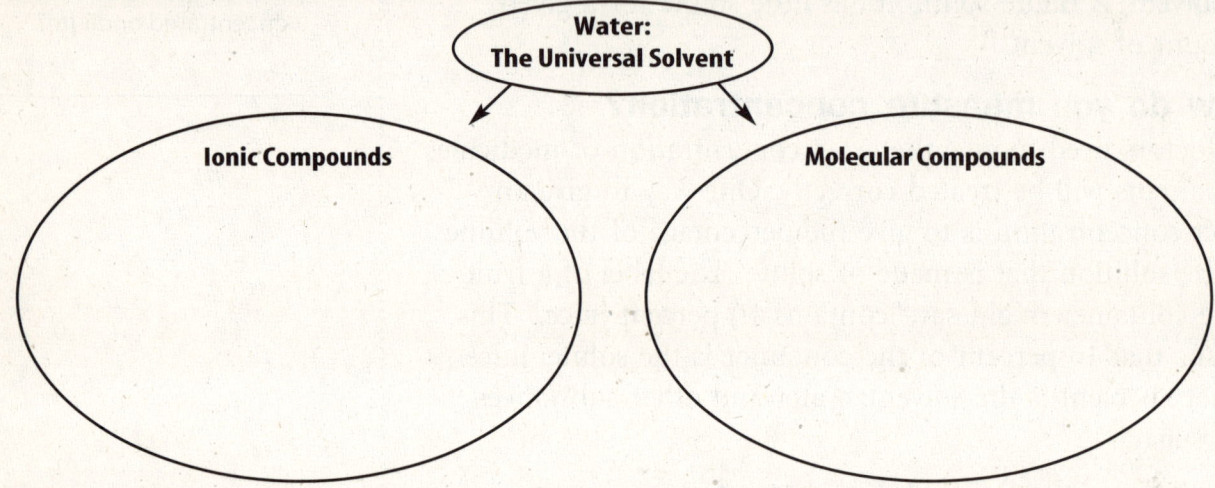

Water: The Universal Solvent

Ionic Compounds

Molecular Compounds

3. How could you use lemonade to teach others about the concentrations of solutions?

_____

_____

_____

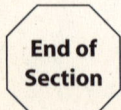

End of Section

Science Online Visit **ips.msscience.com** to access your textbook, interactive games, and projects to help you learn more about weathering.

Substances, Mixtures, and Solubility

section ❸ Acidic and Basic Solutions

● Before You Read

Name some sour foods that you like.

_____

● Read to Learn

Acids

If you like sour foods like dill pickles and lemons, you like foods that have acids. An **acid** is a substance that releases positively charged hydrogen ions (H+) in water. When an acid mixes with water, it dissolves, releasing hydrogen ions. The hydrogen ions then join with water molecules to form hydronium ions. A **hydronium ion** is a positively charged ion that has the formula H_3O^+. The figure shows how a hydronium ion is made.

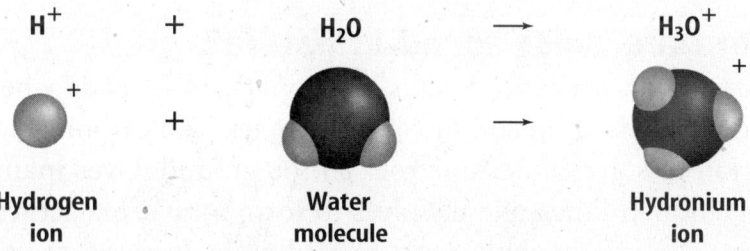

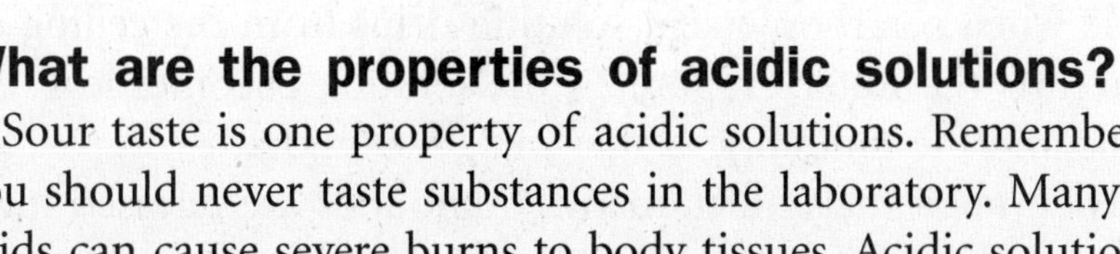

H^+ + H_2O → H_3O^+

Hydrogen ion + Water molecule → Hydronium ion

What are the properties of acidic solutions?

Sour taste is one property of acidic solutions. Remember, you should never taste substances in the laboratory. Many acids can cause severe burns to body tissues. Acidic solutions also can conduct electricity. Hydronium ions are good carriers of electric charges in an electric current. This is why some batteries contain acids.

What You'll Learn

■ about acids, bases, and their properties
■ uses of acids and bases
■ pH of acids and bases

Study Coach

Outline Make an outline of this section as you read. When you finish reading, look over your outline to make sure you understand what you have written down.

FOLDABLES

ⓓ Compare and Contrast
Make the following Foldable to show how acids and bases are alike and different.

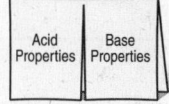

Acid Properties | Base Properties

Many acids are corrosive. This means they can break down certain substances. Many acids can corrode fabric, skin, and paper. Some acids react strongly with metals. When these acids are put on metal, metal compounds and hydrogen gas form, leaving holes in the metal.

What are some uses of acids?

Vinegar contains acetic acid. It is used in salad dressings. Citrus fruits like lemons, limes, and oranges taste sour because they contain citric acid. Your body needs vitamin C, which is ascorbic acid. Ants that sting inject formic acid into their victims. The figure shows products that are made with acids.

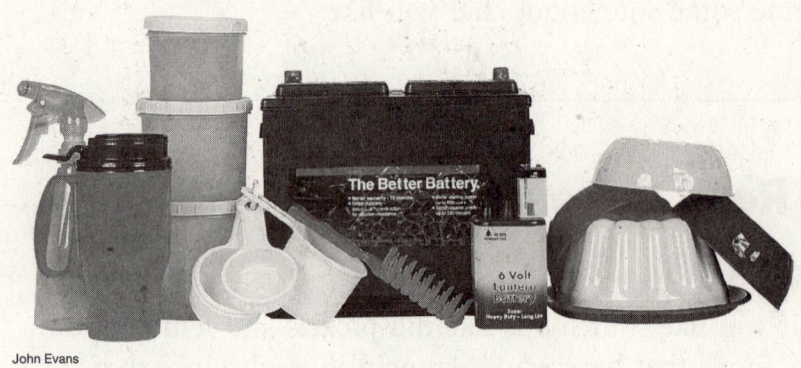

John Evans

Sulfuric acid is used to make fertilizers, steel, paints, and plastics. It is also called battery acid because it is used in many batteries, such as car batteries. Hydrochloric acid is also called muriatic acid. It is used to remove impurities from the surfaces of metals. Hydrochloric acid also can be used to clean mortar from brick walls. Nitric acid is used to make fertilizers, dyes, and plastics.

Where are acids found in nature?

Caves form because of acids. Carbonic acid is made when carbon dioxide from soil dissolves in water. The carbonic acid solution dissolves limestone rock in the ground. Over many years, enough limestone dissolves to form a cave. Stalactites and stalagmites, hanging rocks and columns in caves, are also made when a carbonic acid solution drips from the ceiling of a cave. As the water evaporates, the solution becomes less acidic and the limestone comes out of solution.

When fossil fuels burn, many compounds are released into the air. Some of these compounds form nitric acid and sulfuric acid. These strong acids mix with water vapor and fall back to Earth as rain, sleet, snow, or fog. The acid rain can corrode stone statues, damage forests, and make people sick. ☑

Picture This

1. **List** the products in the figure that you have seen or used.

_____

_____

_____

Reading Check

2. **Name** three acids found in nature.

_____

_____

_____

Bases

Many window and floor cleaners contain an ammonia solution. Ammonia contains a base. A **base** is a substance that can accept hydrogen ions. When bases dissolve in water, they release a hydroxide ion (OH^-). For example, when sodium hydroxide (NaOH) dissolves in water, it separates into sodium ions (Na^+) and hydroxide ions (OH^-). Ammonia (NH_3) is different. When it dissolves in water, it pulls a hydrogen atom away from water. This leaves a hydroxide ion. Look at the figure below.

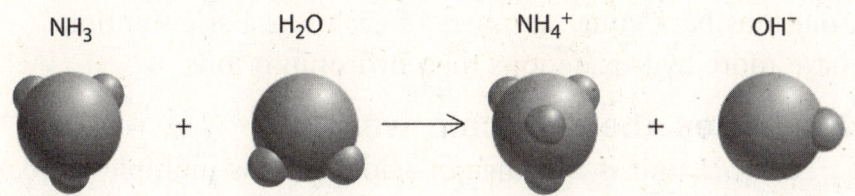

$$NH_3 \quad + \quad H_2O \quad \longrightarrow \quad NH_4^+ \quad + \quad OH^-$$

Picture This
3. Draw and Label Circle the hydroxide ion in this reaction.

What are the properties and uses of bases?

Properties Most soaps are bases. How does soap feel? Basic solutions, like soap, feel slippery. Bases are corrosive like acids. They can cause burns and damage body tissue. That's why you should never touch, smell, or taste a substance to find out if it is a base or an acid. Bases can conduct electricity like acids. They are not as corrosive to metals as acids.

Uses Many uses for bases are shown in the figure below. Bases are used in plastics, soap, ammonia, and other cleaning products. Hydroxide ions can react with dirt and grease to wash them away. Calcium hydroxide, often called lime, is used to mark lines on athletic fields. It also can make soil less acidic. Sodium hydroxide is a base called lye. Lye is a strong base that can cause burns and other health problems. It is used to make soap, clean ovens, and unclog drains.

FOLDABLES™
E Organize Information Use quarter-sheets of paper to help you organize and list information about pH.

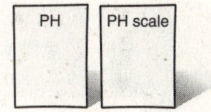

John Evans

Picture This
4. Circle the base in the figure that is used in classrooms every day.

What is pH?

<u>pH</u> is a way to measure how acidic or basic a solution is. Perhaps you've seen someone check the pH of a swimming pool. The pH scale ranges from 0 to 14. Acids have a pH below 7. Bases have a pH above 7. Solutions with a pH of 7 are called neutral. They are neither acids nor bases. Strong acids, like hydrochloric acid, have a pH of 0. Strong bases have a pH of 14.

The pH of a solution depends on its concentration of hydronium ions (H_3O^+) and hydroxide ions (OH^-). Acids have more hydronium ions than hydroxide ions. Neutral solutions have equal numbers of each ion. Basic solutions have more hydroxide ions than hydronium ions.

How does the pH scale work?

Each pH unit is a change in acidity that is multiple of 10. The lower the number, the more acidic a solution is. An acid with a pH of 2 is 10 times stronger than an acid with a pH of 3 and 100 times stronger than an acid with a pH of 4. A base with a pH of 13 is 10 times stronger than a base with a pH of 12 and 100 times stronger than a base with a pH of 11.

Applying Math

5. **Calculate** Look at the pH scale. How many times more acidic is an acid with a pH of 2 than an acid with a pH of 5?

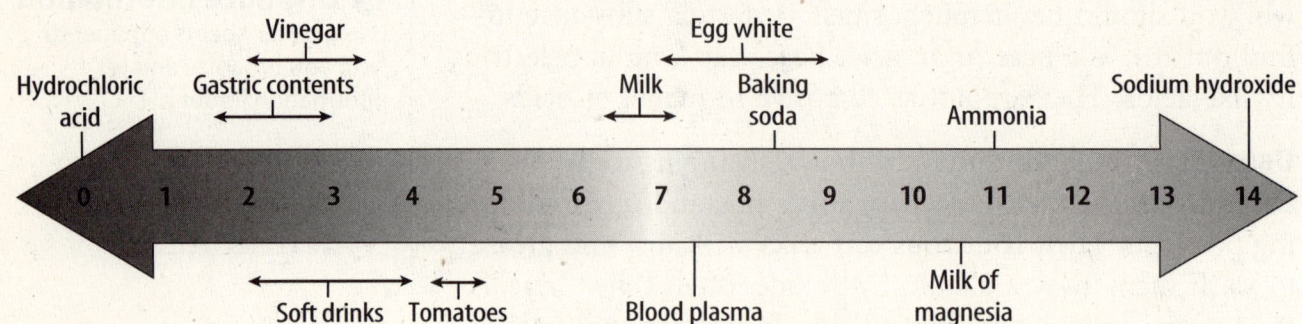

Picture This

6. **Label** In the figure above, write the labels "Acids," "Bases," and "Neutral" at the correct places.

What makes a strong acid or a strong base?

Some acids give foods a sour taste, and some other acids are so strong that they can cause burns. Vinegar, or acetic acid, makes pickles sour, but you can eat pickles because the acid is weak. Hydrochloric acid, a strong acid, would dangerously burn your mouth. What makes these acids different? The ions of strong acids break apart in water more easily than the ions of weak acids. Strong acids form many more hydronium ions than weak acids. More hydronium ions give a lower pH, which is more acidic. Strong bases form many more hydroxide ions than weak bases. More hydroxide ions give a higher pH, which is more basic.

Indicators

Is there a safe way to find out how acidic or basic a solution is? Yes, you can use an indicator. An **indicator** is a compound that turns a certain color in acidic or basic solutions, depending on the pH. An example of an indicator is litmus. This compound is soaked into paper strips. You place the paper strips in a solution and look at the color. Litmus paper turns red in acids and blue in bases.

Neutralization

Have you ever heard of heartburn? Someone with heartburn might take an antacid tablet. The prefix *ant-* means "opposite of." Heartburn is caused by having too much hydrochloric acid in the stomach. An antacid tablet neutralizes the extra acid. How does an antacid tablet work?

An antacid is made from a base that neutralizes the extra acid in the stomach. **Neutralization** (new truh luh ZAY shun) is the reaction of an acid with a base. It is called this because properties of both the acid and the base are reduced, or neutralized. When a base and acid are mixed, they usually form water and a salt. Because of the reaction, there are fewer hydronium and hydroxide ions in the solution. This makes the pH of the solution more neutral. So, when a base such as magnesium hydroxide, $Mg(OH)_2$, in an antacid reacts with hydrochloric acid in the stomach, some of the acid is neutralized. The figure shows the relative amounts of hydronium and hydroxide ions between pH 0 and pH 14.

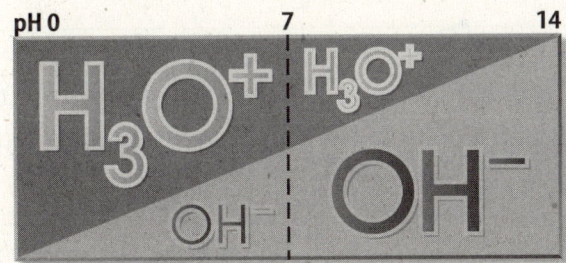

How does neutralization occur?

When a solution is neutralized, hydronium and hydroxide ions react with each other. During neutralization, equal numbers of hydronium ions and hydroxide ions react to produce water molecules. Pure water has a pH of 7, which means it is neutral.

Think it Over

7. **Apply** To neutralize a solution that contains lye, what would be added?

_____

_____

Picture This

8. **Compare** At pH 7, how does the amount of hydronium ions compare to the amount of hydroxide ions?

_____

_____

● After You Read

Mini Glossary

acid: a substance that releases positively charged hydrogen ions (H^+) in water

base: a substance that can accept hydrogen ions

hydronium ion: a positively charged ion that has the formula H_3O^+

indicator: a compound that turns a certain color in acidic or basic solutions, depending on the pH

neutralization: reaction of an acid with a base

pH: a measure of how acidic or basic a solution is

1. Review the terms and their definitions in the Mini Glossary. Write one or two sentences to explain the pH of pure water.

_____

_____

_____

2. Label the location of pure water on the pH scale. Label the acidic side of the scale in red and the basic side in blue.

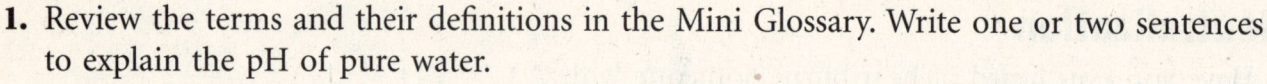

```
pH  1   2   3   4   5   6   7   8   9   10  11  12  13  14
```

3. How much stronger is a pH of 9 than a pH of 4?

_____

4. Suppose you add water to solutions to make acids of different strengths. You add 100 mL of water to an acid to make an acid with pH of 6. How much water would you add to make an acid with pH of 5? Explain.

_____

_____

_____

End of Section

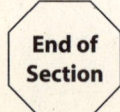  Science Online Visit **ips.msscience.com** to access your textbook, interactive games, and projects to help you learn more about acidic and basic solutions.

134 Substances, Mixtures, and Solubility

Carbon Chemistry

section ❶ Simple Organic Compounds

● Before You Read

Many people believe that if something is called organic it must be good for you. Write what you think *organic* means.

_____

_____

Copyright © Glencoe/McGraw-Hill, a division of The McGraw-Hill Companies, Inc.

● Read to Learn

Organic Compounds

All living things on Earth are made of compounds that have carbon. Carbon is an important element. It forms bonds easily. Each carbon atom has four electrons in its outer energy level. Each of these electrons can form a covalent bond. A covalent bond is a bond where the atoms share electrons. So, each carbon atom can form four covalent bonds with four other atoms. Carbon compounds that have four covalent bonds are very stable. They have eight electrons in their outer energy level. Carbon and hydrogen often bond together in carbon compounds.

Matter can be divided into two groups. It comes from either living things or nonliving things. Most matter that comes from living things contains carbon and hydrogen. Materials from living things were called organic compounds. But, in 1828 scientists discovered that living organisms are not necessary to form organic compounds. Today, scientists still use the term **organic compound** to describe most compounds that contain carbon.

Hydrocarbons

Many compounds are made of only carbon and hydrogen. A **hydrocarbon** is a compound that contains only carbon and hydrogen atoms.

What You'll Learn

- why carbon can form many compounds
- how saturated and unsaturated hydrocarbons differ
- what isomers of organic compounds are

Mark the Text

Identify Main Ideas
Highlight the main idea in each paragraph.

FOLDABLES

Ⓐ Organize Information
Use a half-sheet of notebook paper to organize information about organic compounds.

Organic
Compounds

Methane The simplest hydrocarbon is methane. Each molecule has only one carbon atom and four hydrogen atoms. Methane is found in natural gas. It is also used as the fuel in gas stoves and gas furnaces.

The figure below shows a model of a methane molecule. It contains one carbon atom covalently bonded to four hydrogen atoms. The chemical formula for methane is CH_4.

The model in the middle is a structural formula. In a structural formula, a line between atoms shows a pair of electrons shared between the two atoms. This electron pair forms a single bond. Methane has four single bonds. The model on the right in the figure is an electron dot diagram. It uses dots to show the electrons shared between the atoms. Each dot is one electron.

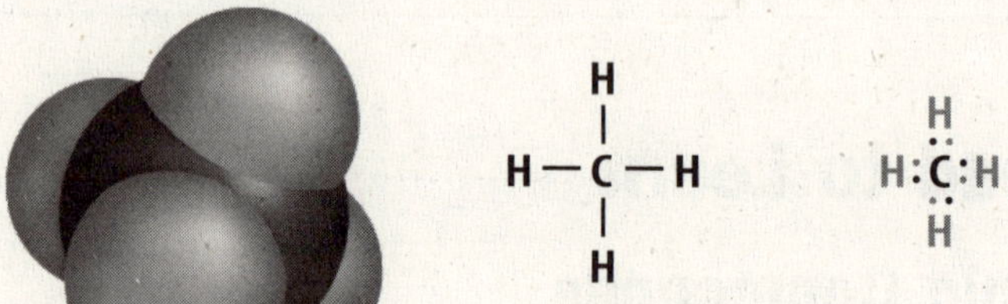

CH_4

Methyl Group The figure below shows one way that a molecule with two carbon atoms forms. Take away one hydrogen atom from a methane molecule. You are left with a carbon atom bonded to three hydrogen atoms. This is called a methyl group. The formula for a methyl group is $-CH_3$.

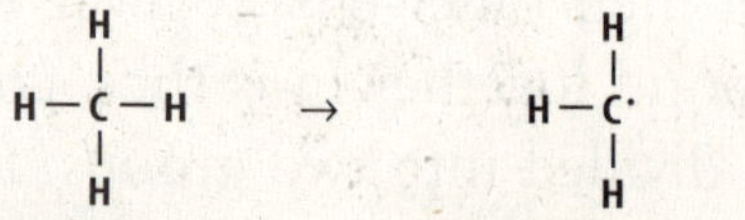

Methane
CH_4

Methyl group
$CH_3 -$

A methyl group can form a single bond with another methyl group, as shown below. When two methyl groups bond together, they form a molecule with two carbon atoms. It is called ethane. Its chemical formula is C_2H_6.

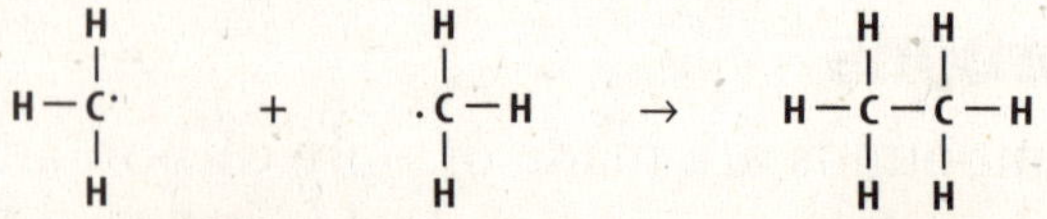

Methyl groups
$CH_3 -$

Ethane
C_2H_6

Picture This

1. **Label** the carbon atom and all the hydrogen atoms in the model of methane on the left figure.

Picture This

2. **Circle** each of the two methyl groups in ethane.

What are saturated hydrocarbons?

Methane and ethane are made of only carbon and hydrogen atoms connected by single covalent bonds. There are many hydrocarbons made of molecules that contain carbon and hydrogen joined by single covalent bonds. A **saturated hydrocarbon** is a hydrocarbon molecule that contains only single bonds. It is called saturated because each carbon atom forms all the single covalent bonds it possibly can with hydrogen. No more hydrogen atoms can be added to a saturated hydrocarbon molecule. Another name for saturated hydrocarbons is alkanes.

How do larger hydrocarbons form?

Larger hydrocarbons form in a way similar to the way ethane forms. A hydrogen atom is taken away from ethane. A $-CH_3$ group replaces the hydrogen atom. Propane has three carbon atoms and eight hydrogen atoms. Its chemical formula is C_3H_8. Propane is the third member of the saturated hydrocarbon series. Butane is the fourth member. It has four carbon atoms and ten hydrogen atoms. The chemical formula for butane is C_4H_{10}. The table below lists the names and formulas of the first four saturated hydrocarbons, or alkanes.

💡 **Think it Over**

3. **Describe** How can a molecule of propane form from a molecule of ethane?

_____

_____

_____

_____

The Structures of Hydrocarbons		
Name	**Structural Formula**	**Chemical Formula**
Methane	H | H—C—H | H	CH_4
Ethane	H H | | H—C—C—H | | H H	C_2H_6
Propane	H H H | | | H—C—C—C—H | | | H H H	C_3H_8
Butane	H H H H | | | | H—C—C—C—C—H | | | | H H H H	C_4H_{10}

Picture This

4. **Calculate** How many more hydrogen atoms does a molecule of butane have than a molecule of methane?

_____

5. Explain Why are methane, propane, and butane used as fuels?

_____

_____

_____

6. Evaluate How many electrons are shared in a double bond?

_____

_____

Picture This

7. Classify Highlight all the single bonds in one color in the three structural formulas. Highlight all the double bonds in a different color.

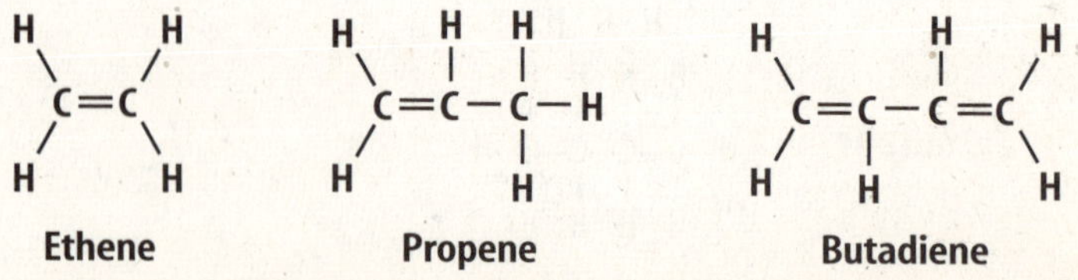

How are hydrocarbons used?

The short hydrocarbon chains have low boiling points. That means they evaporate and burn easily. This makes them good to use as fuels. You learned that methane is burned in stoves and furnaces. Propane is a fuel used in gas grills and lanterns. Butane is used in camp stoves and lighters. Carbon can also form long chains that have hundreds or even thousands of carbon atoms. These long chains make up many of the plastics you use. ✔

What are unsaturated hydrocarbons?

Carbon atoms also form hydrocarbons with double bonds and triple bonds. Remember that each pair of shared electrons forms one bond. In a double bond, two atoms share two pairs of electrons. In a triple bond, two atoms share three pairs of electrons. An **unsaturated hydrocarbon** is a hydrocarbon with one or more double or triple bonds. They are called unsaturated because each carbon atom is bonded to less than four other hydrogen atoms. The figure below shows the formulas of three members of the series of double-bonded unsaturated hydrocarbons. ✔

Ethene The simplest unsaturated hydrocarbon with a double bond is ethene. An ethene molecule has two carbon atoms joined by a double bond. Ethene helps ripen fruits and vegetables. It is also used to make plastic milk and soft-drink bottles.

Propene An unsaturated hydrocarbon molecule with three carbon atoms and one double bond is propene. Look at the structural formula of propene in the figure below. There is one double bond between two of the carbon atoms. Propene is used to make a strong plastic called polypropylene.

Butadiene Some unsaturated hydrocarbon molecules have more than one double bond. Butadiene (byew tuh DI een) has four carbon atoms and two double bonds. It is used to make synthetic rubber.

Ethene   **Propene**   **Butadiene**

Alkenes The names of the compounds in the figure on the previous page end in -*ene*. The names of all unsaturated hydrocarbons with at least one double bond end in -*ene*. These compounds are called alkenes.

What are unsaturated hydrocarbons that have triple bonds called?

Unsaturated hydrocarbons also can have triple bonds. Unsaturated hydrocarbons with one or more triple bonds are called alkynes. Ethyne (EH thine) is the first member of the alkyne series. An ethyne molecule has two carbon atoms joined by a triple bond. You can see the structural formula of ethyne in the figure. Ethyne is also called acetylene (uh SE tuh leen). Acetylene is a gas used for welding because it produces high heat as it burns.

$$H - C \equiv C - H$$

Ethyne or Acetylene

C_2H_2

Picture This

8. **Use Models** Circle the triple bond in the ethyne molecule in the figure.

What are hydrocarbon isomers?

Suppose you had ten blocks you could snap together to make different arrangements. You could make many different arrangements from the same ten blocks. Organic molecules can have different arrangements and still have the same chemical formula. <u>Isomers</u> (I suh murz) are compounds that have different arrangements but the same chemical formula. The figure below shows two isomers, butane and isobutane. You can see that butane's atoms are arranged in a straight chain. The isobutane molecule is not straight. It has a branch.

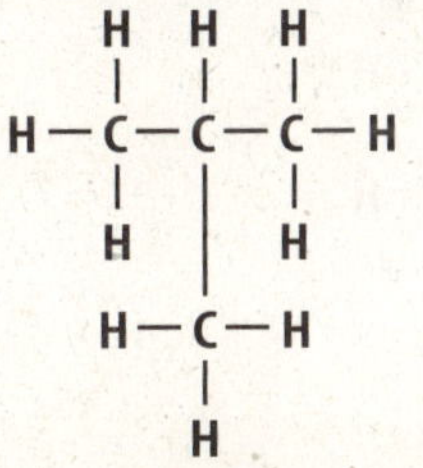

Isobutane
C_4H_{10}

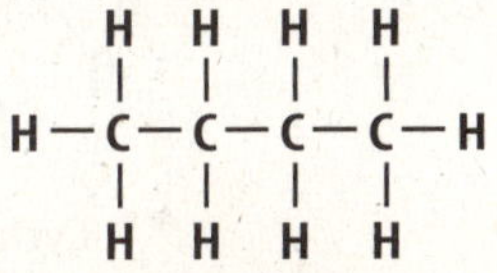

Butane
C_4H_{10}

Picture This

9. **Identify** Highlight the branch on the carbon chain in the structural formula of isobutane.

How are butane and isobutane alike?

Both the butane and the isobutane molecules have four carbon atoms and ten hydrogen atoms. They have the same chemical formula, C_4H_{10}. But, butane and isobutane have different chemical and physical properties because each molecule has a different shape. This is true for all hydrocarbon isomers.

Are all hydrocarbon molecules either straight chains or branched?

By now, you might think that all hydrocarbon molecules are made by carbon atoms in a chain that has a beginning and an end. Some molecules form rings. Look at the two molecules in the figure below. Hexane has six carbon atoms bonded to 14 hydrogen atoms. They form a straight chain.

Cyclohexane Remove one hydrogen atom from each end of a hexane molecule. The carbon atoms at each end can now form a single bond with each other. The new molecule has six carbon atoms in the shape of a ring, as shown in the second figure. Notice that each carbon atom still has four single bonds. The new molecule is called cyclohexane. The prefix *cyclo-* tells you that the molecule is cyclic, or ring-shaped.

Picture This

10. Observe How is the shape of a molecule of cyclohexane different than a molecule of hexane?

_____

_____

_____

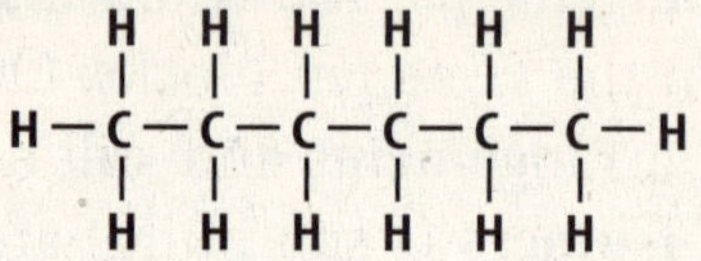

Hexane
C_6H_{14}

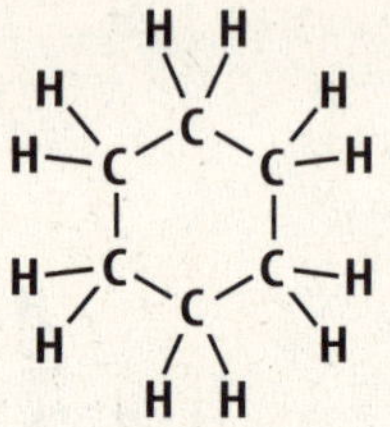

Cyclohexane
C_6H_{12}

What are ring-shaped molecules?

Ring-shaped molecules can have one or more double bonds. There are many chemical compounds that have ring structures. Molecules of the substances fructose, glucose, and sucrose are all ring structures.

● After You Read

Mini Glossary

hydrocarbon: a compound that has only carbon and hydrogen atoms

isomers: compounds that have different arrangements but the same chemical formula

organic compound: most compounds that contain carbon

saturated hydrocarbon: a hydrocarbon molecule that has only single bonds

unsaturated hydrocarbon: a hydrocarbon with one or more double or triple bonds

1. Review the terms and their definitions in the Mini Glossary. Write a sentence that explains why methane is a hydrocarbon.

   _____

   _____

2. Complete the concept web by filling in the name of the type of organic compound you learned about in this section and the names of the three types of that organic compound.

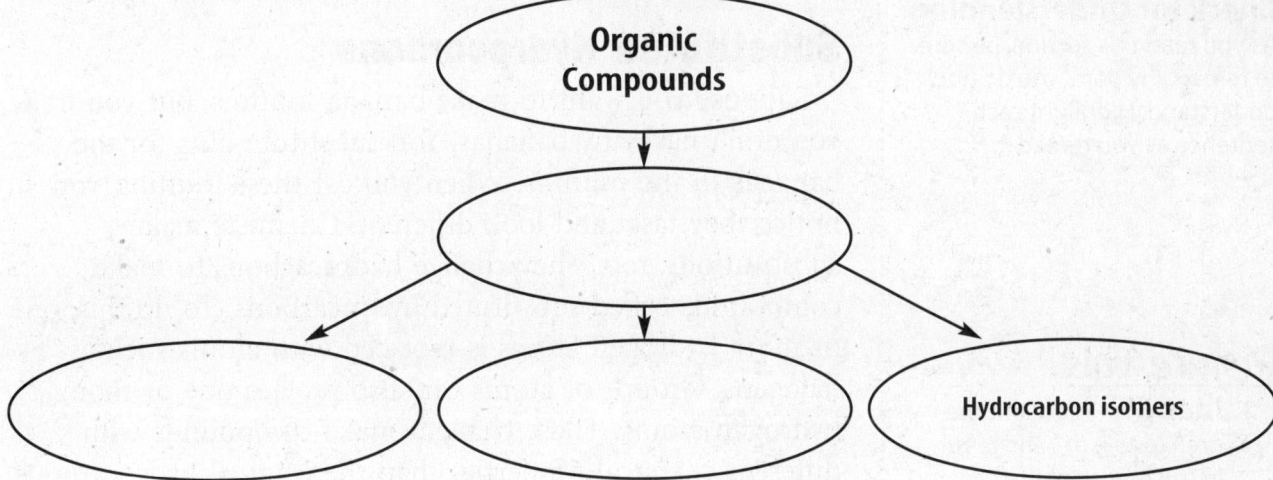

3. You highlighted the main idea in each paragraph. How did you decide what to highlight?

   _____

   _____

   _____

Science Online Visit **ips.msscience.com** to access your textbook, interactive games, and projects to help you learn more about simple organic compounds.

End of Section

Carbon Chemistry

section ❷ Other Organic Compounds

What You'll Learn

■ how new compounds are formed by substituting hydrogens in hydrocarbons
■ what kinds of compounds can be made from substitution

Check for Understanding
As you read this section, be sure to reread any parts you do not understand. Highlight each sentence as you reread it.

Picture This

1. **Identify** Highlight the prefixes *di-*, *tri-*, and *tetra-* in the names of the hydrocarbons. Then circle the chlorine atoms in each structural formula. Notice how each prefix matches the number of chlorine atoms.

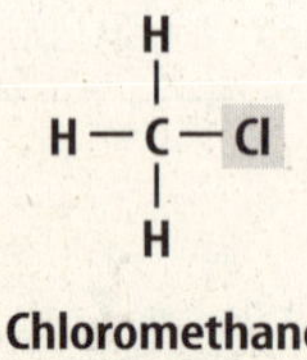

● Before You Read

What is a substitution? Give example of an object and another object that you could use as a substitution for the first object. Explain.

_____

_____

● Read to Learn

Substituted Hydrocarbons

Suppose you want to make banana muffins, but you find you don't have any bananas. You substitute nuts for the bananas in the muffins. When you eat these muffins, you notice they taste and look different. Chemists make substitutions, too. They change hydrocarbons to make compounds called substituted hydrocarbons. To do this, one or more hydrogen atoms is replaced with atoms such as halogens. Groups of atoms can also replace one or more hydrogen atoms. These changes make compounds with different chemical properties than the original hydrocarbons.

The figure below shows some compounds that can form from methane. One or more chlorine atoms have been added in place of the hydrogen atoms. Notice that each molecule has one more chlorine atom and one less hydrogen atom than the molecule before it.

H | H—C—Cl | H	Cl | H—C—Cl | H	Cl | H—C—Cl | Cl	Cl | Cl—C—Cl | Cl
Chloromethane CH_3Cl	**Dichloromethane** CH_2Cl_2	**Trichloromethane** $CHCl_3$	**Carbon tetrachloride** CCl_4

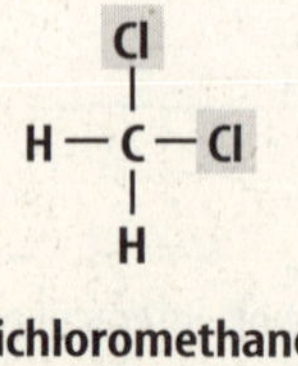

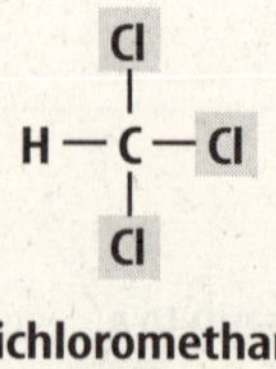

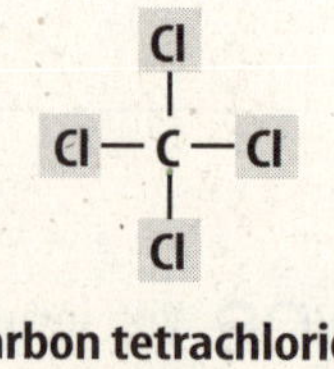

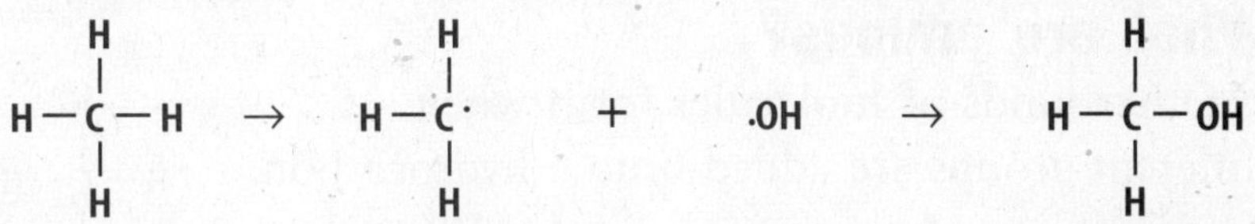

$$H-\overset{\overset{\displaystyle H}{|}}{\underset{\underset{\displaystyle H}{|}}{C}}-H \quad \rightarrow \quad H-\overset{\overset{\displaystyle H}{|}}{\underset{\underset{\displaystyle H}{|}}{C}}\cdot \quad + \quad \cdot OH \quad \rightarrow \quad H-\overset{\overset{\displaystyle H}{|}}{\underset{\underset{\displaystyle H}{|}}{C}}-OH$$

| CH_4 | CH_3- | $-OH$ | CH_3OH |
| Methane | Methyl group | Hydroxy group | Methanol |

What are alcohols?

Chemists also can add groups of atoms to hydrocarbons to make different organic compounds. The figure above shows how the alcohol methanol is formed. First, a hydrogen atom is removed from a methane molecule. Then, a hydroxyl (hi DROK sul) group is added. A **hydroxyl group** is made up of an oxygen atom and a hydrogen atom joined by a covalent bond. The formula for the hydroxyl group is $-OH$. Alcohols form when a hydroxyl group takes the place of a hydrogen atom in a hydrocarbon. Larger alcohol molecules are made by adding more carbon atoms. The table below shows some common alcohols. You have probably used at least one of them.

Common Alcohols			
Uses	Methanol	Ethanol	Isopropyl Alcohol
Fuel	yes	yes	no
Cleaner	yes	yes	yes
Disinfectant	no	yes	yes

What are carboxylic acids?

Have you tasted vinegar? It is a solution of acetic acid and water. The figure on the right shows the structural formula of acetic acid. Notice that it looks like a methane molecule with one hydrogen atom removed. In its place is a carboxyl (car BOK sul) group. A **carboxyl group** is made of a carbon atom that has a double bond with one oxygen atom and a single bond with a hydroxyl group. Its formula is $-COOH$. A carboxylic acid forms when a carboxyl group is substituted in a hydrocarbon molecule. Many carboxylic acids are in foods. For example, citric acid is in citrus fruits such as oranges and lemons. When milk turns sour, lactic acid forms.

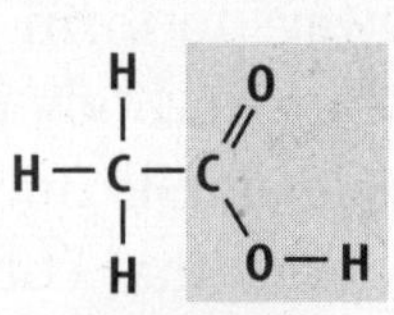

$$H-\overset{\overset{\displaystyle H}{|}}{\underset{\underset{\displaystyle H}{|}}{C}}-\overset{\displaystyle O}{\underset{\displaystyle O-H}{C}}$$

Acetic acid
CH_3COOH

2. Identify Circle the hydroxyl groups in the figure above.

Picture This

3. Use Tables Which alcohols in the table are used as fuel?

_____

_____

FOLDABLES™

B Main Ideas Make the following Foldable to help you identify main ideas about organic compounds.

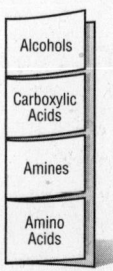

Alcohols

Carboxylic Acids

Amines

Amino Acids

Picture This

4. Identify Circle the amino group in the methylamine molecule.

Think it Over

5. Infer If the shaded single hydrogen atom in the figure on the right were replaced with CH_3, would the molecule still be glycine? Explain your answer.

____

____

____

What are amines?

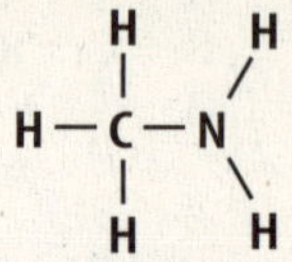

Methylamine
CH_3NH_2

Other kinds of molecules form when different groups are added onto a hydrocarbon. Amines form when an amino (uh MEE noh) group takes the place of a hydrogen atom in a hydrocarbon molecule. An **amino group** is a nitrogen atom joined by covalent bonds to two hydrogen atoms. The formula for an amino group is $-NH_2$. Amino groups are important because they are part of many biological compounds necessary for life. The figure above shows the structural and chemical formulas for an amine called methylamine. Methylamine forms when an amino group takes the place of one of the hydrogen atoms in a methane molecule.

What are amino acids?

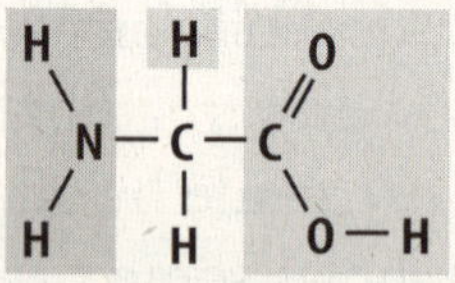

Glycine

You have learned that a group of atoms can take the place of a hydrogen atom at one end of a carbon chain. It is also possible to add groups onto both ends of a chain of carbon atoms. A group of atoms can also take the place of hydrogen atoms in the middle of a chain. The figure above shows an amino acid called glycine. An **amino acid** forms when both an amino group ($-NH_2$) and a carboxylic acid group ($-COOH$) take the place of hydrogen atoms on the same carbon atom. Amino acids are necessary for human life.

Proteins Amino acids are the building blocks of proteins. Proteins are large biological molecules that living cells need. Twenty different amino acids bond together in different ways to form the proteins needed by the human body. The human body makes some of the amino acids it needs. Glycine is one of these amino acids. You do not have to eat foods that contain the amino acids made by your body.

There are nine amino acids your body needs that it cannot make. They are called essential amino acids. You must eat foods that contain protein so your body gets all the amino acids it needs. Some foods that contain protein are meat, eggs, and milk.

● After You Read

Mini Glossary

amino acid: forms when both an amino group ($-NH_2$) and a carboxylic acid group ($-COOH$) take the place of hydrogen atoms on the same carbon atom

amino group: a nitrogen atom joined by covalent bonds to two hydrogen atoms

carboxyl group: made of a carbon atom that has a double bond with one oxygen atom and a single bond with a hydroxyl group, $-OH$

hydroxyl group: made up of an oxygen atom and a hydrogen atom joined by a covalent bond

1. Review the terms and their definitions in the Mini Glossary. Write a sentence about how an alcohol is formed.

_____

_____

2. Match the terms with the correct molecule or group of atoms. Put the letter of the term in Column 2 on the line in front of the diagram it matches in Column 1.

Column 1

_____ 1. $-O-H$

_____ 2. 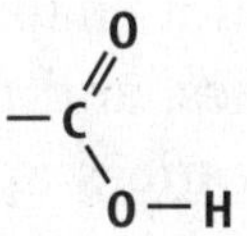 $-C\overset{\displaystyle O}{\underset{\displaystyle O-H}{\big\|}}$

_____ 3. 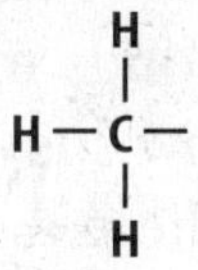 $H-\overset{\displaystyle H}{\underset{\displaystyle H}{C}}-$

_____ 4. 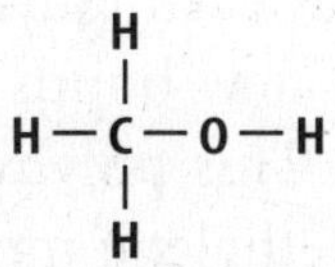 $H-\overset{\displaystyle H}{\underset{\displaystyle H}{C}}-O-H$

_____ 5. 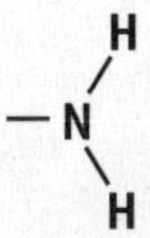 $-N\overset{\displaystyle H}{\underset{\displaystyle H}{\diagdown}}$

Column 2

a. amino group
b. alcohol
c. hydroxyl group
d. carboxyl group
e. methyl group

Science Online Visit **ips.msscience.com** to access your textbook, interactive games, and projects to help you learn more about other organic compounds.

End of Section

Carbon Chemistry

section ❸ Biological Compounds

What You'll Learn

- how large organic molecules are made
- what organic molecules do in the body

Mark the Text

Locate Terms Highlight the definition of each word that appears in bold.

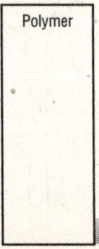

FOLDABLES™

C Take Notes As you read this section, write down important information about polymers on a half-sheet of notebook paper.

Polymer

● Before You Read

List three foods you think are good for your body.

_____

● Read to Learn

What is a polymer?

You have learned about some simple organic molecules. Now you will learn about larger molecules made from these simple molecules. One type of large molecule is a polymer (PAH luh mur). A **polymer** is a large molecule made up of many small organic molecules that are connected by covalent bonds to form a chain. You use polymers every day. Plastics, synthetic fabrics, and nonstick surfaces on cookware are all polymers. **Monomers** are the small organic molecules that link together to form a polymer.

Polymerization (puh lih muh ruh ZAY shun) is the chemical reaction that bonds monomers together to form a polymer. The figure below shows the polymerization reaction that produces the polymer polyethylene from molecules of ethylene. The double bond breaks between the two carbon atoms in each ethylene molecule. Then, the carbon atoms form new bonds with carbon atoms in other ethylene molecules. This polymerization reaction repeats many times. A polyethylene molecule can contain 10,000 ethylene monomers.

Ethylene   Ethylene   Polyethylene

Proteins Are Polymers

You have learned some about proteins. A **protein** is a polymer made of a chain of amino acids linked together. Both ends of an amino acid can link with other amino acids. The amino acids keep linking together until the protein chain is complete. Some proteins speed up chemical reactions in cells. Your hair, fingernails, and muscles are made of proteins. All the cells in your body contain proteins. ☑

Your body uses different proteins for different things. Recall that your body makes many amino acids. It cannot make essential amino acids. They must come from the food you eat such as the foods in the table below.

Approximate Protein Content of Some Foods	
Foods	**Protein Content (g)**
Chicken breast (113 g)	28
Eggs (2)	12
Whole milk (240 mL)	8
Peanut butter (30 g)	8
Kidney beans (127 g)	8

Carbohydrates

The table below lists some foods and the amount of carbohydrates in them. A **carbohydrate** is an organic compound that contains only carbon, hydrogen, and oxygen. There are usually twice as many hydrogen atoms as there are oxygen atoms in a carbohydrate. The different kinds of carbohydrates are divided into three groups. They are sugars, starches, and cellulose.

Your body breaks down carbohydrates into simple sugar molecules that it can use for energy. The day before a big race, athletes often eat a lot of foods with sugars and starches, like pasta, bread, potatoes, and fruit.

Approximate Carbohydrates in Some Foods	
Foods	**Carbohydrate Content (g)**
Apple (1)	21
White rice ($\frac{1}{2}$ cup)	17
Baked potato ($\frac{1}{2}$ cup)	15
Wheat bread (1 slice)	13
Milk (240 mL)	12

✔ **Reading Check**

1. **Identify** What are proteins made of?

_____

_____

Applying Math

2. **Calculate** About how many grams of protein are in one egg?

_____

_____

FOLDABLES

Ⓓ **Classify** Use quarter-sheets of paper to classify foods. Name the foods that contain the biological compounds.

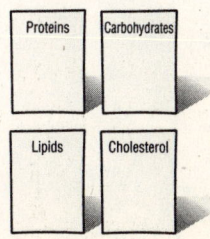

What are simple sugars?

If you like cookies or ice cream, then you know something about sugars. Sugars make fresh fruit and desserts sweet. Simple **sugars** are carbohydrates that have five, six, or seven carbon atoms arranged in a ring. Two common simple sugars are glucose and fructose. ☑

What are glucose and fructose?

The figure below shows a glucose molecule and a fructose molecule. Glucose forms a six-carbon ring. Glucose is found in naturally sweet foods like grapes and bananas. Ripe fruit, honey, and corn syrup contain fructose. Fructose is added to many foods to make them sweet.

✔ **Reading Check**

3. **Describe** the shape of a simple sugar molecule.

_____

_____

Picture This

4. **Determine** How many carbon atoms are in glucose? How many are in fructose?

_____

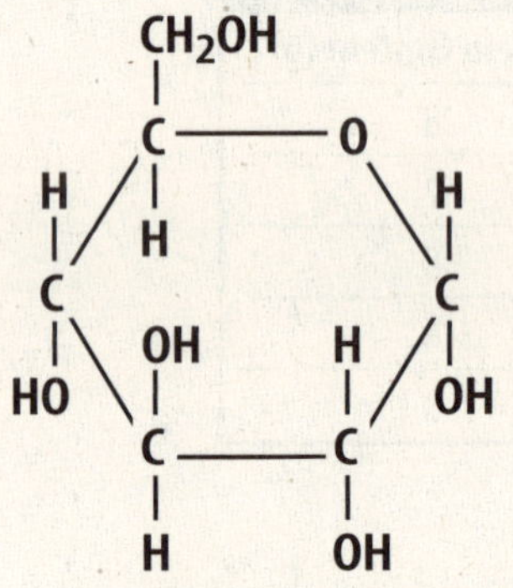

Glucose 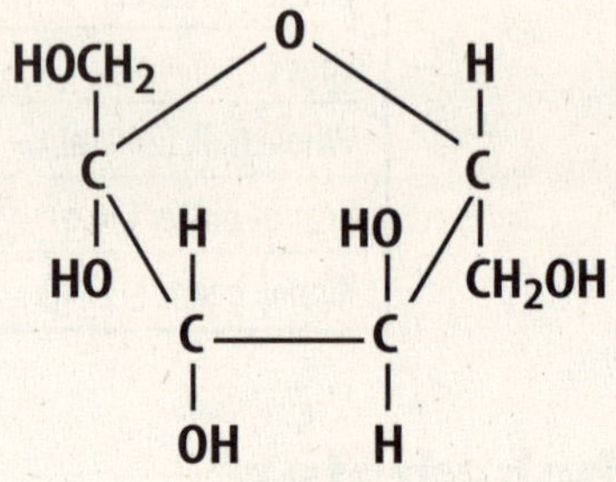   Fructose

What is sucrose?

The sugar you probably are most familiar with is sucrose. That is the sugar used in most baked goods. The sugar you probably have in your sugar bowl is sucrose. The figure below shows a molecule of sucrose. Notice that sucrose is a combination of the sugars glucose and fructose.

Picture This

5. **Identify** Circle the glucose in the molecule of sucrose.

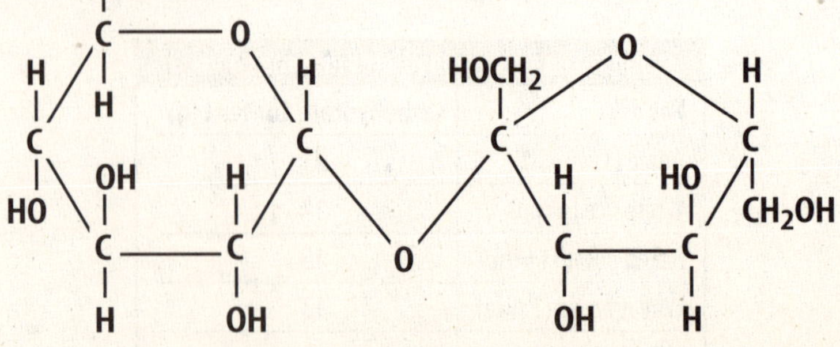

Sucrose

How does your body use sugars?

Your body cannot use sucrose for energy because sucrose cannot move through cell membranes. Your body breaks down sucrose into glucose and fructose. These smaller molecules can move into your cells. Inside the cells, glucose and fructose are broken down even further and used for energy.

What are starches?

Starches are carbohydrates found in foods like rice, wheat, corn, potatoes, lima beans, and peas. **Starches** are polymers made of hundreds or even thousands of glucose molecules joined together. Each sugar molecule in a starch gives off energy when it is broken apart. Because there are so many sugar molecules in starches, a large amount of energy is released. That means starches are a good source of energy. ☑

What other polymers are made of glucose?

Cellulose and glycogen are two other important polymers made of glucose monomers. A cellulose molecule is made of long chains of glucose molecules linked together. The long, stiff fibers that make up the walls of plant cells are made of cellulose. You can see these fibers in the long strands you pull off a stalk of celery.

A glycogen molecule is a polymer made of chains of glucose molecules. But, they are not straight chains. Glycogen molecules have many branches. Animals make glycogen and store it in their muscles and liver. They can keep it there until their bodies need energy.

Humans cannot use cellulose for energy. The human digestive system cannot break cellulose into sugars. Cows and some other animals have special digestive systems that can break down cellulose into sugars. ☑

Lipids

A **lipid** is an organic compound that contains the same elements as carbohydrates—carbon, oxygen, and hydrogen. But lipids have different amounts of these elements. Lipids are made up of two parts. One part, glycerol, has three −OH groups. The other part is three molecules of carboxylic acid. Lipids are found in many foods such as oils, butter, and cheese. Lipids are called fats and oils. Lipids also are found in greases and waxes such as beeswax. Bees release wax from a gland in their abdomens to make beeswax. Beeswax is part of the honeycomb bees make.

✔ **Reading Check**

6. **Identify** What kind of molecule makes up a starch?

_____

_____

✔ **Reading Check**

7. **Explain** Why are humans not able to use cellulose for energy?

_____

_____

_____

Where do lipids store energy?

If you eat more food than your body needs for energy, your body stores the energy by making lipids. The chemical reaction that makes lipids is endothermic. This is a reaction in which energy is absorbed. That means energy is stored in the chemical bonds of lipids. When the bonds are broken, the energy is released. Once it is released, your body can use the energy. ✓

Your body stores lipids in case you need extra energy for some activity. It also stores lipids in case you might not be able to eat for awhile. But, if you regularly eat more food than you need, your body produces large amounts of lipids. Your body stores these lipids as fat.

Are all lipids the same?

You learned the difference between saturated and unsaturated hydrocarbons. Remember, saturated hydrocarbon molecules have only single bonds between carbon atoms. Unsaturated hydrocarbon molecules have one or more double or triple bonds between carbon atoms. Lipid molecules also can be saturated or unsaturated.

Stearic Acid The figure on the right shows a molecule of stearic acid. Bacon and butter contain stearic acid. Stearic acid is a saturated lipid because there are only single bonds between the carbon atoms. When a lipid contains only single bonds, the molecule is a straight chain. Molecules with straight chains of atoms can pack together tightly. Molecules that pack together tightly usually form solids. Other solids that are made of saturated lipids are margarine and shortening. Another name for a saturated lipid is saturated fat. Most saturated fats come from animals.

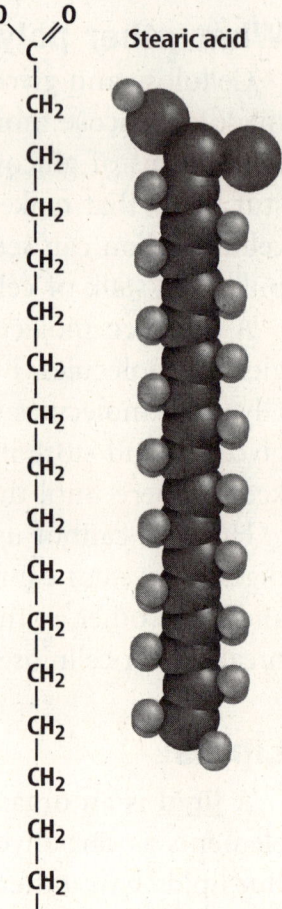

Stearic acid

Copyright © Glencoe/McGraw-Hill, a division of The McGraw-Hill Companies, Inc.

Reading Check

8. Determine Where do lipids store energy?

_____

_____

Picture This

9. Describe the shape of a saturated fat.

_____

_____

Oleic Acid The figure on the right shows a molecule of oleic acid. Notice that oleic acid has a double bond between two of the carbon atoms in its chain. That makes oleic acid an unsaturated lipid. Unsaturated molecules bend wherever there is a double bond. Bent molecules cannot pack as tightly together as straight molecules can. Unsaturated lipids usually are liquid oils, like olive oil. Another name for an unsaturated lipid is unsaturated fat. Corn oil is another unsaturated fat. Unsaturated fats come from plants.

Doctors have observed that people who eat a lot of saturated fats are more likely to have problems like heart disease. There are many foods with unsaturated fats. Making wise choices in the foods you eat can help keep you healthy.

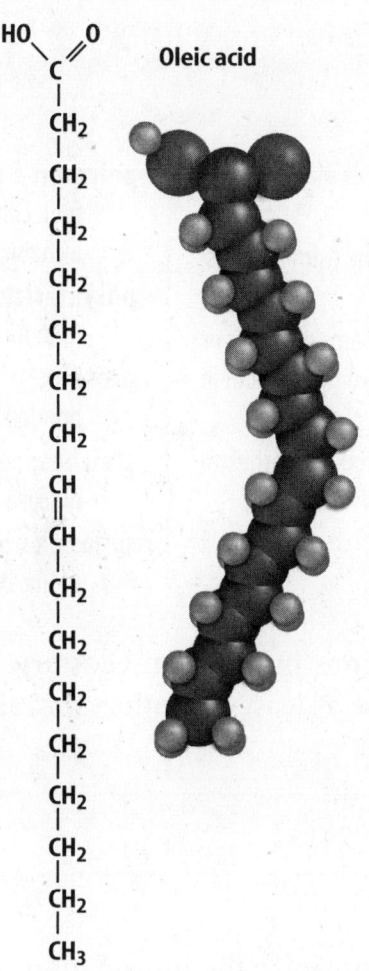

Oleic acid

Cholesterol

Cholesterol is a lipid found in foods that come from animals, such as meat, butter, eggs, and cheese. Even if you don't eat foods that contain cholesterol, your body makes its own. It can change plant oils like corn oil into cholesterol. Your body needs some cholesterol to build cell membranes.

Atherosclerosis Too much cholesterol can be harmful. Cholesterol can collect on the inside walls of arteries. This is called atherosclerosis (ath uhr oh skluh ROH sis). This blocks the blood flow through the arteries. When the blood cannot flow easily through the arteries, it causes high blood pressure. High blood pressure can lead to heart disease. If you eat fewer foods that have saturated fats and cholesterol, you might lower your cholesterol. This can help you reduce your risk of having heart problems.

Picture This

10. **Understand Scientific Illustrations** Highlight the carbon-carbon double bond in the molecule of oleic acid.

Think it Over

11. **Draw Conclusions** Why might a diet that contained a lot of meat, butter, and cheese be harmful to your health?

_____

_____

_____

● After You Read

Mini Glossary

carbohydrate: an organic compound that is made of only carbon, hydrogen, and oxygen

cholesterol: a lipid found in foods that come from animals, like meat, butter, eggs, and cheese

lipid: an organic compound that contains the same elements as carbohydrates—carbon, oxygen, and hydrogen—but in different amounts

monomer: small organic molecules that links together to form a polymer

polymer: a large molecule made up of many small organic molecules that are connected by covalent bonds to form a chain

polymerization: the chemical reaction that bonds monomers together to form a polymer

protein: a polymer made of a chain of amino acids bonded together

starches: polymers made of hundreds or even thousands of glucose molecules joined together

sugars: carbohydrates that have five, six, or seven carbon atoms bonded together in a ring

1. Review the terms and definitions in the Mini Glossary. Write a sentence that explains how a polymer is formed. Use at least two other vocabulary terms in your explanation.

_____

_____

2. Fill in the chart to help you organize the information you learned about biological compounds.

```
                          ┌─────────────────────┐
                          │ Biological Polymers │
                          └─────────────────────┘
```

_____	Carbohydrates	_____	Cholesterol
What your body uses it for:	What your body uses it for:	What your body uses it for:	What your body uses it for:
1. Speeds up	1.	1. Store energy	1.
chemical reactions			
2.			

End of Section

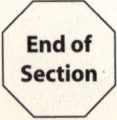Science Online Visit **ips.msscience.com** to access your textbook, interactive games, and projects to help you learn more about biological compounds.

152 Carbon Chemistry

Motion and Momentum

section ❶ What is motion?

● Before You Read

When you move from place to place, how do you know you have moved? Write what you think on the lines below.

_____

_____

● Read to Learn

Matter and Motion

When you are sitting quietly in a chair, are you in motion? It may surprise you to know that all matter in the universe is always in motion. Think about it. In the chair, your heart beats and you breathe. Your blood circulates through your veins. Electrons move around the nuclei of every atom in your body.

Changing Position

How do you know if something is in motion? Something is in motion if it is changing position. Changing position means moving from one place to another. Imagine runners in a 100-meter race. They sprint from the start line to the finish line. Their positions change, so they are in motion.

What is relative motion?

To find out if something changes position, you need a reference point to compare it to. An object changes position if it moves when compared to a reference point. Imagine you are competing in the 100-meter race. You begin just behind the start line. When you pass the finish line, you are 100 m from the start line. If you use the start line as your reference point, then your position has changed by 100 m when compared to the start line. You were in motion. ☑

What You'll Learn

- what distance, speed, and velocity are
- how to graph motion

Mark the Text

Underline As you read, underline material you do not understand the first time you read it. Reread the information until you understand it. Ask your teacher if you still do not understand it after rereading it.

✔ Reading Check

1. **Explain** What do you compare an object to when determining the object's motion?

_____

_____

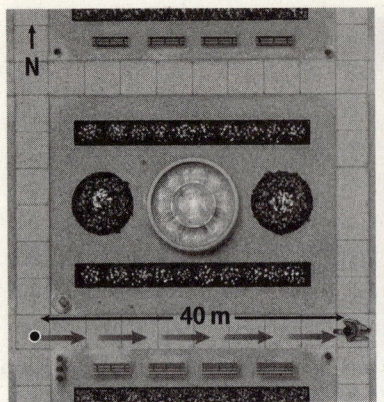

Distance: 40 m
Displacement: 40 m east

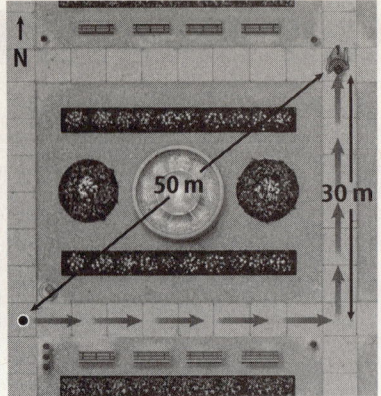

Distance: 70 m
Displacement: 50 m northeast

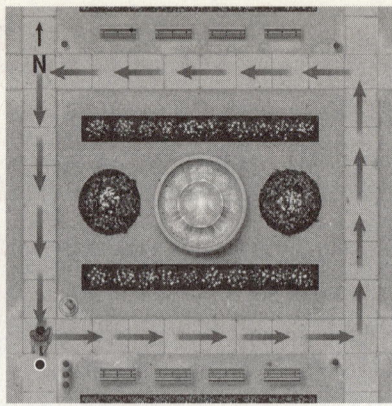

Distance: 140 m
Displacement: 0 m

Picture This

2. **Explain** Why is the displacement in the third figure zero?

_____

_____

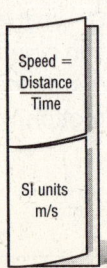
What are distance and displacement?

Suppose you walk from your house to the park around the block. How far away is it? That depends on whether you are talking about distance or displacement. Distance is the length of the route you travel.

Suppose you travel 200 m from your house to the park. How would you describe your location now? You could say you are 200 m from your house. But where you are depends on both the distance you travel and direction. To describe exactly where you are, you need to tell the direction from your house. Displacement includes the distance between your starting and ending points and the direction in which you travel. The figure above shows the difference between distance and displacement.

Speed

When you describe motion, you usually want to say how fast something is moving. The faster something is moving, the less time it takes to travel a certain distance. The slower something is moving, the more time it takes to travel a certain distance. **Speed** is the distance traveled divided by the time it takes to travel that distance. Speed can be calculated with this equation:

$$\textbf{speed} \text{ (in meters/second)} = \frac{\textbf{distance} \text{ (in meters)}}{\textbf{time} \text{ (in seconds)}}$$

$$s = \frac{d}{t}$$

In SI units, distance is measured in m and time is measured in s. The SI measurement for speed is meters per second (m/s). This is the SI distance unit divided by the SI time unit.

What is average speed?

A car in city traffic might have to speed up and slow down many times. How could you describe its speed? One way is to determine the car's average speed between where it starts and stops. The speed equation can be used to find average speed. <u>Average speed</u> is the total distance traveled divided by the total time taken to travel the distance.

What is instantaneous speed?

Have you ever watched the speedometer when you are riding in a car? If the speedometer reads 50 km/h, the car is traveling at 50 km/h at that instant. <u>Instantaneous speed</u> is the speed of an object at one instant of time. ☑

How do average and instantaneous speed differ?

If it takes two hours to travel 200 km in a car, the average speed would be 100 km/h. But the car probably was not moving at this speed the whole time. It might have gone faster on the freeway and stopped at stoplights. There your speed was 0 km/h. If the car were able to travel 100 km/h the whole time, you would have moved at a constant speed.

For another example, see the diagram of the two balls below. Both balls have the same average speed because they both travel 3 m in 4 s. The top ball is moving at a constant speed. In each second, it moves the same distance. The bottom ball is moving at different speeds. Its instantaneous speed is fast between 0 s and 1 s, slower between 2 s and 3 s, and even slower between 3 s and 4 s.

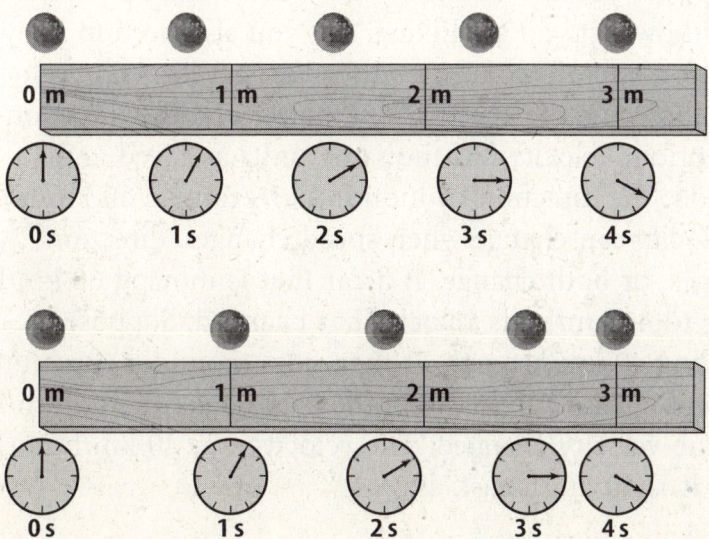

✔ Reading Check

3. Identify What type of speed does the speedometer in a car show?

_____

_____

Picture This

4. Calculate What is the average speed of both balls in the diagram? Show all your work.

_____

_____

_____

Graphing Motion

You can show the motion of an object with a distance-time graph. In a distance-time graph, time is plotted on the horizontal axis. Distance is plotted on the vertical axis.

How do distance-time graphs compare speed?

The graph below is a distance-time graph that shows the motion of two students walking. According to the graph, after 1 s student A traveled 1 m. Her average speed is 1 m/1 s, or 1 m/s. Student B traveled only 0.5 m in 1 s. His average speed is 0.5 m/1 s, or 0.5 m/s. So student A traveled faster than student B. Now compare the steepness of the lines in the graph. The line for student A is steeper than the line for student B. A steeper line shows a faster speed. If the line is horizontal, no change in position happens. A horizontal line means a speed of zero.

Applying Math

5. Calculate Look at the graph. How much farther has student A walked in 2 seconds than student B?

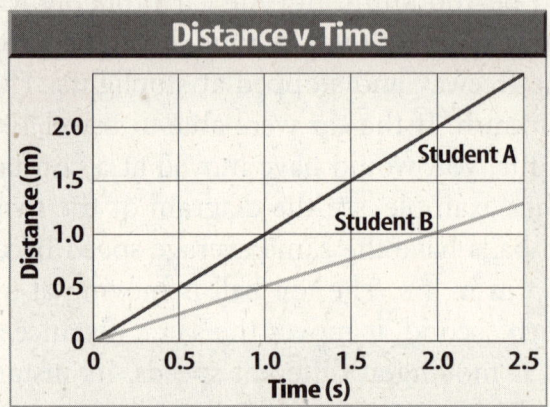

Velocity

Suppose you are hiking in the woods. You may want to know how fast you are hiking. But you also need to know the direction you are going or you might get lost. The **velocity** of an object is the speed of the object and the direction of its motion. Velocity has the same units as speed and includes the direction of motion, for example 20 km/h east.

Velocity can change when speed changes, direction changes, or both change. If a car that is moving 60 km/h slows to 40 km/h, its velocity has changed. Suppose a car is traveling 40 km/h north. It then goes around a curve until it is heading east. All the time, the car's speed was 40 km/h. But the velocity changed. The velocity was 40 km/h north. Now it is 40 km/h east. ✔

✔ **Reading Check**

6. Explain When the car's motion changed from 40 km/h north to 40 km/h east, what changed?

● After You Read

Mini Glossary

average speed: equals the total distance traveled divided by the total time taken to travel the distance

instantaneous speed: the speed of an object at one instant of time

speed: equals the distance traveled divided by the time it takes to travel that distance

velocity: the speed of an object and the direction of its motion

1. Review the terms and their definitions in the Mini Glossary. Ramona divided the distance from her house to school by the time it took her to walk that distance. What quantity did Ramona find? Explain your answer in a complete sentence.

_____

_____

2. The distance-time graph below is for a bicyclist in a bicycle race.

a. What was the bicyclist's average speed after two hours?

_____

b. What happened to her speed during the race?

_____

_____

c. How can you tell?

_____

_____

d. What was her average speed for the entire race?

_____

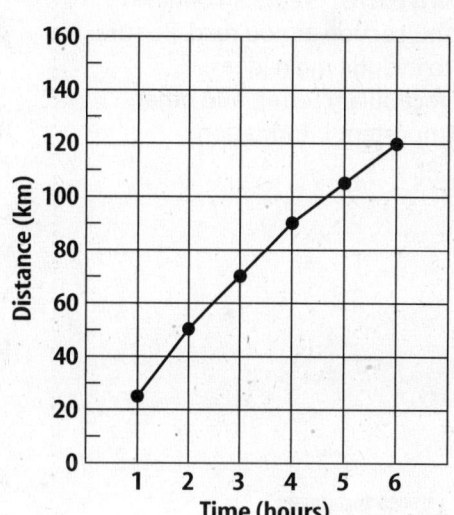

 **Science** nline Visit **ips.msscience.com** to access your textbook, interactive games, and projects to help you learn more about motion.

End of Section

Reading Essentials   **157**

Motion and Momentum

section ❷ Acceleration

What You'll Learn
- what acceleration is
- to predict how acceleration affects motion

Study Coach

Outline Create an outline of this section as you read. Be sure to include main ideas, vocabulary terms, and other important information.

FOLDABLES™

B Classify Make the following three-tab Foldable to help you classify and understand the different types of acceleration.

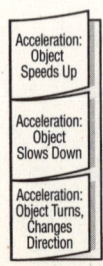

Acceleration: Object Speeds Up

Acceleration: Object Slows Down

Acceleration: Object Turns, Changes Direction

● Before You Read

Have you ever been in a foot race? What kinds of things are measured in a foot race?

_____

_____

● Read to Learn

Acceleration and Motion

Have you ever seen a rocket launch? When the rocket first lifts off, it seems to move very slowly. But very soon the rocket is moving at a fast speed. How can you describe the change in the rocket's motion? When an object changes its motion, it is accelerating. <u>Acceleration</u> is the change in velocity divided by the time it takes for the change to happen.

How is speeding up acceleration?

When you first get on a bike, it is not moving. When you start pedaling, the bike moves faster and faster. This is acceleration. An object that is already moving can accelerate too. Imagine you are biking along a level path. When you start to pedal harder, your speed increases. When the speed of an object increases, the object is accelerating.

How is slowing down acceleration?

Suppose you are biking at a speed of 4 m/s. If you brake, you will slow down. It might sound odd, but when you slow down you are accelerating. Any change in velocity is acceleration. Acceleration happens when an object speeds up or slows down.

When an object is speeding up, its acceleration is in the same direction as its motion. When an object is slowing down, its acceleration is in the opposite direction of its motion.

How is changing direction acceleration?

Remember that acceleration is a change in velocity. A change in velocity can be a change in speed, direction, or both. So, when an object changes direction, it accelerates. Think of yourself on a bicycle. When you turn the handlebars, you and the bicycle turn. The direction of the bike's motion changes, so the bike accelerates. Its acceleration is in the new direction that the bike travels.

Imagine throwing a ball straight up into the air. The ball starts out moving upward. After a while the ball stops moving upward and begins to come back down. The ball has changed its direction of motion. The ball is now accelerating downward.

Calculating Acceleration

If an object is moving in only one direction, its acceleration can be calculated with this equation.

$$\text{acceleration (m/s}^2) = \frac{\textbf{final speed (m/s)} - \textbf{initial speed (m/s)}}{\textbf{time (seconds)}}$$

$$a = \frac{(s_f - s_i)}{t}$$

In this equation, time is the length of time it takes for the motion to change. Initial speed is the starting speed. Acceleration has units of meters per second squared (m/s^2).

What are positive and negative acceleration?

Suppose you are riding your bike in a straight line. You speed up from 2 m/s to 8 m/s in 6 seconds.

$$a = \frac{(s_f - s_i)}{t}$$
$$= \frac{(8 \text{ m/s} - 2 \text{ m/s})}{6s} = \frac{6 \text{ m/s}}{6s} = +1 \text{ m/s}^2$$

So your acceleration is +1 m/s^2. Now suppose you slow down from 8 m/s to 2 m/s in 6 s.

$$a = \frac{(s_f - s_i)}{t}$$
$$= \frac{(2 \text{ m/s} - 8 \text{ m/s})}{6s} = \frac{-6 \text{ m/s}}{6s} = -1 \text{ m/s}^2$$

Your acceleration is −1 m/s^2.

1. **Explain** how an object accelerates when it changes direction.

_____

_____

_____

Applying Math

2. **Calculate** A sports car accelerates from zero to 28 m/s in 4 seconds. What is its acceleration?

_____

What does negative acceleration mean?

When you speed up, your acceleration is positive. When you slow down, your acceleration is negative. That is because when you slow down, your final speed is less than your initial speed. This gives you a negative value in the equation and a negative acceleration. ☑

How do you graph accelerated motion?

You can show the motion of an object moving in one direction on a graph. For this type of graph, speed is plotted on the vertical axis. Time is plotted on the horizontal axis. The graph below is an example.

Positive Acceleration In section A of the graph, speed increases from 0 m/s to 10 m/s during the first 2 seconds. Acceleration is 5 m/s². An object that is speeding up will have a line that slopes up on a speed-time graph.

Zero Acceleration In section B of the graph, the speed does not change. If speed does not change, the object is not accelerating. A horizontal line on a speed-time graph means zero acceleration.

Negative Acceleration In section C of the graph, the object goes from 10 m/s to 4 m/s in 2 s. Acceleration is −3 m/s². You can see that the line on the graph slopes downward as an object slows down.

Reading Check

3. Identify What type of acceleration do you have if you are slowing down?

Picture This

4. Interpret Data For how many seconds does the object in the speed-time graph have an acceleration of zero?

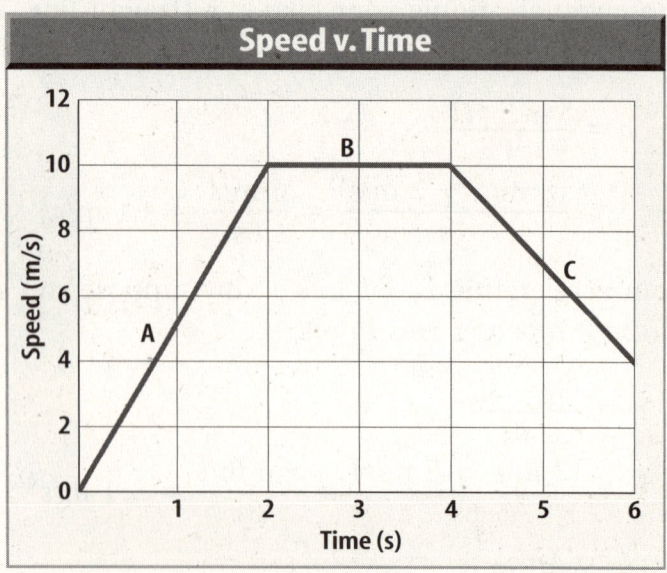

Speed v. Time

● After You Read

Mini Glossary

acceleration: the change in velocity divided by the time it
takes for the change to happen; occurs when an object
speeds up, slows down, or turns

1. Review the term and its definition in the Mini Glossary. Describe the term *acceleration* in your own words.

   _____

   _____

2. Fill in the chart with the different ways an object can accelerate.

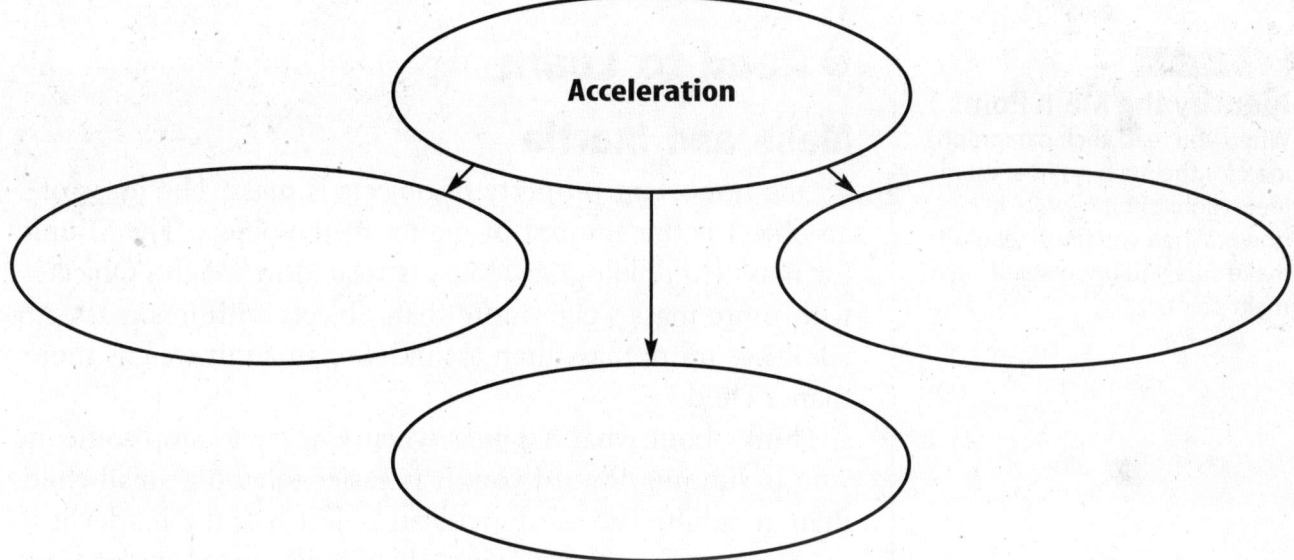

3. Why do you think that slowing down is sometimes called deceleration instead of acceleration?

   _____

   _____

   _____

Science Online Visit **ips.msscience.com** to access your textbook, interactive games, and projects to help you learn more about acceleration.

End of
Section

Motion and Momentum

section ❸ Momentum

What You'll Learn
- how mass and inertia are related
- what momentum is
- to use the law of conservation of momentum to predict motion

 Study Coach

Identify the Main Point
When you read each paragraph, look for the main point or main idea. Highlight it or write it down. When you finish reading, make sure you understand each main point.

💡 Think it Over

1. **Determine** Which has more inertia, a soccer ball or a bowling ball?

_____

⬤ Before You Read

What happens if you are riding in a car and the driver slams on the brakes? Explain on the lines below.

_____

_____

⬤ Read to Learn

Mass and Inertia

One important property of objects is mass. The **mass** of an object is the amount of matter in the object. The SI unit for mass is the kilogram. Mass is related to weight. Objects with more mass weigh more than objects with less mass. An adult has more mass than a child. So, an adult weighs more than a child.

Think about what happens when you try to stop someone who is running toward you. It is easier to stop a small child than an adult. The more mass an object has, the harder it is to start moving, stop moving, slow down, speed up, or turn. **Inertia** is the tendency of an object to resist a change in its motion. The more inertia an object has, the harder it is to change its motion.

Momentum

You know that the faster a bicycle moves, the harder it is to stop. The **momentum** of an object is the measure of how hard it is to stop the object. It depends on the object's mass and velocity. Momentum is usually symbolized by p.

$$\text{momentum (in kg} \cdot \text{m/s)} = \text{mass (in kg)} \times \text{velocity (in m/s)}$$
$$p = mv$$

Mass is measured in kilograms. Velocity is measured in meters per second. So, the unit of momentum is kilograms multiplied by meters per second (kg • m/s). Momentum has a direction that is the same as the direction of the velocity.

Conservation of Momentum

When you play billiards, you knock the cue ball into other balls. When a cue ball hits another ball, the motion of both balls changes. The cue ball slows down and may change direction. So its momentum decreases. The other ball starts moving. So its momentum increases.

What happens to lost momentum?

The momentum lost by the cue ball is gained by the other ball. This means that the total momentum of the two balls is the same before and after the collision. This is true for any collision, but only when no outside forces, like friction that slows down objects, act on the objects. The **law of conservation of momentum** states that the total momentum of objects that collide is the same before and after the collision. This is true for the collision of the billiard balls. It is also true for collisions of atoms, cars, football players, or any other matter. ☑

Using Momentum Conservation

Outside forces are almost always acting on objects that are colliding. These are forces like friction and gravity. But sometimes, these forces are very small and can be ignored. Then the law of conservation of mass can be used to predict how the motions of objects will change after a collision.

What happens after objects collide?

There are many ways that collisions can happen. Sometimes the objects that collide will bounce off each other. In another type of collision, objects stick to each other after they collide.

Bounce Off What happens when you knock down bowling pins with a bowling ball? Picture a bowling ball rolling down the alley and hitting some bowling pins. The bowling ball and pins bounce off each other. When the ball hits the pins, some of the ball's momentum is transferred to the pins. The ball slows down and the pins speed up. The speeds change, but the total momentum does not. Momentum is conserved.

Applying Math

2. Use Formulas
Calculate the momentum of a 14-kg bicycle traveling north at 2 m/s. Show all your work.

_____

_____

✔ Reading Check

3. Identify The law of conservation of momentum affects objects that

a. rotate.
b. turn.
c. collide.
d. roll.

FOLDABLES™

ⓒ Organize Information
Make the following Foldable to help you organize information about how momentum is transferred and the law of conservation of momentum.

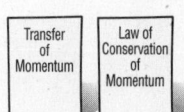

Imagine you are standing still on a pair of skates. You are not moving. Then someone standing in front of you throws you a backpack. You catch the backpack and begin to move backwards. You and the backpack move in the same direction that the backpack was moving before the collision.

You can use the law of conservation of momentum to find your velocity after you catch the backpack. Suppose the backpack has a mass of 2 kg and is tossed at a velocity of 5 m/s. Your mass is 48 kg and you have no velocity because you are standing still. So, your velocity before the collision is 0 m/s.

First, find the total momentum of you and the backpack. Remember, momentum equals mass times velocity.

$$
\begin{aligned}
\text{total momentum} &= \text{your momentum} + \text{backpack momentum} \\
&= (48 \text{ kg} \times 0 \text{ m/s}) + (2 \text{ kg} \times 5 \text{ m/s}) \\
&= 0 \text{ kg} \cdot \text{m/s} + 10 \text{ kg} \cdot \text{m/s} \\
&= 10 \text{ kg} \cdot \text{m/s}
\end{aligned}
$$

The law of conservation of momentum tells you that the total momentum before the collision is the same as the total momentum after the collision. After the collision, the total momentum does not change. You and the backpack have become one object and are moving at the same velocity. You can use the equation for momentum to find the final velocity.

$$
\begin{aligned}
\text{total momentum} &= (\text{mass of backpack} + \text{your mass}) \times \text{velocity} \\
10 \text{ kg} \cdot \text{m/s east} &= (2 \text{ kg} + 48 \text{ kg}) \times \text{velocity} \\
10 \text{ kg} \cdot \text{m/s east} &= (50 \text{ kg}) \times \text{velocity} \\
\frac{10 \text{ kg} \cdot \text{m/s east}}{(50 \text{ kg})} &= \text{velocity} \\
0.2 \text{ m/s} &= \text{velocity}
\end{aligned}
$$

Your velocity right after you catch the backpack is 0.2 m/s.

Think it Over

4. **Predict** Will the velocity of the student and the backpack together be faster or slower than the velocity of the backpack by itself?

_____

_____

Applying Math

5. **Calculate** Find the velocity of the student and the backpack if the backpack's mass is 3 kg, it was tossed at a velocity of 4 m/s, and the mass of the student is 57 kg. Show all your work.

_____

_____

_____

_____

_____

Stopping Friction between your skates and the ground will slow you down as you move on your skates. The momentum of you and the backpack will continue to decrease until you stop because of friction.

How can mass predict motion after collisions?

You can use the law of conservation of momentum to predict collisions between two objects. What happens when one marble hits another marble that is at rest? It depends on the masses of the marbles that collide. The figure shows a marble with a smaller mass hitting a marble with a larger mass. The larger marble is at rest. After the collision, the marble with a smaller mass bounces off in the opposite direction. The larger marble moves in the same direction that the small marble was moving.

What if the larger marble hits a smaller marble that is not moving? Both marbles will move in the same direction. But the marble with the smaller mass always moves faster than the marble with the greater mass.

How does bouncing affect momentum?

Two objects can also bounce off of each other. The two marbles in the figure have the same mass and are moving at the same speed. They bounce off each other when they collide. Before the collision, the momentum of each marble was the same but in opposite directions. So the total momentum was zero. That means that the total momentum after the collision has to be zero too. The two marbles must move in opposite directions with the same speed after the collision. Then the total momentum is zero again.

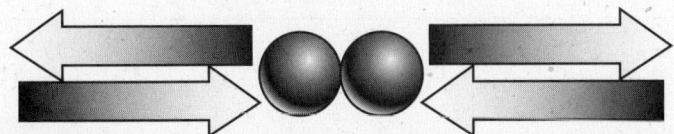

Picture This

6. Describe From which marble to which marble was momentum moved?

_____

_____

Applying Math

7. Analyze Would the total momentum still be zero if one marble had greater mass than the other marble?

_____

● After You Read

Mini Glossary

inertia: tendency of an object to resist a change in motion.

law of conservation of momentum: states that the total momentum of objects that collide is the same before and after the collision

mass: amount of matter in an object

momentum: the measure of how hard it is to stop an object

1. Review the terms and their definitions in the Mini Glossary. Explain in complete sentences what affects the inertia of an object.

_____

_____

2. The sketch below shows two marbles. The arrows show the size and the direction of the momentum of the two marbles. Draw arrows in the space below that show what will happen to these two marbles because of the law of conservation of momentum when they collide.

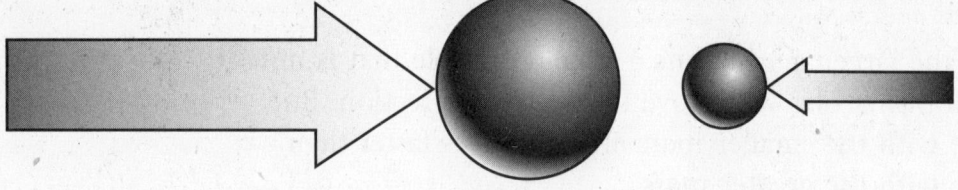

3. How can a football game be used to explain inertia and momentum?

_____

_____

_____

End of Section

Science ●nline Visit **ips.msscience.com** to access your textbook, interactive games, and projects to help you learn more about momentum.

Force and Newton's Laws

section ➊ Newton's First Law

● Before You Read

What do you have to do to move an object like a shopping cart? What causes motion?

_____

_____

● Read to Learn

Force

To make a soccer ball move, you kick it. You can pick up a book from your desk. If you hold the book in the air and then let it go, gravity pulls it to the floor. The motion of the soccer ball and the book was changed by something pushing or pulling on each of them.

A **force** is a push or a pull. When you throw a ball, your hand exerts, or puts, a force on the ball. Then, gravity puts another force on the ball. Gravity pulls it to the ground. When the ball hits the ground, the ground exerts a force on the ball to stop it from moving.

Forces can act on objects in different ways. For example, you can pick up a paper clip with a magnet. The magnet puts a force on the paper clip. Or, you can put a force on the paper clip with your hand to pick it up. If you let go of the paper clip, Earth's gravity exerts a force on the paper clip and it falls to the ground.

How can forces be combined?

More than one force can act on an object at the same time. Imagine holding a paper clip near a magnet. You, the magnet, and Earth's gravity are all putting forces on the paper clip. The **net force** is the combination of all forces acting on an object.

What You'll Learn

- the difference between balanced and net forces
- Newton's first law of motion
- how friction affects motion

Study Coach

Make Flash Cards As you read, write main ideas and vocabulary terms on note cards. When you finish reading, use your flash cards to make sure you understand the main ideas and terms.

FOLDABLES™

Ⓐ Compare and Contrast Make the following two-tab Foldable to organize important information about balanced and unbalanced forces.

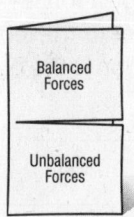

Balanced Forces

Unbalanced Forces

How does net force determine motion?

When more than one force is acting on an object, the net force determines the motion of the object. If a paper clip near a magnet is not moving, then the net force on the paper clip is zero.

How do forces combine to form the net force? If the forces are in the same direction, they add together. If two forces are in opposite directions, the net force is the difference between the two forces. If one of the forces is greater than the other, the motion of the object is in the direction of the greater force.

What are balanced forces?

Suppose you and a friend push on opposite ends of a wagon. You both push with the same force, and the wagon does not move. Your forces cancel each other because they are equal and in opposite directions. **Balanced forces** are two or more forces acting on an object that cancel each other and do not change the object's motion. The net force is zero if the forces acting on an object are balanced. The figure below shows balanced forces.

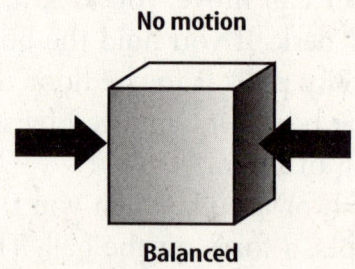

No motion

Balanced

What are unbalanced forces?

Unbalanced forces are forces that don't cancel each other. When unbalanced forces act on an object, the net force is not zero. The net force causes the motion of the object to change. The figure below shows how unbalanced forces change the motion of an object.

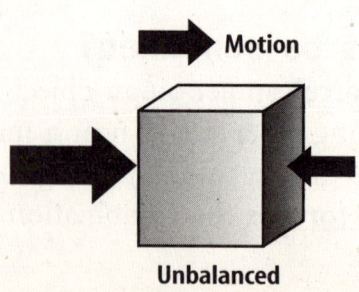

Motion

Unbalanced

Think it Over

1. **Infer** Imagine two people pushing on a door. One person pushes the door to close it. The other person pushes on the other side of the door to open it. If both people are pushing with the same force, what will happen to the door?

_____

_____

Picture This

2. **Identify** Look at the box with unbalanced forces. In which direction is the strongest force—to the right or to the left? In which direction is the box moving?

_____

_____

Newton's First Law of Motion

When you stand on a skateboard, you don't move. If someone gives you a push, you and the skateboard move. You and the skateboard were objects at rest until someone pushed you. An object at rest stays at rest unless an unbalanced force acts on it and causes it to move.

If someone pushes you on a skateboard, do you keep going? You probably would roll for a while, even after the person stops pushing you. An object can be moving without a net force acting on it.

One of the first to understand that objects could be moving without a force acting on them was Galileo Galilei. He was an Italian scientist who lived from 1564 to 1642. Galileo's ideas helped Isaac Newton understand motion better. Newton was able to explain the motion of objects in three rules. These rules are called Newton's laws of motion.

Newton's first law of motion describes how an object moves when the net force acting on it is zero. **Newton's first law of motion** states that if the net force acting on an object is zero, the object stays at rest or, if the object is already moving, it continues to move in a straight line with the same, or constant, speed.

Friction

Galileo knew that the motion of an object doesn't change unless an unbalanced force acts on it. So, why does a book stop sliding across a desktop just after you push it? There is a force acting on the sliding book. **Friction** is the force that resists sliding motion between two touching surfaces.

Friction also acts on objects moving through air or water. If two objects are touching each other, friction always will try to keep them from sliding past each other. Friction always will slow an object down.

What is static friction?

Have you ever tried to push something heavy, like a refrigerator or a sofa? At first heavy objects don't move. As you push harder and harder, the object will start to move. When you first push, the friction between the object and the floor is opposite to the force you are putting on it. So, the net force is zero. The object does not move. Static friction is the type of friction that prevents an object from moving when a force is applied.

B Classify Make the following table Foldable to help you organize Newton's laws of motion with examples from your own life. Write about Newton's first law as you read this section. You can complete your Foldable as you read Sections 2 and 3.

Force	Example in Your Life
First Law	

💡 **Think it Over**

3. **Infer** Think about Newton's first law. What would happen to a moving object if there were no friction?

_____

_____

_____

What causes static friction?

Static friction is caused by the attraction between the atoms of two surfaces that are touching each other. This makes the two surfaces stick together. The force of static friction is greater when the object is heavy or if the surfaces are rough.

What is sliding friction?

Static friction keeps an object at rest. Sliding friction slows down an object that slides. If you push a box across a floor, you have to keep pushing to overcome the force of sliding friction. Sliding friction is caused by the roughness of the surfaces that are sliding. A force must be applied to move the rough areas of one surface past the rough areas of the other. Sliding friction slows down the sliding baseball player in the figure.

Picture This

4. Identify Draw an arrow below the sliding baseball player to show the direction of the force due to friction.

What is rolling friction?

Rolling friction is what makes a wheel turn. There is rolling friction between the ground and the part of the wheel touching the ground. Rolling friction keeps the wheel from slipping on the ground. If a wheel is rolling forward, rolling friction exerts a force on the wheel that pushes the wheel forward. ☑

It is usually easier to pull a load on a wagon that has wheels than it is to drag the load along the ground. This is because the rolling friction between the wheels and the ground is less than the sliding friction between the load and the ground.

☑ **Reading Check**

5. Explain If a wheel is rolling forward, what type of friction pushes the wheel forward?

● After You Read

Mini Glossary

balanced forces: two or more forces acting on an object that cancel each other and do not change the motion of the object

force: a push or a pull

friction: the force that resists sliding motion between two touching surfaces

net force: the combination of all forces acting on an object

Newton's first law of motion: if the net force acting on an object is zero, the object stays at rest; or, if the object is already moving, it continues to move in a straight line with constant speed

unbalanced forces: forces that don't cancel each other

1. Review the terms and their definitions in the Mini Glossary. When you push a skateboard on a flat surface, why does it stop after a while? Use at least one term in your answer.

_____

_____

2. Complete the table below to show how Newton's first law of motion affects objects at rest and objects that are moving. Name the types of friction that could affect objects at rest and moving objects.

	How is the object affected by Newton's first law?	Which type or types of friction affect it?
Object at rest		
Object in motion		

3. At the beginning of the section, you were asked to make flash cards. Did your flash cards help you learn about Newton's first law? Why or why not?

_____

_____

 Visit **ips.msscience.com** to access your textbook, interactive games, and projects to help you learn more about Newton's first law of motion.

End of Section

chapter 11

Force and Newton's Laws

section 2 Newton's Second Law

What You'll Learn

- Newton's second law of motion
- why the direction of force is important

Before You Read

If someone told you that a car was accelerating, what would that mean to you?

____

____

Mark the Text

Underline As you read, underline the main ideas under each heading. After you finish reading, review the main ideas that you have underlined.

FOLDABLES™

B Classify As you read this section, use your table Foldable to write about Newton's second law.

Force	Example in Your Life
First Law	
Second Law	

Read to Learn

Force and Acceleration

You know that it takes force to make a heavy shopping cart go faster. You must push harder and harder to make the cart speed up. When the heavy cart is moving, what do you have to do to slow it down? You have to use force to pull on the cart to make it slow down or stop. You also have to use force to turn a cart that is already moving. When the motion of an object changes, the object is accelerating. Speeding up, slowing down, and changing directions are all examples of acceleration.

Newton's second law of motion states that when a force acts on an object, the object accelerates in the direction of the force. You can calculate acceleration by using the equation below.

$$\textbf{acceleration} \text{ (in meters/second}^2) = \frac{\textbf{net force} \text{ (in newtons)}}{\textbf{mass} \text{ (in kilograms)}}$$

$$a = \frac{F_{net}}{m}$$

In this equation, a is acceleration, m is the mass of the object, and F_{net} is the net force. You can multiply both sides of the equation by the mass, and write the equation this way:

$$F_{net} = ma$$

What are the units of force?

Force is measured in newtons (N). The newton is an SI measurement. So, if you are calculating force, the mass must be measured in kilograms (kg). The acceleration must be measured in meters per second squared (m/s^2). One N is equal to 1 kg $\cdot$ m/s^2. ☑

Gravity

One force that you may already know about is gravity. Gravity is the force that pulls you downward when you jump into a pool or coast down a hill on a bike. Gravity also keeps Earth in orbit around the Sun and the Moon in orbit around Earth.

What is gravity?

Gravity is a force that exists between any two objects that have mass. It pulls two objects toward each other. Gravity depends on the mass of the objects and the distance between them. The force of gravity becomes weaker as objects move away from each other or as the mass of objects gets smaller. Large objects like Earth and the Sun have great gravitational forces. Objects with less mass like you or a pencil have weak gravitational forces.

There is a gravitational force between you and the Sun. There is also a gravitational force between you and Earth. Why doesn't the Sun's gravity pull you off of Earth? The gravitational force between you and the Sun is very weak because the Sun is so far away. Only Earth is close enough and massive enough to exert a noticeable gravitational force on you. Earth's gravitational force on you is 1,650 times greater than the Sun's gravitational force on you.

What is weight?

Earth's gravity causes all objects to fall toward Earth with an acceleration of 9.8 m/s^2. You can use the equation of Newton's second law to find the force of Earth's gravity on any object near Earth's surface:

$$F = ma = m \times (9.8 \text{ m/s}^2)$$

<u>Weight</u> is the amount of gravitational force on an object. Your weight on another planet would be different from your weight on Earth. That's because the gravitational force on other planets is different. Other planets have masses different from Earth's. So, your weight would be different on other planets.

✔ Reading Check

1. **Explain** What units are used when force is measured?

_____

_____

Applying Math

2. **Calculate** Jamie has a mass of 35 kg. What is her weight on Earth, in newtons? Use the formula for gravitational force. Show your work.

_____

_____

3. Use Definitions If you were on Mars instead of on Earth, which would be different—your weight or your mass?

_____

_____

Applying Math

4. Explain By what number do you multiply 588 to get your weight in newtons on Pluto?

_____

_____

How do weight and mass differ?

Weight and mass are different. Weight is the amount of gravitational force on an object. Your bathroom scale measures how much Earth's gravity pulls you down. Mass is the amount of matter in an object. Gravity doesn't affect the amount of matter in an object. Mass is always the same, even on different planets. A person with a mass of 60 kg has a mass of 60 kg on Earth or on Mars. But, the weight of the person on Earth and Mars would be different, as shown in the table. That's because the force of gravity on each planet is different. ✔

Weight of 60-kg Person on Different Planets		
Place	Weight in Newtons If Your Mass Were 60 kg	Percent of Your Weight on Earth
Mars	223	38
Earth	588	100
Jupiter	1,388	236
Pluto	4	0.7

Using Newton's Second Law

Newton's second law tells how to calculate the acceleration of an object. You must know the object's mass and the forces acting on the object. Remember that velocity is how fast an object is moving and in what direction. Acceleration tells how velocity changes.

How is speeding up acceleration?

When an object speeds up, it accelerates. Think about a soccer ball sitting on the ground. If you kick the ball, it starts moving. You exert a force on the ball. The ball accelerates only while your foot is in contact with the ball. While something is speeding up, something is pushing or pulling the object in the direction it is moving. The direction of the push or pull is the direction of the force. It is also the direction of the acceleration.

How is slowing down acceleration?

If the net force on an object is in the direction opposite to the object's velocity, the object slows down. As shown below, the force of sliding friction becomes larger when the boy puts his feet in the snow. The net force on the sled is the combination of gravity and sliding friction. When the sliding friction force becomes large enough, the net force is opposite to the sled's velocity. This causes the sled to slow down.

Picture This

5. **Label** In the figure, label one arrow "Force due to friction" and the other arrow "Direction of motion."

How do you calculate acceleration?

Calculate acceleration using the equation from Newton's second law of motion. For example, suppose you pull a 10-kg sled with a net force of 5 N. You can find the acceleration as follows:

$$a = \frac{F_{net}}{m} = \frac{5\ N}{10\ kg} = 0.5\ m/s^2$$

The sled keeps accelerating as long as you keep pulling on it. The acceleration does not depend on how fast the sled is moving. It depends only on the net force and the mass of the sled.

Applying Math

6. **Calculate** Suppose you kick a 2-kg ball with a force of 14 N. What is the acceleration of the ball? Show your work.

_____

_____

_____

Copyright © Glencoe/McGraw-Hill, a division of The McGraw-Hill Companies, Inc.

7. **Draw** Imagine that you throw a basketball and it goes through a basketball hoop. In the space below, sketch the path that the ball would follow.

How do objects turn?

Forces and motion don't always happen in a straight line. If a net force acts at an angle to the direction an object is moving, the object will follow a curved path. Imagine shooting a basketball. When the ball leaves your hands, it doesn't continue to move in a straight line. Instead, it starts to curve downward due to gravity. The curved path of the ball is a combination of its original motion and the downward motion caused by gravity.

Circular Motion

You move in a circle when you ride on a merry-go-round. This motion is called circular motion. In circular motion, your direction of motion is constantly changing. This means you are constantly accelerating. There is a force acting on you the whole time. That's why you have to hold on tightly—to keep the force from causing you to fall off.

Imagine a ball on a string moving in a circle. The string pulls on the ball and keeps it moving in a circle. The force exerted by the string is called centripetal (cen TRIP eh tal) force. The centripetal force points to the center of the circle. Centripetal force is always perpendicular to the motion. The figure shows the direction of motion, centripetal force, and acceleration of a ball traveling in a circle on a string.

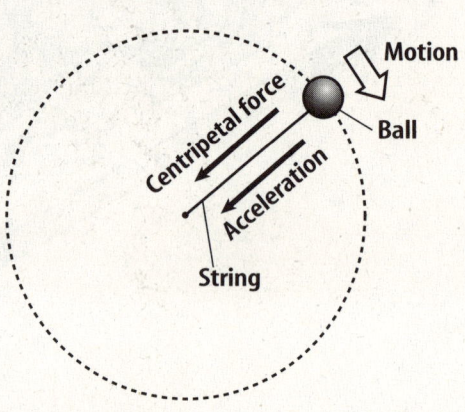

Picture This

8. **Evaluate** Look at the figure. Suppose the ball was traveling in the opposite direction around the circle. What would be the direction of the centripetal force? Why?

_____

_____

_____

How do satellites stay in orbit?

Satellites are objects that orbit Earth. They go around Earth in nearly circular orbits. The centripetal force acting on a satellite is gravity. But why doesn't a satellite fall to Earth like a baseball? Actually, satellites do fall toward Earth.

When you throw a baseball, its path curves until it hits Earth. If you throw the baseball faster, it goes a little farther before it hits Earth. If you could throw the ball fast enough, its curved path would follow the curve of Earth's surface. The baseball would never hit the ground. It would keep traveling around Earth.

How fast must a satellite travel?

The speed at which a satellite must travel to stay in orbit near Earth's surface is 8 km/s, or about 29,000 km/h.

Air Resistance

Have you ever run against the wind? If so, you have felt the force of air resistance. When an object moves through air, there is friction between the object and the air. This friction, or air resistance, slows down the object. Air resistance is a force that gets larger as an object moves faster. Air resistance also depends on the shape of an object. Think about two pieces of paper. One piece is crumpled into a ball and the other piece is flat. The paper that is crumpled into a ball will fall faster than the flat piece of paper falls.

When an object falls it speeds up as gravity pulls it downward. At the same time, the force of air resistance pushing up on the object is increasing as the object moves faster. Finally, the upward force of air resistance becomes large enough to equal the downward force of gravity.

When the air resistance force equals the weight of an object, the net force on the object is zero. Newton's second law explains that the object's acceleration then is zero. Its speed no longer increases. When air resistance balances the force of gravity, the object falls at a constant speed. This constant speed is called the terminal velocity.

Center of Mass

Imagine throwing a stick. The stick spins while it flies through the air. Even though the stick spins, there is one point on the stick, the center of mass, that moves in a smooth path. The **center of mass** is the point in an object that moves as if all the object's mass was concentrated at that point. For a symmetrical object, such as a ball, the center of mass is the center of the object. ☑

Think it Over

9. **Infer** Imagine that you could throw a baseball at a speed of 29,000 km/h. What would happen to the ball if you threw it that fast?

_____

_____

_____

Reading Check

10. **Identify** Where is the center of mass of a ball?

_____

_____

● After You Read

Mini Glossary

center of mass: the point in an object that moves as if all the object's mass was concentrated at that point

Newton's second law of motion: when a force acts on an object, the object accelerates in the direction of the force

weight: the amount of gravitational force on an object

1. Review the terms and their definitions in the Mini Glossary. What are three ways an object can accelerate? Answer in complete sentences.

   _____

   _____

2. Look at the figures below. For each object, draw and label an arrow to show the direction of the motion. Then draw and label an arrow to show the direction of acceleration.

3. You were asked to underline the main ideas as you read this section, then review what you underlined. Why do you think you were asked to review what you underlined?

   _____

   _____

   _____

End of Section

 Visit **ips.msscience.com** to access your textbook, interactive games, and projects to help you learn more about Newton's second law of motion.

Force and Newton's Laws

section ❸ Newton's Third Law

● Before You Read

Imagine stepping out of a canoe onto the shore of a lake. What happens to the canoe when you step out?

_____

_____

● Read to Learn

Action and Reaction

Newton's first two laws of motion explain how the motion of one object changes. You have learned that if balanced forces act on an object, the object will remain at rest or stay in motion with constant velocity. If the forces are unbalanced, the object will accelerate in the direction of the net force.

Another of Newton's laws describes something else that happens when one object exerts a force on another object. When you push on a wall, did you know that the wall also pushes on you? **Newton's third law of motion** states that forces always act in equal but opposite pairs. When you push on a wall, you apply a force to the wall. But, the wall also applies a force equal in strength to you. When one object applies a force on another object, the second object exerts the same size force on the first object.

Why don't action and reaction forces cancel?

The forces that two objects put on each other are called an action-reaction force pair. The forces in a force pair are equal in strength, but opposite in direction. The forces in a force pair don't cancel each other out because they act on different objects. Forces can cancel each other only if they act on the same object.

What You'll Learn

■ about forces that objects exert on each other

Study Coach

Outline As you read the section, create an outline using each heading from the text. Under each heading, write the main points or ideas that you read.

FOLDABLES

❸ Classify As you read this section, use your table Foldable to write about Newton's third law.

Force	Example in Your Life
First Law	
Second Law	
Third Law	

Action and Reaction Forces Imagine a bowling ball hitting a bowling pin. The action force from the bowling ball acts on the pin. The pin flies in the direction of the force. The reaction force from the pin acts on the ball. It causes the ball to slow down.

How do action-reaction force pairs work on large and small objects?

When you walk forward, your shoe pushes Earth backward. Earth pushes your shoe forward. So why do you move when Earth does not? Earth has so much mass compared to you that it does not appear to move when you push on it. If you step on a skateboard, the force from your shoe makes the skateboard roll backward. This is because you have more mass than the skateboard. ✔

How do rockets take off?

The launching of a space shuttle is a good example of Newton's third law. When the fuel in the shuttle's engines is ignited, a hot gas is produced. The gas molecules collide with the inside walls of the engines. The walls exert an action force that pushes the gas out of the bottom of the engine. The gas molecules put reaction forces on the walls of the engine. These reaction forces are what push the engine and the rocket forward. The force of the rocket engines is called thrust.

Weightlessness

You may have seen pictures of astronauts floating inside a space shuttle. The astronauts are said to be weightless—as if Earth's gravity were not pulling on them. But, Earth's gravity is what keeps a shuttle in orbit. Newton's laws of motion can explain why the astronauts float as if there weren't any forces acting on them.

How is weight measured?

Think about how you measure your weight. When you stand on a bathroom scale, your weight pushes down on the scale. This causes the scale pointer to show your weight. Newton's third law tells you that the scale pushes back up on you with a force equal to your weight. This force balances the downward pull of gravity on you, as shown in the figure on the left on the next page. ✔

✔ Reading Check

1. **Describe** Why doesn't Earth appear to move when you push down on it with your foot?

_____

_____

_____

✔ Reading Check

2. **Explain** When you stand on a scale, which force balances the downward pull of gravity on you?

_____

_____

_____

How does free fall cause weightlessness?

Imagine standing on a scale in an elevator that is falling, as shown in the figure on the right below. An object is in free fall when the only force acting on it is gravity. The elevator, you, and the scale are all in free fall. In free fall, the scale doesn't push back up on you. That's because the only force acting on you is gravity. According to Newton's third law, you are also not pushing down on the scale. So, the scale pointer stays at zero. You seem to be weightless. However, you are not really weightless. Earth's gravity is still pulling down on you. But, because nothing is pushing up on you, you have no sensation of weight.

Force exerted by scale

Weight of student

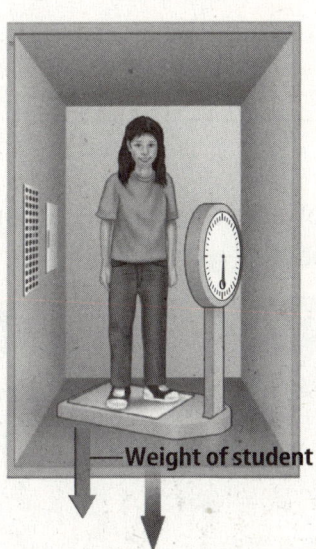

Weight of student

Why are spacecraft in orbit weightless?

Remember that an object will orbit Earth when its path follows the curve of Earth's surface. Gravity keeps pulling the object down. But, the forward motion keeps it from falling straight downward. Objects that orbit the Earth, like satellites and the space shuttle, are in free fall.

Objects inside the shuttle are also in free fall. This makes the shuttle and everything inside it seem weightless, even though gravity is acting on them.

Suppose an astronaut in the shuttle is holding a ball. When she lets go of the ball, it will not move unless she pushes it. The ball does not move because the ball, the astronaut, and the shuttle are all falling at the same speed. If the astronaut pushes the ball forward, it accelerates to a speed that is faster than the shuttle and astronaut. The ball moves forward inside the shuttle.

Think it Over

3. Explain Why isn't an object in free fall really weightless?

_____

_____

_____

Picture This

4. Describe Look at the figure. What is the only force acting on the girl in the elevator on the right?

_____

● After You Read

Mini Glossary

Newton's third law of motion: forces always act in equal
   but opposite pairs

1. Review the term and its definition in the Mini Glossary. What are the action and reaction forces that make a rocket move forward? Answer in complete sentences.

_____

_____

_____

_____

2. On the figure below, draw arrows and label the action and reaction forces that are on the objects as the bat hits the baseball.

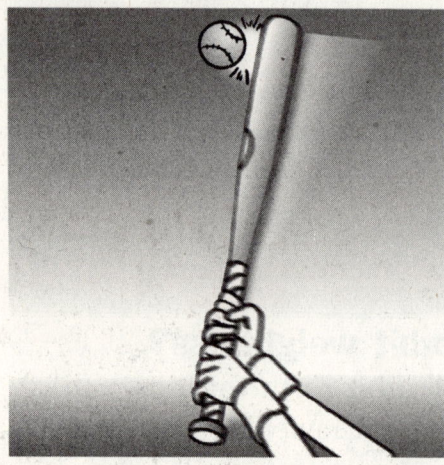

3. How could you use a skateboard to show Newton's third law of motion to a group of elementary school students?

_____

_____

_____

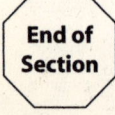
End of Section

 **Science** Online Visit **ips.msscience.com** to access your textbook, interactive games, and projects to help you learn more about Newton's third law of motion.

 **chapter 12**

Forces and Fluids

section ❶ Pressure

● Before You Read

When people say that they are under a lot of pressure, what do they usually mean?

_____

_____

_____

What You'll Learn

■ what pressure is and how to calculate it
■ to model pressure changes in a fluid

● Read to Learn

Pressure

Have you ever walked in deep, soft snow? Your feet sink. It can be difficult to walk. The same thing happens when you walk on dry sand. If you ride a bicycle in deep snow or dry sand, the tires would sink even deeper than your feet.

How deep you sink depends on your weight. It also depends on the area over which your weight is spread. When you stand on two feet, your weight is spread out over the area covered by your feet. If you are wearing snowshoes, the snowshoes spread your weight out over a larger area of snow. You do not sink as far. The area of contact between you and the snow changes.

What is pressure?

When you stand on snow, or any surface, your feet and your weight put, or exert, a downward force on the surface. This force is called pressure. **Pressure** is the force that is applied on a surface per unit area. When you put snowshoes on your feet, the force of your weight is spread out over a larger area. This decreases the pressure you put on the snow. When you change the area of contact, you also change the amount of pressure.

Study Coach

Create a Quiz When you read, write down questions that will help you remember the main ideas and vocabulary words. When you finish reading, make the questions into a quiz. See how many of the questions you can answer correctly.

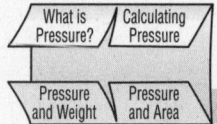

Ⓐ Organize Information Make the following four-tab notebook paper Foldable to help you organize information about pressure.

How do you calculate pressure?

What would happen to the pressure exerted by your feet if your weight increased? You would sink deeper into the snow, so pressure also would increase. Pressure increases if the force exerted increases. Pressure also increases if the area of contact decreases. You can calculate pressure using this formula. ☑

$$\text{Pressure (pascals)} = \frac{\text{force (newtons)}}{\text{area (meters squared)}}$$

$$P = \frac{F}{A}$$

The SI unit for pressure is the pascal (Pa). One pascal equals the force of 1 N applied over an area of 1 m^2, or 1 Pa = 1 N/m^2. The weight of a dollar bill lying flat on a table exerts a pressure of about 1 Pa on the table. Since 1 Pa is such a small unit, you often see pressure given in units of kPa, which is 1,000 Pa.

How are pressure and weight related?

To calculate the pressure that is exerted on a surface, you need to know the force and the area over which it is applied. Often, the force is the weight of an object. For example, you might want to know the pressure that is exerted on your hand when you hold a 2-kg book. Remember, 2 kg is the mass of the book. You first need to find the force that the book is exerting on your hand, or its weight. Use the following equation to find the weight of the book.

$$\text{Weight} = \text{mass} \times \text{acceleration due to gravity}$$
$$W = (2 \text{ kg}) \times (9.8 \text{ m/s}^2)$$
$$W = 19.6 \text{ N}$$

Now, suppose that the contact area between the book and your hand is 0.003 m^2. Calculate the pressure exerted on your hand by the book.

$$P = \frac{F}{A}$$

$$P = \frac{(19.6 \text{ N})}{(0.003 \text{ m}^2)}$$

$$P = 6,533 \text{ Pa} = 6.53 \text{ kPa}$$

✔ Reading Check

1. **Explain** What happens to pressure if the area of contact decreases?

_____

_____

Applying Math

2. **Calculate** the pressure exerted on your hand if the book had a mass of 3 kg. Show your work.

_____

_____

_____

_____

How are pressure and area related?

You already know that wearing showshoes will keep you from sinking deeply into snow. Why? Changing the area over which a force is applied changes the pressure. Increasing the area with snowshoes decreases the pressure. The opposite is also true. Think of driving a nail into a piece of wood, as shown in the figure below. The end of the nail is pointed. All of the force of the hammer is exerted on the wood by the tiny area covered by the point. Because the area is so small, the pressure is large. It is so large that the nail pushes the wood fibers apart. This lets the nail go into the wood.

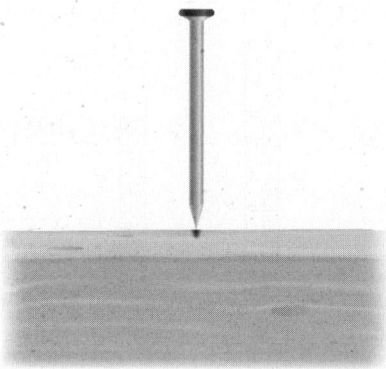

Picture This

3. **Identify** Circle the place on the nail where there is more pressure when a hammer strikes it—on the head or the point of the nail.

Fluids

A **fluid** is any substance that has no definite shape and is able to flow. You probably think of fluids as being liquids, but gases are also fluids. When you are outside on a windy day, you can feel the air flowing around you. Air can flow and has no definite shape. So air is a fluid. Gases, liquids, and plasma are fluids and can flow. Plasma is a state of matter found in the Sun and other stars.

Pressure in a Fluid

Suppose you placed an empty glass on a table. The weight of the glass exerts pressure on the table. If you pour water into the glass, the weight of both the water and the glass exert pressure on the table. So the pressure exerted on the table increases. The water has weight. So it also exerts pressure on the bottom of the glass. The pressure is the weight of the water divided by the area of the bottom of the glass. If you pour more water into the glass, the height of the water increases. The weight also increases, so the pressure exerted by the water increases.

FOLDABLES

B **Compare and Contrast**
Use half-sheets of notebook paper to compare and contrast pressure in fluids and atmospheric pressure.

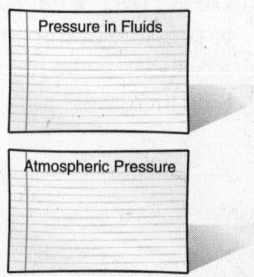

Can the same volume have different pressures?

In the figure below, the graduated cylinders both have the same amount of water. Since the cylinder on the right is narrower, the height of the water in it is greater.

Is the pressure the same at the bottom of each cylinder? You know the weight of the water is the same in each cylinder. But the contact area between the water and the bottom of the narrower cylinder is smaller than the contact area at the bottom of the wider cylinder. You already know that when contact area decreases, the pressure increases. So the pressure at the bottom of the narrower cylinder is greater than at the bottom of the wider cylinder.

Picture This

4. **Interpret Data** How many milliliters of water are in each graduated cylinder?

_____

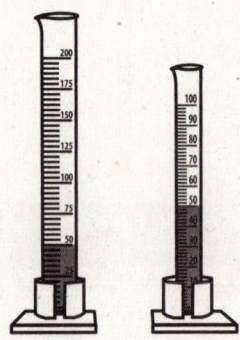

The pressure a fluid exerts on the bottom of a container depends on the height of the fluid. This is always true for any fluid or any container. The greater a fluid's height, the greater the pressure exerted by the fluid. The shape of the container does not matter.

Why does pressure increase with depth?

When you swim, you might feel pressure in your ears when you are underwater. The pressure increases as you swim deeper. The pressure you feel is from the weight of the water above you. As you swim deeper, the height of the water above you increases. When the height of the water increases, so does the weight. The pressure exerted by a fluid increases as the fluid gets deeper. ✓

How is pressure exerted by fluids?

The pressure exerted by a fluid is due to the weight of the fluid. Is the pressure exerted only downward? No. The pressure applied by a fluid on an object is perpendicular to all of the surfaces of the object. So when you dive to the bottom of a swimming pool, the pressure on you is the same on your back as it is on your stomach.

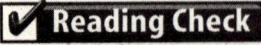

Reading Check

5. **Describe** What happens to pressure at the bottom of a cylinder as the height of a fluid increases?

_____

Atmospheric Pressure

You may not feel it, but you are surrounded by a fluid. It is the atmosphere and it exerts pressure on you all the time. The atmosphere at Earth's surface is about 1,000 times less dense than water. But the thickness of the atmosphere is large enough to exert a large pressure on objects at Earth's surface. The atmospheric pressure at sea level is about 100,000 Pa. The weight of Earth's atmosphere exerts about 100,000 N of force over every square meter on Earth. When you sit down, the force pushing down on your body from the atmospheric pressure can be equal to the weight of several small cars. ☑

Why doesn't atmospheric pressure crush you? Your body is filled with fluids such as blood. The pressure exerted outward by the fluids inside your body balances the pressure applied by the atmosphere. Therefore, atmospheric pressure does not crush you.

How does atmospheric pressure change?

When you go higher in the atmosphere, atmospheric pressure decreases as the amount of air above you decreases. The same thing happens in water. Water pressure is highest at the ocean floor and decreases as you go upward.

What is a barometer?

A barometer, like in the figure below, is an instrument used to measure atmospheric pressure. A barometer is made of a tube that is closed at the top and open at the bottom. The space at the top of the tube is a vacuum. Atmospheric pressure pushes liquid up the tube. When the pressure at the bottom of the column of liquid equals the atmospheric pressure, the liquid stops going up the tube. The force pushing on the surface of the liquid changes as the atmospheric pressure changes. So the height of the liquid in the tube increases as the atmospheric pressure increases.

☑ Reading Check

6. Identify What surrounds you and constantly exerts pressure on you?

_____

_____

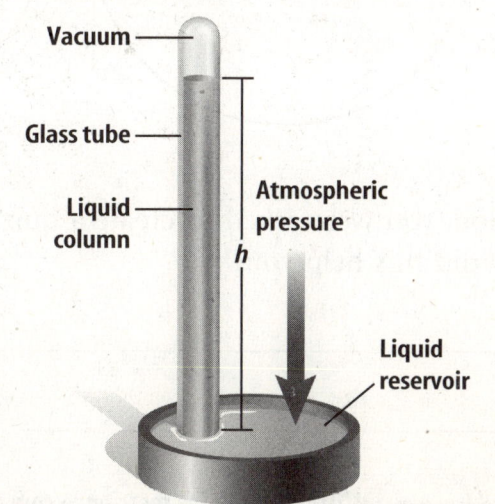

Vacuum

Glass tube

Liquid column

Atmospheric pressure

h

Liquid reservoir

Picture This

7. Apply What happens to the height of the liquid in the tube when atmospheric pressure decreases?

_____

_____

● After You Read

Mini Glossary

fluid: any substance that has no definite shape and is able to flow

pressure: the force that is applied on a surface per unit area

1. Review the terms and their definitions in the Mini Glossary. How does a fluid exert pressure on an object that is in the fluid?

_____

_____

2. Fill in the graphic organizer below to explain pressure, how to calculate pressure, and how the height of a fluid affects pressure.

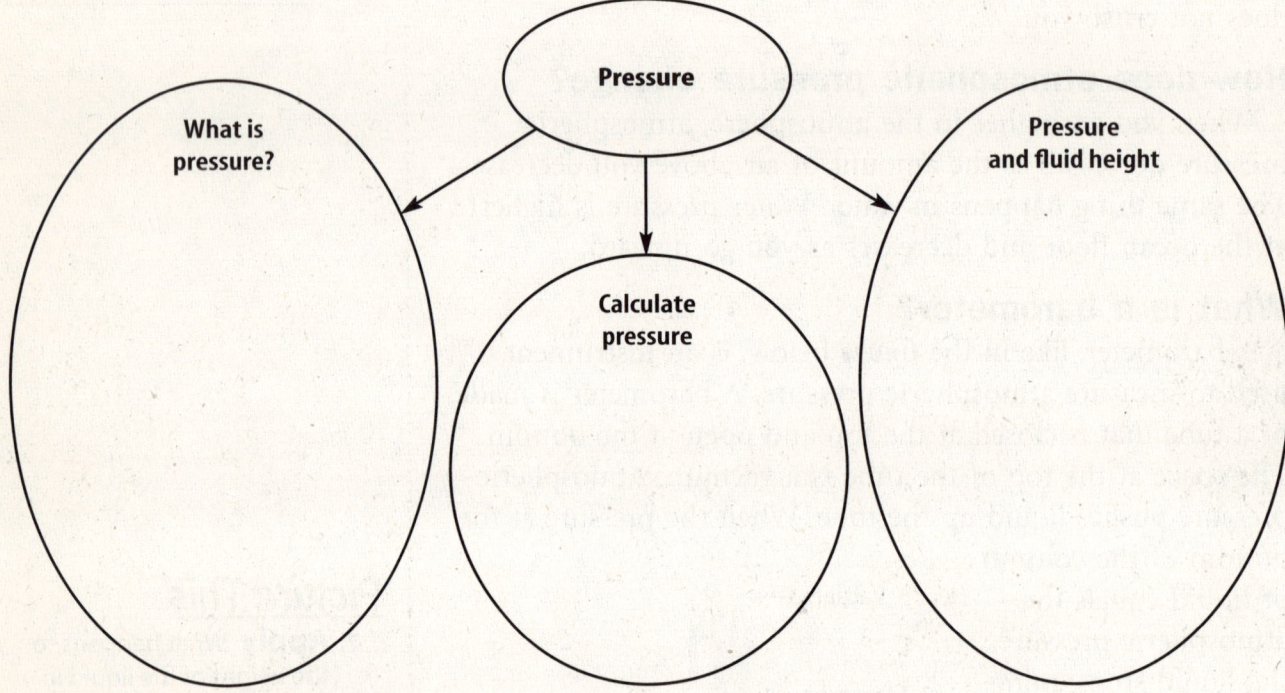

3. At the beginning of the section, you were asked to create a quiz to help you learn the material in the section. How did this help you?

_____

_____

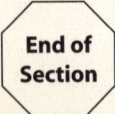
End of Section

Science ⊙ **nline** Visit **ips.msscience.com** to access your textbook, interactive games, and projects to help you learn more about pressure.

Forces and Fluids

section ❷ Why do objects float?

● Before You Read

What happens when you jump into a pool? Describe on the lines below how your body moves through the water.

_____

_____

_____

● Read to Learn

The Buoyant Force

Can you float? Think about the forces that act on you when you float in a pool. You are not moving when you float. So, according to Newton's second law of motion, the forces acting on you must be balanced. You know that Earth's gravity is pulling you downward. So what balances gravity? It is a force called the buoyant force. The **buoyant force** is an upward force that is exerted by a fluid on any object in the fluid. A balance between gravity and the buoyant force is shown in the figure below.

Ryan McVay/PhotoDisc

What You'll Learn

- how the pressure in a fluid makes a buoyant force
- what density is
- to explain floating and sinking using Archimedes' principle.

Mark the Text

Highlight Highlight words or sentences that you do not understand. When you finish reading, ask your teacher to help you understand the things that you highlighted.

Picture This

1. **Identify** Label the arrows in the figure as *Buoyant force* and *Gravity*.

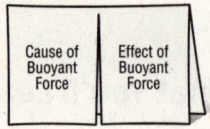
Picture This

2. Explain In the figure, why is the pressure pushing up on the bottom of the cube greater than the pressure pushing down on the top of the cube?

_____

_____

What causes the buoyant force?

The buoyant force is caused by the pressure exerted by a fluid on an object in the fluid. The figure below shows a cube-shaped object in a fluid. The fluid exerts a pressure everywhere on the surface of the cube. Remember that the direction of the pressure on a surface in a fluid is always perpendicular to the surface. Recall also that pressure exerted by a fluid increases as you go deeper into the fluid.

How does pressure affect buoyant force?

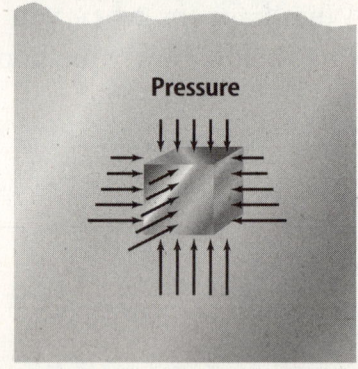

Pressure

The pressure exerted by the water on the cube in the figure is greater toward the bottom of the cube. This is because the bottom of the cube is deeper in the water. The pressure exerted on the cube from the bottom is greater than the pressure pushing down on the top of the cube. Since these forces are not balanced, there is a net force pushing upward on the cube. This upward force is the buoyant force. A buoyant force acts on all objects that are in a fluid. It does not matter whether the objects float or sink.

Sinking and Floating

When you toss a stone into a pond, it sinks. If you toss a stick into a pond, it floats. A buoyant force acts on both the stone and the stick. Why does one sink and the other float?

Remember, gravity always pulls objects downward. If the weight of an object is greater than the buoyant force pushing upward, the object sinks. If the buoyant force is equal to the weight of an object, it floats.

Changing the Buoyant Force

Whether an object floats or sinks depends on whether the buoyant force is less than its weight. The weight of an object depends only on the mass of the object. Remember, mass is the amount of matter in an object. The weight of an object does not change if its shape changes. Think about a ball of clay. The clay has the same amount of matter whether it is in a ball or pressed flat.

How does shape affect buoyant force?

Buoyant force does depend on the shape of an object. Remember, fluid exerts upward pressure on the entire lower surface of an object that is in contact with the fluid. If the surface is made larger, more upward pressure is exerted on the object. This increases the buoyant force. If a piece of aluminum foil is folded, the buoyant force is less than the weight, so it sinks. If the aluminum is flattened into a thin sheet, the area of the bottom of the aluminum increases. On this piece, the buoyant force is large enough that the sheet floats.

Shape is why large metal ships float. If the metal of a ship were crushed into a cube, it is heavy enough to sink. But ships are made with curved bottoms called hulls. A ship's hull has a large area in contact with the water. This increases the buoyant force enough so the ship floats. ☑

Why doesn't buoyant force change with depth?

You know that the pressure exerted on an object by a fluid increases with depth. You might think that a rock will only sink to a depth where the buoyant force on the rock balances its weight. But this does not happen. As a rock sinks in a pond, the pressure pushing up on the bottom surface does increase. But so does the pressure pushing down on the top surface. The difference between these pressures is always the same. So the buoyant force acting on the rock is always the same, no matter how deep it goes.

Archimedes' Principle

The Greek mathematician Archimedes (ar kuh MEE deez) figured out how to find buoyant force more than 2,200 years ago. **Archimedes' principle** states that the buoyant force on an object is equal to the weight of the fluid it displaces, or moves.

Think about dropping an ice cube in a glass that is full to the top with water. When you drop the ice cube in, it takes the place of some of the water and causes this water to overflow. Suppose you catch the water that spills out of the glass. If you weighed this water, its weight would be equal to the buoyant force on the ice cube. Because the ice cube is floating, you know the buoyant force is balanced by the weight of the ice cube. So the water that is displaced by the ice cube, or the buoyant force, is equal to the weight of the ice cube. ☑

☑ Reading Check

3. **Explain** why a ship's hull is curved.

_____

_____

_____

☑ Reading Check

4. **Describe** According to Archimedes' principle, what is the weight of the displaced fluid equal to?

_____

What is density?

Whether an object floats or sinks depends on the density of the fluid and the density of the object. **Density** is the mass of an object divided by the volume it takes up. You can find density with the following formula:

$$\text{density (in g/cm}^3) = \frac{\text{mass (in g)}}{\text{volume (in cm}^3)}$$

$$D = \frac{m}{V}$$

For example, water has a density of 1.0 g/cm³. If you multiply both sides of the equation above by the volume, you can find the mass for any volume of a substance. This gives you the equation mass = density × volume. If you know the density and volume of a material, you can find its mass.

How does density affect sinking and floating?

Suppose you place a solid plastic block in a container of fluid. Whether the block sinks or floats depends on the density of the plastic and the density of the fluid. Suppose the density of the block is less than the density of the fluid. In this case, the block will float. Now suppose the density of the block is greater than the density of the fluid. In this case, the block will sink.

Boats

Suppose you have a steel boat and a steel cube, like those in the figure. Both have the same mass. The boat is shaped so it takes up a large volume—much larger than the cube. So the boat displaces more water than the cube. The boat displaces so much water that the water it 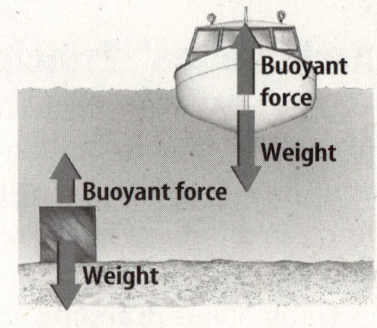 displaces weighs more than the boat. According to Archimedes' principle, increasing the weight of the water that is displaced increases the buoyant force. By making the volume of the boat large enough, enough water can be displaced to balance the weight of the boat. So the boat floats.

The cube and the boat have the same mass, but the boat has a greater volume. So the boat is less dense. The boat floats because its density is less than the density of water.

Applying Math

5. **Calculate** What is the density of an object that has a mass of 42 g and a volume of 7 cm³? Show your work.

_____

_____

_____

Picture This

6. **Identify** Circle the object in the figure that is less dense. Explain how you know.

_____

_____

_____

● After You Read

Mini Glossary

Archimedes' principle: the buoyant force on an object is equal to the weight of the fluid it displaces, or moves

buoyant force: an upward force that is exerted by a fluid on any object in the fluid

density: the mass of an object divided by the volume it takes up

1. Review the terms and their definitions in the Mini Glossary. Use the word density in a sentence to explain why an object floats or sinks.

   _____

   _____

2. Fill in the table below by describing how each concept affects whether objects will float or sink.

Concept	Floating	Sinking
Buoyant force	The buoyant force is greater than the weight of the object.	
Archimedes' principle		The fluid moved weighs less than the object.

3. Why do you think that you go deeper into a pool when you dive than when you do a belly flop?

   _____

   _____

Science Online Visit **ips.msscience.com** to access your textbook, interactive games, and projects to help you learn more about why objects float.

End of Section

Forces and Fluids

section ❸ Doing Work with Fluids

What You'll Learn

- how forces are moved through fluids
- how a hydraulic system increases force
- about Bernoulli's principle

◉ Before You Read

Have you ever jumped on one side of an air mattress when someone was lying on the other side? On the lines below, explain what happens.

_____

_____

_____

Study Coach

Make Flash Cards Make a flash card for each question heading in this section. On the back of each flash card, write the answer to the question. When you're finished reading, review your flash cards.

◉ Read to Learn

Using Fluid Forces

Fluids can be made to exert forces that do work, such as making cars stop, making airplanes fly, and pumping water. How are these forces made by fluids?

What happens when you push on a fluid?

The pressure in a fluid can be increased by pushing on the fluid. The figure shows a container of fluid with a movable cover. The cover is called a piston. If you push down on the piston, the fluid cannot escape around the piston. The fluid does not move because it cannot go anywhere. The force on the bottom of the container is the weight of the fluid plus the downward force on the piston. The force exerted by the fluid at the bottom of the container has increased. The pressure exerted by the fluid also has increased. When you push on the brake pedal of a car, a rod pushes a piston into a fluid, as in the figure.

Picture This

1. **Identify** Circle the area of the figure where the most force is.

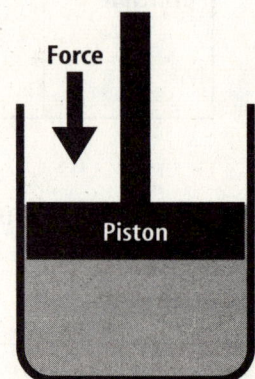

Pascal's Principle

Have you ever had a drink that comes in a box? You have to poke a hole in the box with a straw to drink the liquid inside. What happens if you squeeze the container? The drink comes squirting out through the straw. When you squeeze the container, you apply a force on the fluid. This increases the pressure in the fluid. The increased pressure pushes the fluid out of the straw.

Suppose you poke the straw in the container on the side instead of the top. Would the liquid still squirt out when you squeeze the container? Yes, it would. When you squeeze, the force you exert on the fluid by squeezing is moved to all parts of the container. This is an example of Pascal's principle. **Pascal's principle** states that when a force is applied to a fluid in a closed container, the pressure in the fluid increases everywhere by the same amount.

Hydraulic Systems

Pascal's principle is used in hydraulic systems like the ones used to lift cars. A **hydraulic system** uses a fluid to increase an input force. The fluid in a hydraulic system transfers pressure from one piston to another.

Pressure Transfer An example of a hydraulic system is shown in the figure below. An input force pushes down on the small piston. This increases the pressure in the fluid. The pressure increase moves through the fluid and acts on the large piston. The force the fluid exerts on the large piston is the pressure in the fluid times the area of the piston. The area of the large piston is greater than the area of the small piston. So the output force exerted on the large piston is greater than the input force exerted on the small piston.

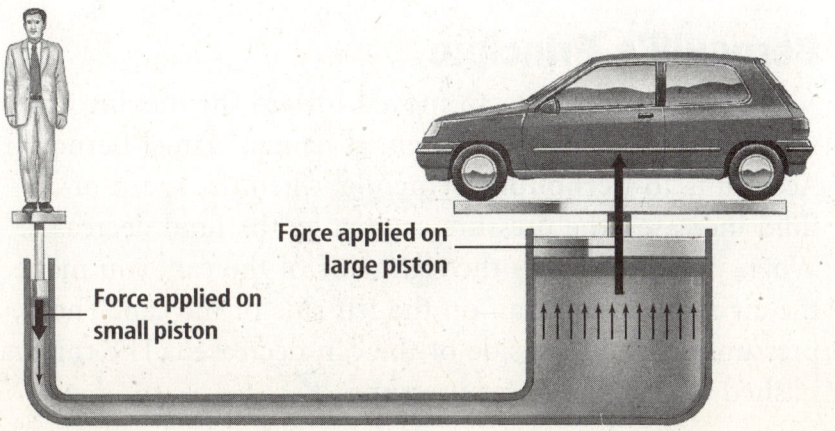

Force applied on large piston

Force applied on small piston

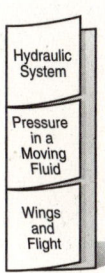

FOLDABLES™

D Organize Information
Make the following Foldable to help you organize information about doing work with fluids.

Hydraulic System

Pressure in a Moving Fluid

Wings and Flight

Picture This

2. Infer How would the force on the large piston change if its area decreased?

_____

_____

How do hydraulic systems increase force?

How do you find the amount of force pushing up on the large piston in a hydraulic system? Suppose that the area of the small piston is 1 m² and the area of the large piston is 2 m². If you push on the small piston with a force of 10 N, the increase in pressure at the bottom of the small piston is

$$P = F/A$$
$$= (10 \text{ N})/(1 \text{ m}^2)$$
$$= 10 \text{ Pa}$$

Pascal's principle states that the increase in pressure happens throughout the fluid. This means the pressure exerted by the fluid on the large piston is 10 Pa. The increase in force on the large piston can be found by multiplying both sides of the formula by A.

$$F = P \times A$$
$$= 10 \text{ Pa} \times 2 \text{ m}^2$$
$$= 20 \text{ N}$$

The force pushing on the larger piston is twice as large as the force pushing on the smaller piston. If the area of the large piston increases, the force pushing up on the piston increases. So a small force can lift a very heavy object.

Pressure in a Moving Fluid

What happens to the pressure in a fluid if the fluid is moving? If you place an empty soda can on your desktop and blow to the right of the can, the can moves to the right. It moves toward the moving air. The moving air lowers the air pressure on the right side of the can. This makes the force exerted on the left side of the can by air pressure greater. The can is pushed to the right by this greater pressure.

Bernoulli's Principle

The reason why the can moved toward the moving air was discovered by a Swiss scientist named Daniel Bernoulli. According to **Bernoulli's principle**, when the speed of a fluid increases, the pressure exerted by the fluid decreases. When you blew across the right side of the can, you made the air move faster than on the left side of the can. The pressure on the right side of the can decreased. The can was pushed toward the lower pressure. ✔

Applying Math

3. **Calculate** What would be the force pushing on the larger piston if the area of the larger piston was 5 m² instead of 2m²? Show your work.

✔ Reading Check

4. **Describe** According to Bernoulli's principle, what happens to the pressure exerted by a fluid when its speed increases?
   a. The pressure increases.
   b. The pressure decreases.
   c. The pressure stays the same.
   d. The pressure moves objects.

How are chimneys affected by Bernoulli's principle?

Hot air is less dense than cold air. Hot air above a fire is pushed up a chimney by the cooler, denser air in the room. Wind outside increases the rate at which smoke rises. Look at the figure below. Air moving across the top of the chimney decreases the air pressure above the chimney. This follows Bernoulli's principle. The decreased pressure causes more smoke to be pushed up by the higher pressure of the air in the room.

Picture This

5. **Identify** On the figure, label the areas where the air pressure is low, and where the air pressure is high.

How do high winds cause damage?

Bernoulli's principle also applies to high winds. Hurricanes are storms that cause very high winds. High winds from a hurricane blowing across a house cause the pressure outside of the house to be less than the pressure inside. The difference in pressure between outside and inside can be large enough to cause windows to be pushed out and to shatter. High winds sometimes can blow roofs off of houses. When winds blow across a roof, the pressure above the roof decreases. If the winds are blowing fast enough, the outside pressure can become so low that the higher pressure inside the house can push the roof off of the house. ☑

✔ **Reading Check**

6. **Explain** What principle explains why high winds can cause windows to blow out and roofs to blow off of houses?

_____

_____

Wings and Flight

Have you ever stuck your hand out the window of a moving car? If so, you have felt a push on your hand from the air moving by. If you angle your hand upward, what happens? You feel your hand being pushed upward. If you angle your hand up even more, the upward push is stronger. Did you know that your hand was behaving like an airplane wing? The force that lifts your hand is made by a fluid—the air. Remember that fluids are liquids, gases, and plasma.

How do wings produce lift?

Air Flow A jet airplane's engine pushes the plane forward through the air. A propeller airplane's engine pulls the plane forward through the air. Air flows over the wings as the plane moves. The wings are tilted upward into the airflow. This is just like your hand that was tilted outside the car window. The figure below shows how the tilt of a wing causes air flowing around the wing's upper and lower surfaces to be directed downward.

Lift Making the air flow downward creates lift. Remember that air is made of molecules. The wing exerts a force on the molecules, pushing them downward. Recall Newton's third law of motion: for every action force there is an equal but opposite reaction force. When the wing exerts a downward action force on the air molecules, the air molecules exert an upward reaction force on the wing. This reaction force pushes the wing upward. This upward force is called lift. ☑

☑ Reading Check

7. Explain what happens when a wing pushes down on air molecules.

_____

_____

_____

_____

Picture This

8. Identify What provides the action force and what provides the reaction force shown in the figure?

_____

_____

_____

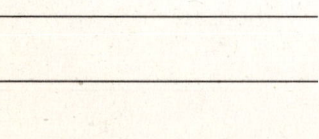

Action force

Reaction force

Why are there different types of wings?

Not all wings look the same. The wing shape of an airplane depends on how the airplane is used. Lift depends on the amount of air the wing pushes downward and how fast that air is moving. Lift can be increased by increasing the surface area of a wing. A larger wing is able to push more air downward. The amount of lift also depends on how fast the plane moves. Fast planes, such as jet fighters, can have small wings. Large planes, such as cargo planes, that are heavy and fly at slow speeds need large wings with more surface area to provide more lift. ☑

How do bird wings work?

A bird's wing provides lift the same way an airplane's wing does. But birds wings have two functions. Not only do they provide lift, but they also act as propellers. When a bird flaps its wings, they pull the bird forward.

Why do bird wings have different shapes?

Bird wings have different shapes, depending on the way the bird usually flies. Seabirds have long, narrow wings, like the wings of a glider. These wings help seabirds glide long distances. Forest and field birds, like pheasants, have short, rounded wings. These wings help the birds take off quickly. They also allow the birds to make sharp turns. Some birds, like swallows, swifts, and falcons fly at high speeds. These birds have small, narrow, tapered wings. Their wings look somewhat like those of a jet fighter. The figure below shows some examples of different types of bird wings.

Seagull   **Sparrow**   **Swift**

☑ **Reading Check**

9. **Describe** What happens to lift as the area of a wing increases?

_____

_____

Picture This

10. **Identify** Which of these birds has wings that are designed for fast flight? How can you tell?

_____

_____

_____

● After You Read

Mini Glossary

Bernoulli's principle: when the speed of a fluid increases, the pressure exerted by the fluid decreases

hydraulic system: uses a fluid to increase an input force

Pascal's principle: when a force is applied to a fluid in a closed container, the pressure in the fluid increases everywhere by the same amount

1. Review the terms and their definitions in the Mini Glossary. Explain which principle allows a hydraulic system to work and how it works.

_____

_____

_____

2. Write the name of the principle that explains each example on the line next to the example.

   a. A roof is blown off of a house in a hurricane. _____

   b. A hydraulic system is used to lift a car. _____

   c. A soda can moves when you blow air past it. _____

   d. Juice squirts out of a plastic bottle when you poke a hole in it

   and squeeze. _____

3. You were asked to make a flash card for each heading as you read the section. How could you use your flashcards to study for a test?

_____

_____

_____

End of Section

Science Online Visit **ips.msscience.com** to access your textbook, interactive games, and projects to help you learn more about doing work with fluids.

200 Forces and Fluids

Energy and Energy Resources

section ❶ What is energy?

◉ Before You Read

What does the phrase "She has a lot of energy" mean to you?

_____

_____

What You'll Learn
- what energy is
- the difference between kinetic energy and potential energy
- the different forms of energy

◉ Read to Learn

The Nature of Energy

__Energy__ is the ability to cause change. An object that has energy can make things happen. Look around you. Changes are happening. Someone might be walking by. Sunshine might be warming your desk. Maybe you can see the wind move the leaves on a tree. What changes are happening?

When is energy noticed?

You have a lot of energy. So does everything around you. But you only notice this energy when a change takes place. When a change happens, energy moves from one object to another. Energy from sunlight moves to the spot on the desktop and makes it warm. Energy from the wind moves to leaves. All objects, including desktops and leaves, have energy.

Energy of Motion

Things that move can cause change. Suppose a bowling ball rolls down the alley and knocks down some bowling pins. Does this involve energy? A change happens when the pins fall over. The bowling ball causes this change. Since energy is the ability to cause change, the bowling ball has energy. The energy in the movement of the bowling ball makes the pins fall. The energy an object has because of its motion is __kinetic energy__. So as a bowling ball moves, it has kinetic energy. If an object is not moving, it does not have kinetic energy.

Mark the Text

Highlight Forms of Energy As you read this section, highlight the different forms of energy. Then write an example of each type of energy next to the places you highlighted.

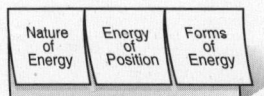

Ⓐ Organize Information Make the following Foldable to organize information about the nature of energy, the energy of position, and the different forms of energy.

Nature of Energy	Energy of Position	Forms of Energy

How are kinetic energy and speed related?

What would happen to the bowling pins if the bowling ball rolls faster? More of the pins might fall down or they might move farther. A faster bowling ball causes more change to happen than a slower bowling ball. The faster the bowling ball goes, the more kinetic energy it has. This is true for all moving objects. Kinetic energy increases as an object moves faster. ✔

How are kinetic energy and mass related?

Suppose you roll a volleyball down the alley at the same speed as a bowling ball. Will the volleyball move the pins as far as the bowling ball will? The answer is no. The volleyball might not knock down any pins.

How are the volleyball and the bowling ball different? They are moving at the same speed, but the volleyball has less mass. The volleyball has less kinetic energy than the bowling ball because it has less mass. Kinetic energy increases as the mass of an object increases.

Energy of Position

An object can have energy even if it is not moving. Look at the vase on top of the bookcase. The vase does not have any kinetic energy because it is not moving. What if it accidentally falls to the floor? Changes happen. Gravity pulls the vase downward. The vase has kinetic energy as it falls. Where did this energy come from?

When the vase was sitting on the shelf, it had potential (puh TEN chul) energy. **Potential energy** is the energy stored in an object because of its position. The position of the vase is its height above the floor. As the vase falls, the potential energy is transformed, or changed, from one form to another. It is transformed into kinetic energy. A vase has more potential energy if it is higher above the floor. Potential energy also depends on mass. The more mass an object has, the more potential energy it has. The objects in the figure have different amounts of potential energy.

✔ **Reading Check**

1. **Apply** Does a slower-moving object have more or less kinetic energy than a faster-moving object?

_____

FOLDABLES™

B Compare and Contrast
Make the following Foldable to compare and contrast kinetic energy and potential energy.

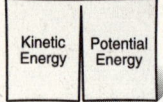
Kinetic Energy | Potential Energy

Picture This

2. **Determine** Which vase on the shelves has the most potential energy?

_____

Forms of Energy

Food, sunlight, and wind have energy. But they have different kinds of energy. The energy in food and sunlight is different from the kinetic energy in the wind. The warmth you feel from sunlight is different from kinetic energy or potential energy.

What is thermal energy?

When you sit near a sunny window, you get warm. The feeling of warmth is a sign that you are getting more thermal energy. **Thermal energy** is energy of an object that increases as the object's temperature increases. All objects have thermal energy. In the figure below, a cup of hot chocolate has more thermal energy than a bottle of cold water. The bottle of cold water has more thermal energy than a block of ice with the same mass.

  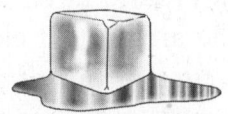

Your body makes thermal energy all the time. Chemical reactions that happen inside your cells make thermal energy. Where does this energy come from? Thermal energy is released by chemical reactions. Thermal energy comes from another kind of energy called chemical energy.

What is chemical energy?

Chemical energy is the energy stored in chemical bonds. Some of this energy is released when chemicals are broken apart and new chemicals are made.

For example, food has chemical energy that your body uses to help you think, move, and grow. Food has chemicals, such as sugar. The chemicals are made of atoms that are bonded together. Energy is stored in the bonds between atoms. These chemical bonds can be broken down in your body to release energy. ☑

Also, the flame of a candle comes from chemical energy stored in wax. When the wax burns, chemical energy changes into thermal energy and light energy.

Picture This

3. **Identify** Circle the object with the greatest thermal energy. Put a box around the object with the least thermal energy.

☑ Reading Check

4. **Explain** What has stored chemical energy that your body uses?

_____

What is radiant energy?

Light from the candle flame travels very fast through the air. It moves at a speed of 300,000 km/s. This is fast enough to circle Earth almost eight times in 1 s. When light hits an object, three things can happen. The light can be absorbed by the object, reflected by the object, or be passed through the object. When an object absorbs light energy, the object can get warmer. The light energy changes into thermal energy. You can feel this happening if you wear a black shirt outside on a sunny day. ✔

The energy carried by light is **radiant energy**. You can use electrical energy to make radiant energy. Imagine a metal heating coil on an electric stovetop. As it is heated, it becomes red hot. The hotter it gets, the more radiant energy it gives off. Electrical energy is being used to make the heating coil warmer.

What is electrical energy?

Electrical energy is used in many ways. **Electrical energy** is carried by the electric current that comes out of batteries and electrical outlets. Electrical lighting uses electrical energy. Look around at all the devices that use electrical energy.

The amount of electrical energy depends on the voltage. The current out of a 120-V electrical outlet can carry more energy than the current out of a 1.5-V battery. Large power plants are needed to make the huge amount of electrical energy people use every day. About 20 percent of the electrical energy made in the United States comes from nuclear power plants.

What is nuclear energy?

Nuclear power plants use the energy stored in the nucleus of an atom to make electricity. **Nuclear energy** is the energy in the nucleus of every atom. Nuclear energy can be transformed into other kinds of energy. Releasing nuclear energy is difficult. Complicated power plants are necessary to produce nuclear energy. Releasing nuclear energy from an atom is very different from releasing chemical energy from wood. To do that, all you need is a lighted match.

5. **Summarize** When does light energy change to thermal energy?

_____

💡 **Think it Over**

6. **Compare** Which of the following can carry the most current?

   a. 9-V battery
   b. 220-V electrical outlet
   c. 12-V battery
   d. 110-V electrical outlet

● After You Read

Mini Glossary

chemical energy: the energy stored in chemical bonds

electrical energy: the energy carried by the electric current that comes out of batteries and electrical outlets

energy: the ability to cause change

kinetic energy: the energy an object has because of its motion

nuclear energy: the energy in the nucleus of every atom.

potential energy: the energy stored in an object because of its position

radiant energy: the energy carried by light

thermal energy: the energy of an object that increases as temperature increases

1. Read the key terms and definitions in the Mini Glossary above. On the lines below, explain the difference between the terms *potential energy* and *kinetic energy*.

_____

_____

2. Match the forms of energy with the correct examples. Write the letter of each example in Column 2 on the line in front of the form of energy it matches in Column 1.

Column 1	Column 2
_____ 1. potential energy	a. the energy that makes a television work
_____ 2. kinetic energy	b. a lamp giving off light
_____ 3. electrical energy	c. the energy in food
_____ 4. thermal energy	d. a ball rolling
_____ 5. chemical energy	e. a book sitting on a shelf
_____ 6. nuclear energy	f. the energy in a cup of hot tea
_____ 7. radiant energy	g. the energy in an atom's nucleus

3. You were asked to highlight the different forms of energy in this section. What do you think would be another way to help you remember the different forms of energy?

_____

Science Online Visit **ips.msscience.com** to access your textbook, interactive games, and projects to help you learn more about energy.

End of Section

Reading Essentials **205**

Energy and Energy Resources

section ❷ Energy Transformations

What You'll Learn
- to apply the law of conservation of energy
- energy changes form
- how electric power plants make energy

Mark the Text

Identify Main Ideas
Highlight the main point in each paragraph as you read this section. Study the main points, then state each point in your own words.

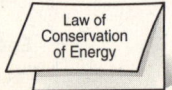

FOLDABLES

C Describe Make the following Foldable. Inside, describe the law of conservation of energy and give examples.

Law of Conservation of Energy

● Before You Read

Explain how you used one kind of energy today.

● Read to Learn

Changing Forms of Energy

Energy can have different forms such as chemical, thermal, radiant, and electrical. All around you, at all times, energy is being transformed. This means it is changing from one form to another. You see some of these transformations when you notice a change in your environment. Forest fires are an example of a change involving energy. They can happen naturally because of lightning strikes. Changes also happen as a mountain biker pedals up a hill.

How can you track energy transformations?

A mountain biker's leg muscles transform chemical energy into kinetic energy as he pedals. The kinetic energy of his leg muscles is transformed into kinetic energy of the bicycle as he pedals. As he moves up the hill, some of this energy is transformed into potential energy.

Some energy also is transformed into thermal energy. His body is warmer because chemical energy is being released. The parts of the bicycle are warmer too because of friction. When energy is transformed, thermal energy usually is made. People exercising, cars running, and living things growing all produce thermal energy.

The Law of Conservation of Energy

The **law of conservation of energy** states that energy is never created or destroyed. The only thing that changes is the form of the energy.

When the biker is resting at the top of the hill, all of his original energy is still around. Some of his energy changed into potential energy. Some changed into thermal energy. No energy is missing. It can all be accounted for.

Changing Kinetic and Potential Energy

The law of conservation of energy can be used to identify energy changes in a system. For example, tossing a ball into the air and catching it is a simple system. As the ball leaves your hand, most of its energy is kinetic. As it rises, it gets slower. It loses kinetic energy. The kinetic energy is changed into potential energy. The amount of kinetic energy that it loses equals the amount of potential energy that it gains. The total amount of energy stays the same.

Energy Changes Form

Energy changes form all the time all around you. Many machines transform energy from one kind to another. For example, an automobile engine transforms the chemical energy in gasoline into kinetic energy. Some of the chemical energy also is transformed into thermal energy, making the engine hot. An engine that converts chemical energy into more kinetic energy and less thermal energy is a more efficient engine. New kinds of cars use an electric motor along with a gasoline engine. These engines are more efficient so the car can travel farther on a gallon of gas.

How is chemical energy transformed?

Chemical energy can be transformed into kinetic energy inside your body. This happens in muscle cells. Chemical reactions take place and cause certain molecules to change shape. Many of these changes make your muscles contract. This makes a part of your body move.

Biomass Biomass is the matter in living organisms. Biomass contains chemical energy. When organisms die, chemical compounds in their biomass break down. Bacteria, fungi, and other organisms help change these chemical compounds into simpler chemicals. These simpler chemicals are used by other living things. ☑

Thermal energy also is released when these biomass breaks down. For example, as a compost pile decomposes, chemical energy is changed into thermal energy. The temperature of a compost pile can reach 60°C.

1. **Draw Conclusions**
At what point does the ball have the most potential energy?

    a. when it reaches its highest point
    b. when it leaves your hand
    c. just before you catch the ball
    d. halfway up in the air

Reading Check

2. **Summarize** What kind of energy does bacteria and fungi help transform?

How is electrical energy transformed?

You use electrical energy every day. When you flip a light switch or turn on a radio, electrical energy is transformed into other forms of energy. You use electrical energy when you plug something into an electrical outlet or use a battery.

Hearing Sounds The figure shows how electrical energy is transformed into other kinds of energy when you listen to a radio. A loudspeaker in the radio changes electrical energy into sound waves. The sound waves travel to your ear. This is energy in motion. The energy carried by the sound waves makes parts of your ear move too. This energy of motion is transformed into chemical and electrical energy in nerve cells. The nerve cells send the energy to your brain. Your brain figures out that the energy is a voice or music. Where does the energy go after the brain? It finally is transformed into thermal energy.

<u>Picture This</u>

3. Identify What kind of energy travels through the air from a radio?

_____

_____

Energy Transformations

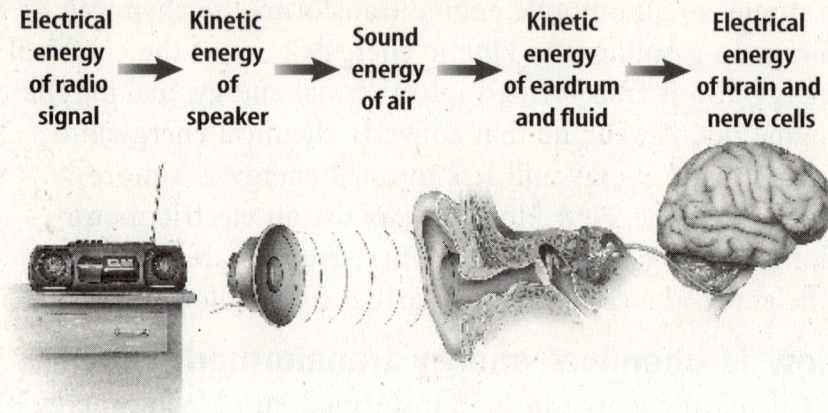

| Electrical energy of radio signal | Kinetic energy of speaker | Sound energy of air | Kinetic energy of eardrum and fluid | Electrical energy of brain and nerve cells |

☑ **Reading Check**

4. Check Understanding How can thermal energy be used to make kinetic energy?

_____

_____

_____

What changes into thermal energy?

Different kinds of energy can be transformed into thermal energy. When something burns, chemical energy changes into thermal energy. Electrical energy changes into thermal energy when a wire that is carrying an electrical current gets hot.

Thermal energy can be used to heat buildings and keep you warm. Thermal energy also can be used to heat water. If you heat water to its boiling point, it changes to steam. Steam can be used by steam engines to make kinetic energy. Steam engines were once used on steam locomotives to pull trains. Thermal energy also can be transformed into radiant energy. This happens when you heat a metal bar until it is so hot that it glows and gives off heat. ☑

How does thermal energy move?

Thermal energy can move from one place to another. A cup of hot chocolate has thermal energy. Its thermal energy moves from the cup to the cooler air around it. Thermal energy only moves from something at a higher temperature to something at a lower temperature.

Generating Electrical Energy

Where does the electrical energy in an electrical outlet come from? It must be made all the time by power plants. In fossil fuel power plants, coal, oil, or natural gas is burned to boil water. Steam from the boiling water rushes through a turbine. A **turbine** is a machine that has a set of fan blades that are close together. The steam pushes on the blades and turns the turbine. The turbine rotates a shaft in the generator A **generator** is a device that changes kinetic energy into electrical energy. All power plants work in a similar way—they use energy to turn a generator. ☑

Are there different kinds of power plants?

Almost 90 percent of the electrical energy in the United States comes from nuclear and fossil fuel power plants. Other kinds of power plants are hydroelectric (hi droh ih LEK trihk) and wind. Hydroelectric power plants use generators to change the kinetic energy of moving water into electrical energy. Wind power plants use generators to change the kinetic energy of wind into electrical energy.

You can diagram the energy transformations in a power plant using arrows. A power plant that burns coal makes energy through the following energy transformations. Nuclear power plants also use energy transformations like the ones below.

chemical   thermal   kinetic   kinetic   electrical
energy $\rightarrow$ energy $\rightarrow$ energy $\rightarrow$ energy of $\rightarrow$ energy out
of coal   of water   of steam   turbine   of generator

How are hydroelectric power plants different?

Hydroelectric power plants do not change water into steam. This is because the water hits the turbine. So the first two steps in the diagram are not needed. The process starts with the kinetic energy of the water.

Think it Over

5. Describe Imagine you are taking a hot pan out of the oven using an oven mitt. Describe where thermal energy moves in this example.

_____

_____

_____

Reading Check

6. Define What machine turns a generator to make electricity?

_____

● After You Read

Mini Glossary

generator: a device that transforms kinetic energy into electrical energy

law of conservation of energy: states that energy is never created or destroyed

turbine: a set of steam-powered fan blades that spins a generator at a power plant

1. Review the terms and their definitions in the Mini Glossary. Write a paragraph about how a turbine and a generator are used to make electrical energy.

_____

_____

_____

_____

2. Fill in the blanks to tell what type of energy is being transformed as a biker rides a bicycle.

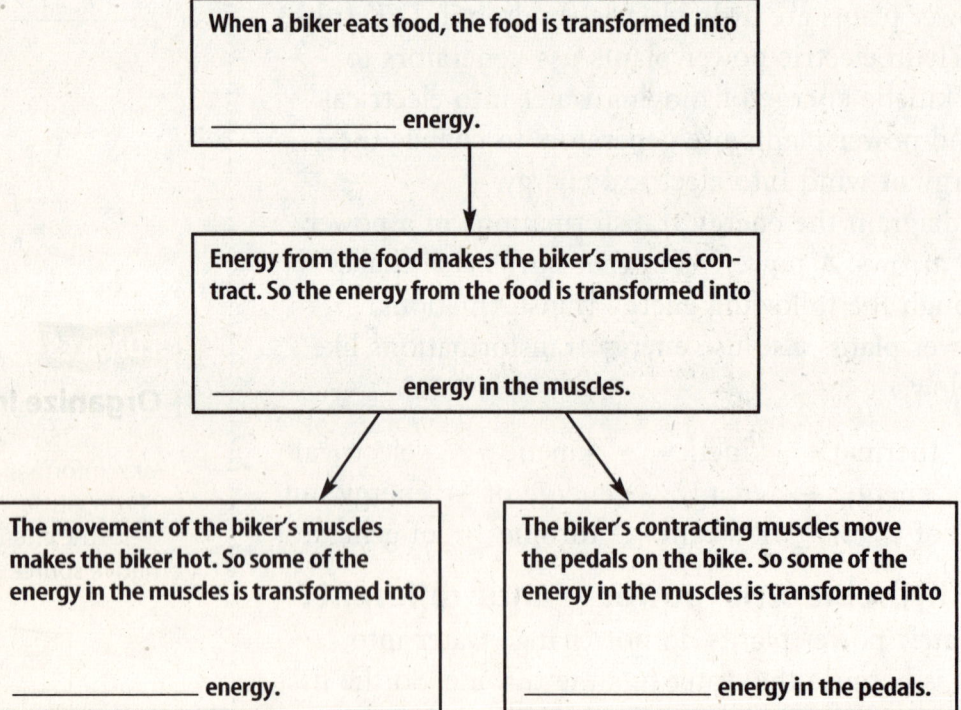

When a biker eats food, the food is transformed into

_____ energy.

Energy from the food makes the biker's muscles contract. So the energy from the food is transformed into

_____ energy in the muscles.

The movement of the biker's muscles makes the biker hot. So some of the energy in the muscles is transformed into

_____ energy.

The biker's contracting muscles move the pedals on the bike. So some of the energy in the muscles is transformed into

_____ energy in the pedals.

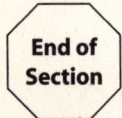
End of Section

Science Online Visit **ips.msscience.com** to access your textbook, interactive games, and projects to help you learn more about energy transformations.

Energy and Energy Resources

section ❸ Sources of Energy

● Before You Read

You must plug in most appliances before they will work. Where does the energy in an electrical outlet come from?

_____

_____

● Read to Learn

Using Energy

Energy is used every day to provide light and heat to homes, schools, and workplaces. The law of conservation of energy states that energy cannot be created or destroyed. It can only change form. If a car or refrigerator cannot create energy, where does the energy come from?

Energy Resources

Energy must come from the natural world. The surface of Earth gets energy from two places. It comes from the Sun and radioactive atoms in Earth's interior. Earth gets far more energy from the Sun than is made in Earth's interior. Almost all the energy you use today can be traced to the Sun. Even the gasoline used to power a car can be traced to the Sun.

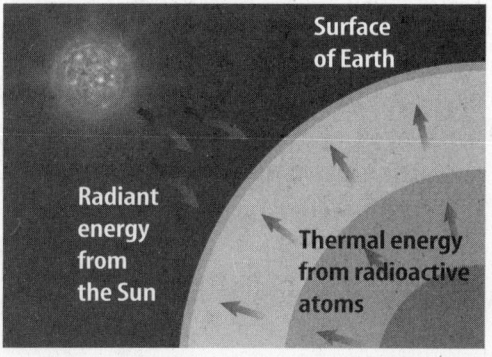

Radiant energy from the Sun

Surface of Earth

Thermal energy from radioactive atoms

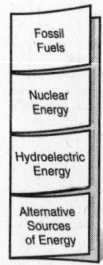

Fossil Fuels

Nuclear Energy

Hydroelectric Energy

Alternative Sources of Energy

Fossil Fuels

Fossil fuels are coal, oil, and natural gas. Oil and natural gas were made from the remains of microscopic organisms. These organisms lived in Earth's oceans millions of years ago. Heat and pressure slowly turned these organisms into oil and natural gas. Coal was formed in a similar way.

As shown in the figures below, coal was made from the remains of plants that once lived on land. Through photosynthesis (foh toh SIHN thuh sus), ancient plants transformed the radiant energy from sunlight into chemical energy. The chemical energy is stored in molecules. Over time, heat and pressure changed these molecules into fossil fuel. Chemical energy stored in fossil fuels is released when the fossil fuels are burned.

Picture This
1. **Identify** What three things changed the plant molecules into coal molecules?

_____

_____

Radiant energy

Time
Heat
Pressure

Coal mine

FOLDABLES

E **Compare and Contrast**
Use three quarter-sheets of notebook paper to help you compare and contrast nonrenewable resources, renewable resources, and inexhaustible energy sources.

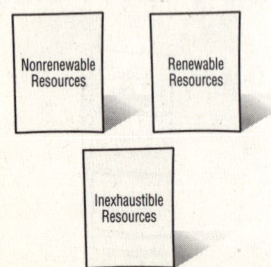

Nonrenewable Resources

Renewable Resources

Inexhaustible Resources

Can fossil fuels be replaced?

Most of the energy you use comes from fossil fuels. It takes millions of years to replace each drop of gasoline and each lump of coal that is burned. This means that the amount of fossil fuels on Earth will keep decreasing as it is used. Fossil fuels are nonrenewable resources. A **nonrenewable resource** is an energy source that is used up much faster than it can be replaced.

Disadvantages of Fossil Fuels Burning fossil fuels also makes chemical compounds that cause pollution. Each year billions of kilograms of air pollutants are made by burning fossil fuels. These pollutants cause respiratory illnesses and acid rain. Carbon dioxide gas is made when fossil fuels are burned. This carbon dioxide gas might cause Earth's climate to warm.

Nuclear Energy

Can you imagine 1 kg of fuel that has almost 3 million times more energy than 1 L of gas? What could have so much energy in so little mass? The answer is the nuclei of uranium atoms. When these nuclei break apart, they release huge amounts of energy. This energy is used to make electricity by heating water. The figure shows this process. The water makes steam that spins an electric generator. The generator makes electricity.

Picture This

2. Identify What type of energy does the steam and the turbine have in a nuclear power plant?

_____

Electrical Energy from Nuclear Energy

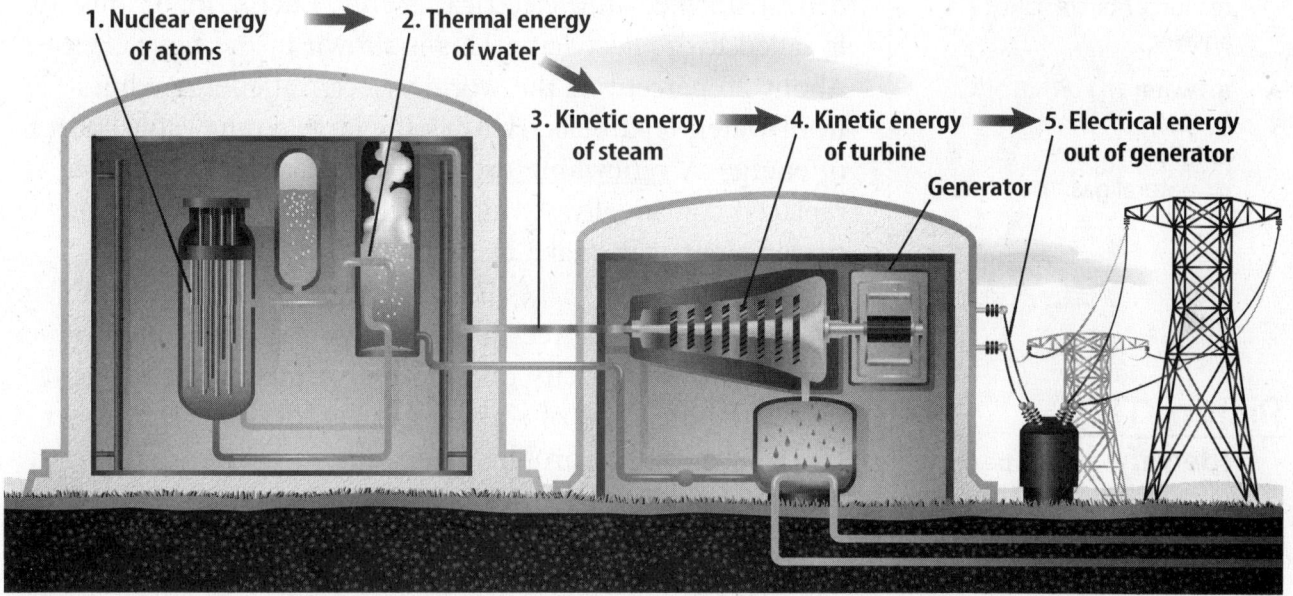

1. Nuclear energy of atoms → 2. Thermal energy of water → 3. Kinetic energy of steam → 4. Kinetic energy of turbine → 5. Electrical energy out of generator

Generator

What are the advantages of nuclear energy?

Making electricity by using nuclear energy helps make the supply of fossil fuels last longer. Nuclear power plants also produce almost no air pollution. In one year, a typical nuclear power plant makes enough energy to supply 600,000 homes with electricity. To do this, it produces only 1 m^3 of waste.

What are the disadvantages of nuclear energy?

One disadvantage of nuclear energy is that uranium is a nonrenewable resource. It comes from Earth's crust. Another disadvantage is that nuclear waste is radioactive and can be dangerous to living things. Some of the materials in nuclear waste will remain radioactive for many thousands of years. This means nuclear waste must be carefully stored so no radioactivity will be released into the environment for a long time. ☑

☑ Reading Check

3. Determine What are two disadvantages of nuclear energy?

_____

_____

How can nuclear waste be stored?

One way to store nuclear waste is to seal it in a ceramic material that is put in protective containers. Then the containers are buried far underground. The place to bury them has to be chosen carefully. It cannot be near underground water supplies. It also has to be safe from earthquakes and other natural disasters. Earthquakes and other natural disasters could cause the radioactive material to leak.

Hydroelectricity

The potential energy of water trapped behind a dam can be transformed into electrical energy. Energy made this way is called hydroelectricity. This is shown in the figure below. About 20 percent of the world's electrical energy comes from water. Hydroelectricity is the largest renewable source of energy. A **renewable resource** is an energy source that is replaced continually. As long as rivers flow, hydroelectric power plants can make electricity.

Hydroelectricity makes little pollution. This is an advantage over some other sources of electricity. However, the production of hydroelectricity does have a major disadvantage. It upsets the life cycle of some animals that live in the water. Dams have caused problems for salmon in the Northwest. Salmon return to the spot where they were hatched to lay their eggs. Many salmon cannot reach these places because of dams. There are plans to remove some dams and build fish ladders to help fish go around other dams.

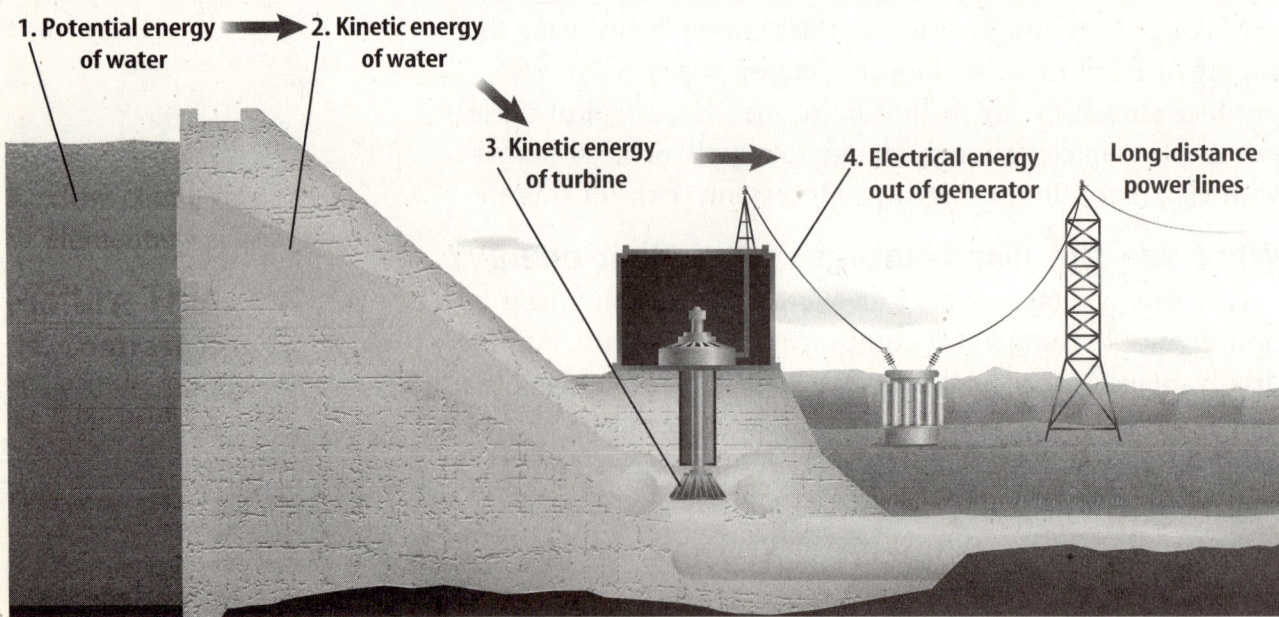

1. Potential energy of water → 2. Kinetic energy of water

3. Kinetic energy of turbine

4. Electrical energy out of generator

Long-distance power lines

Think it Over

4. Infer Which of the following is a renewable resource? Circle your answer.

a. water
b. coal
c. oil
d. natural gas

Picture This

5. Identify What type of energy does the water have when it flows through the dam?

_____

Are energy consumption and production equal?

More energy is being consumed, or used, in the United States than is being produced, or made. You use energy every day—to get to school, to watch TV, and to heat or cool your home. The amount of energy used by an average person has increased. Therefore, more energy must be made. The graph shows energy consumption and production by the United States from 1949 to 1999.

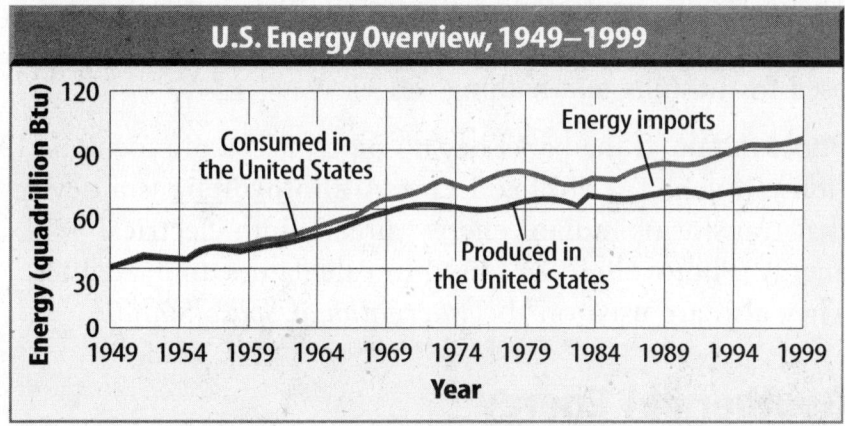

U.S. Energy Overview, 1949–1999

Applying Math

6. Reading Graphs
About what year did the United States start consuming more energy than it produced?

_____

Alternative Sources of Energy

There are many ways to make electrical energy. Each has disadvantages that can affect the environment and humans. Alternative resources are being researched. **Alternative resources** are new sources of energy that are safer and less harmful to the environment. Alterative resources include solar energy, wind energy, and geothermal energy.

Solar Energy

The Sun is an inexhaustible resource. An **inexhaustible resource** is an energy source that cannot be used up by humans. The amount of solar energy that hits the United States in one day is more than the total amount of energy used by the country in one year. But less than 0.1 percent of the energy used in the United States comes directly from solar energy. One reason is that solar energy is more expensive to use than fossil fuels. However, as the supply of fossil fuels decreases, it might become more expensive to find and mine fossil fuels. It might also become more expensive to mine them from Earth. Then, it might be cheaper to use solar energy or other energy sources to make electricity. ☑

Reading Check

7. Identify What type of resource is the Sun?

_____

How is the Sun's energy collected?

Two types of collectors take in the Sun's rays. Have you ever seen large rectangular panels on the roofs of houses or buildings? These are collectors for solar energy.

Thermal Collector If the panels had pipes coming out of them, they were thermal collectors. A thermal collector uses a black surface to absorb the Sun's radiant energy. Black absorbs more radiant energy than any other color. The thermal collector uses the Sun's radiant energy to heat water. The water can be heated to about 70°C. The hot water can be pumped through a house to provide heat. It can also be used for washing and bathing. ☑

Photovoltaic If the panel has no pipes, it is a photovoltaic (foh toh vohl TAY ihk) collector. A **photovoltaic** is a device that transforms radiant energy directly into electrical energy. Photovoltaics are used in calculators and satellites. They also are used on the *International Space Station.*

Geothermal Energy

Imagine you could go to the center of Earth, about 6,400 km below the surface. As you went deeper and deeper, the temperature would increase. After going only about 3 km, the temperature would be warm enough to boil water. At a depth of 100 km, the temperature could be over 900°C.

The heat made inside Earth is called geothermal energy. Some geothermal energy is made when unstable radioactive atoms inside Earth decay. This transforms nuclear energy into thermal energy. At some places deep within Earth, the temperature is hot enough to melt rock. Melted, or molten, rock is called magma. Magma rises up close to the surface through cracks in Earth's crust. Magma reaches the surface when a volcano erupts. In other places, magma gets close to the surface and heats the rock around it.

What are geothermal reservoirs?

In some places, magma is very close to Earth's surface. Rainwater and water from melted snow can seep down to the magma through the cracks and openings in Earth's surface. The magma heats the water and it can become steam. The hot water and steam can be trapped under high pressure in cracks and pockets. These are called geothermal reservoirs. Geothermal reservoirs are sometimes close enough to the surface to make hot springs and geysers. ☑

✔ Reading Check

8. Apply Why are thermal collectors black?

_____

_____

_____

✔ Reading Check

9. Determine What does the magma in geothermic reservoirs turn water into?

_____

How is geothermal power made?

Wells can be drilled to reach geothermal reservoirs in places where the reservoirs are less than several kilometers deep. Hot water and steam from geothermal energy is used by geothermal power plants to make electricity.

Geothermal Power Plant

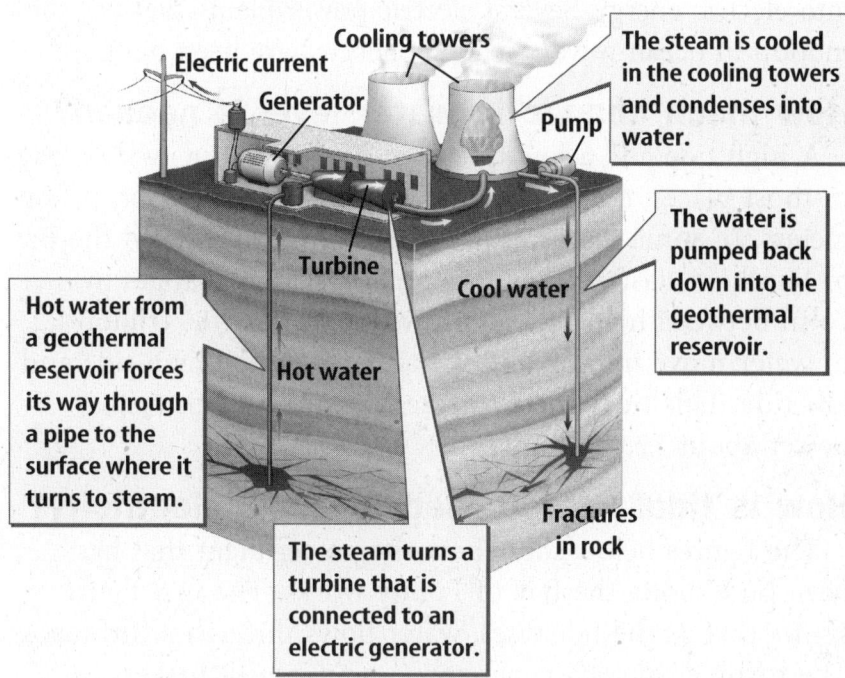

Electric current

Generator

Cooling towers

The steam is cooled in the cooling towers and condenses into water.

Pump

Turbine

Cool water

The water is pumped back down into the geothermal reservoir.

Hot water from a geothermal reservoir forces its way through a pipe to the surface where it turns to steam.

Hot water

Fractures in rock

The steam turns a turbine that is connected to an electric generator.

Picture This

10. **Interpret an Illustration** Which is pumped back down into a geothermic reservoir from a geothermic power plant: steam, hot water, or cool water?

_____

The figure shows how geothermal reservoirs make electricity. Geothermal power is an inexhaustible resource. But geothermal power plants can be built only where geothermal reservoirs are close to Earth's surface, like in the western United States.

How are heat pumps used?

Geothermal heat usually keeps the temperature of the ground that is several meters deep at 10° to 20°C. This constant temperature can be used to heat or cool buildings by using a heat pump.

During the summer, the air is warmer than the ground below. A heat pump sends warm water from the building through the cooler ground. The water cools and then is pumped back to the building to absorb heat. In the winter, the air is cooler than the ground below. Then, the cool water absorbs heat from the ground and releases it from the heat pump into the building.

Think it Over

11. **Explain** Why does the cool water in the building absorb heat in the summer?

_____

_____

12. Explain What kind of resource is the movement of the ocean?

_____

Energy from the Oceans

The ocean is constantly moving. If you have been to the seashore, you have seen the waves roll in. If you spent the day at the beach, you may have also seen the level of the ocean rise and fall. The rise and fall in the ocean level is called a tide. The movement of the ocean is an inexhaustible source of mechanical energy. Mechanical energy can be transformed into electric energy. Several electric power plants that use the motion in ocean waves, or tidal energy, have been built. ✔

How much change in water level is needed?

A high tide and a low tide each happen about twice a day. In most places, the level of the ocean changes by only a few meters. In some places, it changes by much more. In the Bay of Fundy in Eastern Canada, the ocean level changes by 16 m between high tide and low tide. Almost 14 trillion kg of water move into or out of the bay between high tide and low tide. This tidal energy makes enough electricity to power about 12,000 homes.

How is tidal energy used to make electricity?

The figures below show how the power plant that has been built along the Bay of Fundy works. The first figure shows that as the tide rises, water flows through a turbine. The turbine causes a generator to spin, which makes electricity. The water is then trapped behind a dam. The second figure shows that when the tide goes out, the trapped water is released. It flows through the turbine making the generator spin. This makes more electricity. Electric power is made each day for about 10 hours.

Tidal energy is a clean, inexhaustible resource. But, only a few places have a large enough difference between high and low tide to build an electric power plant.

Picture This

13. Highlight Use a highlighter to trace the flow of water into and out of the tidal power plant.

Tidal Power Plant

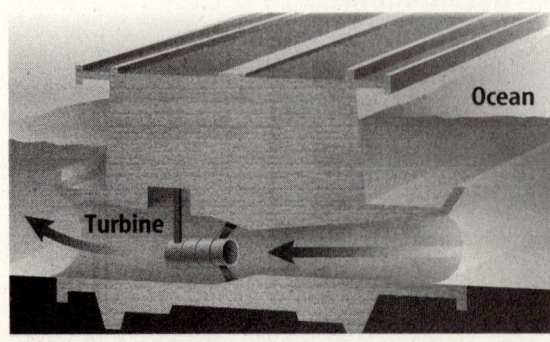
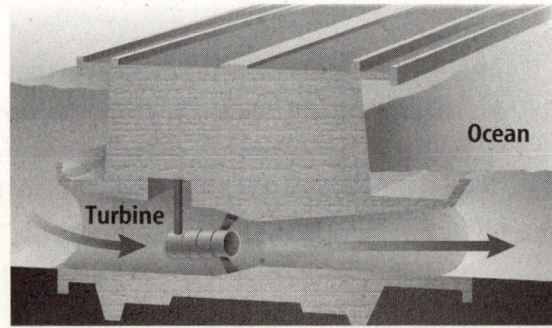

Wind

Wind is another inexhaustible supply of energy. Modern windmills, like the ones in the figure, transform the kinetic energy of the wind into electrical energy. Electrical energy is made when wind spins the propeller. The propeller is connected to a generator, which makes electricity. These windmills produce almost no pollution. But windmills do make a lot of noise. You also need a large area of land to place a lot of windmills. Also, studies have shown that birds sometimes are killed by windmills.

Picture This

14. Infer Why do you think the windmills shown in the figure are placed on top of mountains instead of between hills or mountains?

_____

_____

Conserving Energy

Fossil fuels are a valuable resource. They are burned to provide energy. Oil and coal can be used to make plastics and other materials. To make the supply of fossil fuels last longer, people need to use less energy. Using less energy is called conserving energy.

You can save money by conserving energy. You should turn off appliances like televisions when you are not using them to conserve energy. Keep doors and windows closed tightly when it is hot or cold outside. This will keep heat from leaking out of or into your house. If cars were used less or were made more efficient, they would use less gas and oil, and therefore less energy. You also help conserve energy when you recycle aluminum cans and glass.

Think it Over

15. Describe What is another way you can conserve energy?

_____

_____

_____

● After You Read

Mini Glossary

alternative resources: new renewable or inexhaustible energy sources

inexhaustible resource: an energy source that cannot be used up by humans

nonrenewable resource: an energy source that is used up much faster than it can be replaced

photovoltaic: a device that transforms radiant energy directly into electrical energy

renewable resource: an energy source that is replaced continually

1. Review the terms and their definitions in the Mini Glossary. What is the difference between a renewable resource and a nonrenewable resource?

_____

_____

2. Write as many examples of renewable, nonrenewable, and inexhaustible resources in the chart as you can.

Renewable Resources	Nonrenewable Resources	Inexhaustible Resources

3. You were asked to highlight the text each time you read about an energy source. How did this help you learn about energy sources?

_____

_____

End of Section

 **Science**Online Visit **ips.msscience.com** to access your textbook, interactive games, and projects to help you learn more about sources of energy.

Work and Simple Machines

chapter 14

section ❶ Work and Power

● Before You Read

Describe the work you have done today.

_____

_____

● Read to Learn

What is work?

In science, there is a special definition of work. **Work** is done when a force makes an object move in the same direction as the force that is applied. You do work when you lift your books, turn a doorknob, or write with a pen.

What does motion have to do with work?

Suppose your teacher asks you to move a box of books. You try, but the box will not move. It is too heavy. You are tired because you tried to force the box to move. But you have not done any work. Two things must happen for you to do work. First, you must apply a force on an object. Second, the object must move in the same direction as the force that you applied. Imagine a girl standing still and holding two bags of groceries. Is she doing work? No, she is not moving or causing anything to move.

How does the direction of force affect work?

Your arms apply a force upward when you lift a basket of clothes. Your arms have done work because the basket moved in the same direction as the force applied by your arms. If you walk forward with the basket, your arms are still applying an upward force on the basket. But you and the basket are moving forward. The basket is not moving in the same direction as the upward force applied by your arms. So, no work is done by your arms.

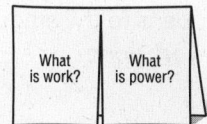

FOLDABLES

Ⓐ Organize Information Make the following Foldable to help you understand work and power.

What is work? What is power?

Does all of a force do work?

Sometimes only part of a force moves an object. Think about what happens when you push a lawn mower. Look at the figure. You push at an angle to the ground. Part of the force is forward. Part of the force is downward. Which part of the force does work? Only the part of the force that is forward does work. It is in the same direction as the motion of the mower.

Picture This

1. **Identify** Which force in the figure does work? Circle the label and arrow.

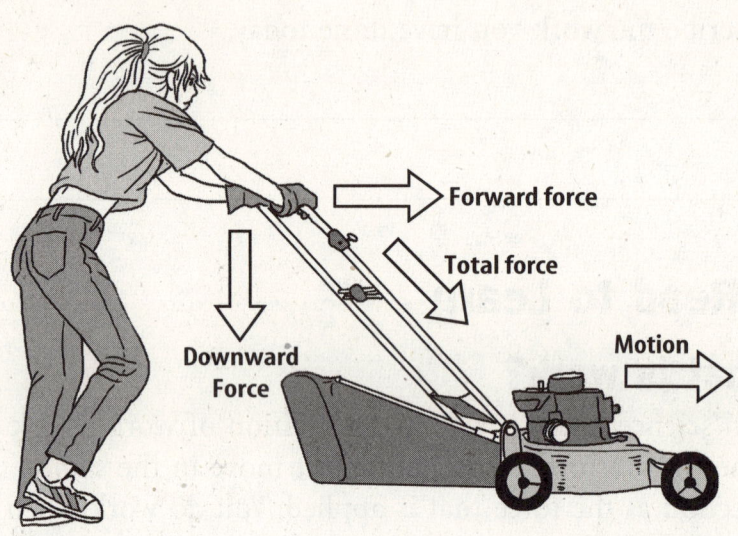

Forward force

Total force

Downward Force

Motion

Applying Math

2. **Calculate** A woman lifted a box with a force of 50 N. She lifted the box 2 m. How much work did she do? Show your work.

_____

_____

Calculating Work

Work is done when a force makes an object move. More work is done if the force is increased or if the object moves farther. You can calculate how much work is done by using the work equation below. The SI unit for work is the joule (JEWL). The joule is named for the nineteenth-century scientist James Prescott Joule.

$$\textbf{work (joules)} = \textbf{force (newtons)} \times \textbf{distance (meters)}$$
$$W = Fd$$

What distance is used for the work equation?

Suppose you give a book a push and it slides across a table. You use the distance an object moves while a force is acting on it to calculate work, not the total distance the object moved. So, the distance in the work equation is the distance the book moved while you were pushing it.

Think it Over

3. **Infer** Suppose the woman in Problem 2 above lifted the box only 1 m. What happens to the amount of work done?

_____

What is power?

What does it mean to be powerful? Imagine two weightlifters lifting the same amount of weight. They lift the weight the same distance above the floor. They both do the same amount of work. But the amount of power they use depends on how long it took to do the work. <u>Power</u> is how quickly work is done. The weightlifter who lifted the weight in less time is more powerful.

How do you calculate power?

You can calculate power by dividing the amount of work done by the time needed to do the work.

$$\text{power (watts)} = \frac{\text{work (joules)}}{\text{time (seconds)}}$$

$$P = \frac{W}{t}$$

The SI unit of power is the watt. The watt is named for James Watt, a nineteenth-century British scientist.

How can doing work change energy?

Remember that when something moves, it has kinetic energy. If you push a chair and make it move, you do work on the chair. You also change the chair's energy. By making the chair move, you increase the chair's kinetic energy.

If you lift an object higher, you also change the energy of the object. The potential energy of an object increases when it is higher above Earth's surface. When you lift an object, you do work on the object and increase its potential energy. ☑

How are power and energy related?

You increase the energy of an object when you do work on it. Energy cannot be created or destroyed. If the object gains energy, you must lose energy. When you do work on an object, you move, or transfer, energy to the object and your energy decreases. The amount of work done is the amount of energy transferred to the object. So, power is also equal to the amount of energy transferred in a certain amount of time.

Sometimes energy can be transferred even when no work is done. This happens when heat flows from a warm object to a cold one. Energy can be transferred in many ways, even when no work is done. Power is always the rate, or speed, at which energy is transferred. The rate is the amount of energy transferred divided by the time needed to transfer it.

4. Calculate A teacher does 140 J of work in 20 s. How much power in watts did he use? Show your work.

_____

_____

☑ **Reading Check**

5. Summarize When you do work on an object, what two kinds of energy can you increase for the object?

_____

_____

● After You Read

Mini Glossary

power: how quickly work is done

work: is done when a force makes an object move in the same direction as the force that is applied

1. Read the key terms and definitions in the Mini Glossary above. Describe work in your own words.

   _____

   _____

   _____

2. Complete the table.

Action	Was work done on the book?	In which direction was work done?	How did the action change the energy of the object?
Lifting your books from the bottom of your locker	yes	up	The books now have potential energy.
Carrying your books from your locker to class			
Pushing your book across your desk for a friend to see			

3. You were asked to make two flash cards for every page of the section. How did this help you learn the material in the section?

   _____

   _____

   _____

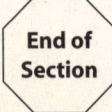
End of Section

Science Online Visit **ips.msscience.com** to access your textbook, interactive games, and projects to help you learn more about work.

Work and Simple Machines

section 2 Using Machines

● Before You Read

Describe a machine you used today and what it does.

_____

_____

● Read to Learn

What is a machine?

Scissors, brooms, and knives are all machines. A machine is a device that makes doing work easier.

Mechanical Advantage

Machines change the way you do work. When you use a machine, you apply a force over a distance. You use force to move a rake. The force that you apply on a machine is **input force.** The work you do on a machine is equal to the input force times the distance over which your force is applied. The work that you do on the machine is the input work.

The machine also does work. It applies a force to move an object over a certain distance. A rake applies a force to move leaves. Sometimes this force is called the resistance force. This means the machine must overcome some resistance. The force that the machine applies is **output force.** The work that the machine does is the output work.

The output work can never be greater than the input work when you use a machine. So why use a machine? A machine can make work easier in three ways because it can:

- change the amount of force you need to apply.
- change the distance over which the force is applied.
- change the direction in which the force is applied.

What You'll Learn

- how a machine makes work easier
- about mechanical advantage and efficiency
- how friction reduces efficiency

Mark the Text

Locate Information Look at the section headings. Highlight each question head as you read. Then, use a different color to highlight the answers to the questions.

FOLDABLES

B Organize Information
Make the following Foldable to help you organize information about machines, mechanical advantage, and efficiency.

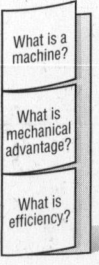

What is a machine?

What is mechanical advantage?

What is efficiency?

1. Calculate Suppose you use a machine to move a large rock. You apply a force of 100 N to the machine. The machine applies a force of 2,000 N to the rock. What is the mechanical advantage of the machine? Circle your answer.

a. 2
b. 10
c. 20
d. 100

Picture This

2. Describe Write the words that make the sentence true on the lines below: When a machine increases **a.**_____, it is applied over a shorter **b.**_____.

a. _____

b. _____

How does changing force make work easier?

Some machines make doing a job easier by reducing the force you need to apply. You need less force to do the work. This kind of machine makes the output force greater than the input force. How much larger the output force is compared to the input force is the **mechanical advantage** (MA) of the machine. You can calculate mechanical advantage using this equation:

$$\text{mechanical advantage} = \frac{\text{output force (newtons)}}{\text{input force (newtons)}}$$

$$MA = \frac{F_{\text{out}}}{F_{\text{in}}}$$

How does changing distance make work easier?

Some machines let you apply force over a shorter distance. In these machines, the output force is less than the input force. A rake is an example of this kind of machine. You move your hands a small distance at the top of the handle. But, the bottom of the rake moves a greater distance. The mechanical advantage of this kind of machine is less than 1. This is because the output force is less than the input force.

How does changing direction make work easier?

Sometimes it is easier to apply a force in a certain direction. Imagine putting a flag up on a flagpole. It is easier to pull down on the rope on the flagpole than to pull up on it. Some machines let you change the direction of the input force. In these machines, the distance and the force do not change. The mechanical advantage of this kind of machine is equal to 1. The output force is equal to the input force. The figures show the three ways machines make doing work easier.

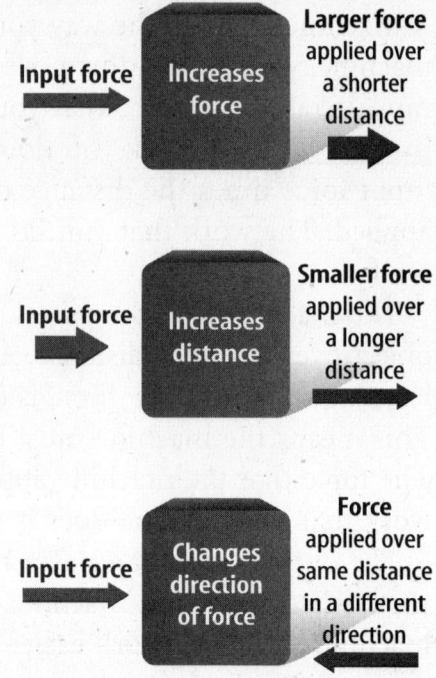

Input force → Increases force → Larger force applied over a shorter distance

Input force → Increases distance → Smaller force applied over a longer distance

Input force → Changes direction of force → Force applied over same distance in a different direction

Efficiency

A machine can't make the output work greater than the input work. In fact, for a real machine, the output work is always less than the input work. There is friction when parts of the machine move. Friction changes some of the input work into heat. So, the output work is less. If friction in the machine decreases, the efficiency, or amount of effort, of the machine increases. The **efficiency** of a machine is the ratio of the output work to the input work. You can find efficiency by using this equation:

$$\text{efficiency (in percent)} = \frac{\textbf{output work (joules)}}{\textbf{input work (joules)}} \times 100\%$$

$$eff = \frac{W_{\text{out}}}{W_{\text{in}}} \times 100\%$$

How does friction affect a machine?

Imagine pushing a heavy box up a ramp. The bottom surface of the box slides across the top surface of the ramp. Neither the box nor the ramp is perfectly smooth. Each surface has high spots and low spots.

As the two surfaces slide past each other, high spots on the two surfaces touch each other. The places that they touch are called contact points. At these contact points, atoms and molecules can bond together. This makes the contact points stick together. The attractive forces between all of the bonds added together is the frictional force. The frictional force tries to keep the two surfaces from sliding past each other. ☑

To keep the box moving, a force must be applied. The force has to break the bonds between the contact points. Even after these bonds are broken and the box moves, new bonds form as different parts of the two surfaces touch.

How can friction be reduced?

One way to reduce friction between two surfaces is to add oil to the surfaces. Oil can fill the gaps between the surfaces. Oil keeps many of the high spots from touching each other. There are fewer contact points between the surfaces. So, the force of friction is less. This means more of the input work is changed to output work by the machine.

3. Calculate Workers use a ramp to load a piano into a truck. The output work, or the amount of work needed to move the piano, is 12,000 J. The workers do 15,000 J of work. What is the efficiency of the ramp? Show your work.

_____

_____

Reading Check

4. Identify What causes friction?

_____

_____

Think it Over

5. Evaluate Which will have more friction, two pieces of sandpaper or two pieces of notebook paper?

_____

● After You Read

Mini Glossary

efficiency: the ratio of the output work to the input work

input force: the force that you apply on a machine

mechanical advantage: the number of times that a machine increases the input force

output force: the force that a machine applies

1. Review the terms and their definitions in the Mini Glossary. Describe how the input force and output force of a machine work together to make work easier.

_____

_____

_____

2. In the figure below, write the way each machine can be useful. Write the terms *increases force, changes direction of force,* and *increases distance* in the correct locations.

How Machines Make Work Easier

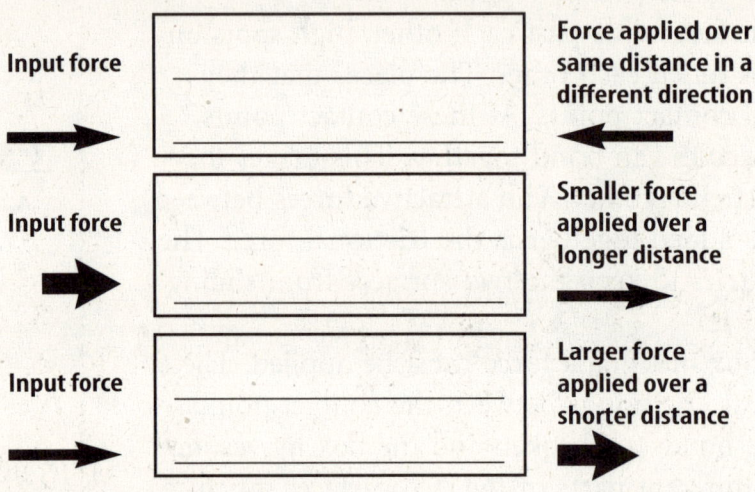

Input force

Force applied over same distance in a different direction

Input force

Smaller force applied over a longer distance

Input force

Larger force applied over a shorter distance

3. How did highlighting the answers to the headings that were questions help you make sure you understood the material in the section?

_____

_____

_____

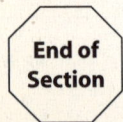
End of Section

Science Online Visit **ips.msscience.com** to access your textbook, interactive games, and projects to help you learn more about using machines.

Work and Simple Machines

section ❸ Simple Machines

● Before You Read

Suppose you need to put a heavy box into a truck. Would you rather push the box up a ramp or lift it straight into the air? Explain.

_____

_____

Copyright © Glencoe/McGraw-Hill, a division of The McGraw-Hill Companies, Inc.

● Read to Learn

What is a simple machine?

In the last section you learned that machines make work easier. Some machines like cars, elevators, or computers are very complicated. But machines can be very simple. A hammer, a shovel, and a ramp are all machines. A **simple machine** is a machine that does work with only one movement. There are six simple machines: an inclined plane, a lever, a wheel and axle, a screw, a wedge, and a pulley. A **compound machine** is a machine made up of more than one simple machine. A bicycle is a compound machine.

Inclined Plane

Ramps have been used for thousands of years. Ancient Egyptians might have used them to build the pyramids. Archaeologists hypothesize that the Egyptians built huge ramps to move limestone blocks. The blocks each weighed more than 1,000 kg. A ramp is a simple machine known as an inclined plane. An **inclined plane** is a flat, sloped surface. You might need a lot of force if you have to lift an object. An inclined plane lets you use less force to move an object from one height to another. The longer an inclined plane is, the less force is needed to move the object.

What You'll Learn

■ what the different simple machines are
■ how to find the mechanical advantage of each simple machine

> **Mark the Text**
>
> **Identify Simple Machines** As you read through this section, underline each kind of simple machine.

FOLDABLES

Ⓒ Compare and Contrast
Make the following Foldable to help you compare and contrast simple and compound machines.

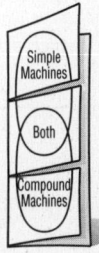

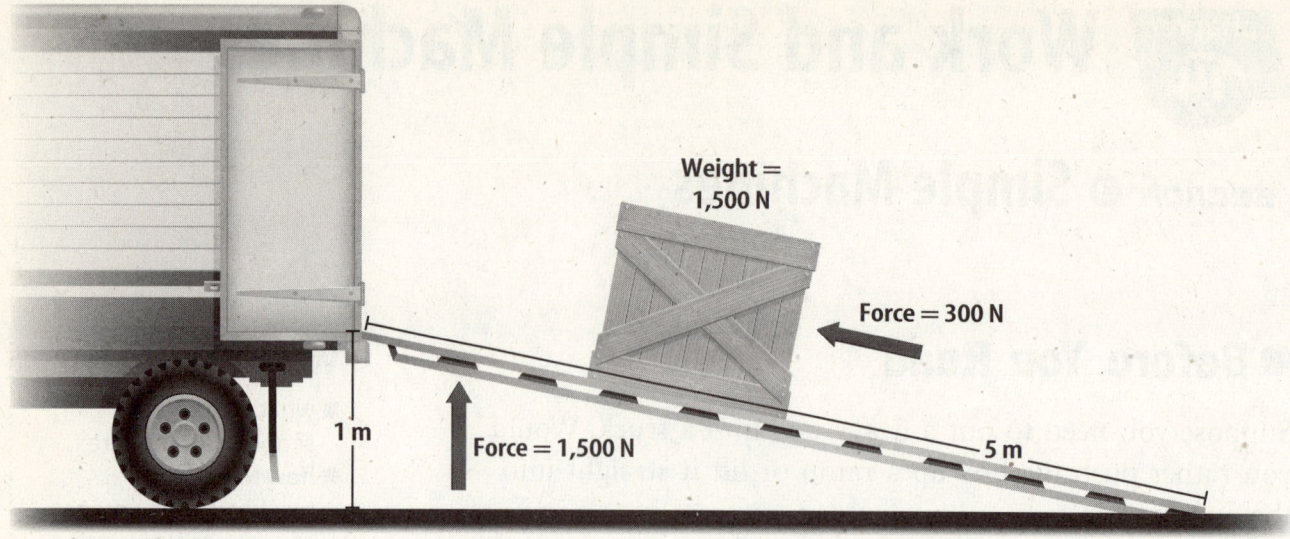

Weight =
1,500 N

Force = 300 N

1 m

Force = 1,500 N

5 m

Picture This

1. Evaluate Circle the force needed to push the box up the ramp. Then, circle the force needed to lift the box straight up. How much more force is needed to lift the box straight up?

_____

_____

Applying Math

2. Calculate You need to lift a different box into the truck. The amount of work it takes is 1,000 J. You use a 2 m ramp. How much force do you need? Show your work.

_____

_____

How are inclined planes used?

Suppose you have to lift a box weighing 1,500 N into the back of the truck that is 1 m off the ground. Could you do that? The force (1,500 N) times the distance (1 m) equals 1,500 J of work. Look at the figure. Suppose you use a 5-m-long ramp to lift the box into the truck. The amount of work you need to do does not change. You still need to do 1,500 J of work. But, the distance over which you apply the force is now 5 m. You can find the force you need by dividing both sides of the work equation by distance.

$$\text{Force} = \frac{\text{work (joules)}}{\text{distance (meters)}}$$

$$\text{Force} = \frac{1,500 \text{ J}}{5 \text{ m}} = 300 \text{ N}$$

When you use the ramp, you need to apply a force of only 300 N. A force of 300 N is much less than a force of 1,500 N. With the ramp, you apply the force over a distance that is five times longer. So, the force is five times less.

The mechanical advantage of an inclined plane is the length of the inclined plane divided by its height. In this example, the ramp has a mechanical advantage of 5.

What is a wedge?

A **wedge** is an inclined plane that moves. A wedge can have one or two sloping sides. A knife is an example of a wedge. An axe and certain kinds of doorstops also are wedges. The mechanical advantage of a wedge increases as it becomes longer and thinner.

Are there wedges in your body?

You have wedges in your body. Think of biting into an apple. The bite marks on the apple show that your front teeth are wedge shaped. A wedge changes the direction of the applied force. The downward force of your bite is changed into a sideways force. The sideways force pushes the skin of the apple apart.

What is a screw?

The screw is another form of inclined plane. A **screw** is an inclined plane wrapped around a cylinder or post. The inclined plane on a screw forms the threads of the screw. A screw changes the direction of the applied force. The applied force pulls the screw into the material. Friction between the threads and the material holds the screw tightly in place. The mechanical advantage of the screw is the length of the inclined plane wrapped around the screw divided by the length of the screw. The more tightly wrapped the threads are, the easier it is to turn the screw. ☑

Lever

A **lever** is any stiff rod or plank that turns around a point. The point that the lever turns around is called a fulcrum. You can find the mechanical advantage of a lever by dividing the distance from the input force to the fulcrum by the distance from the fulcrum to the output force. This is shown in the figure.

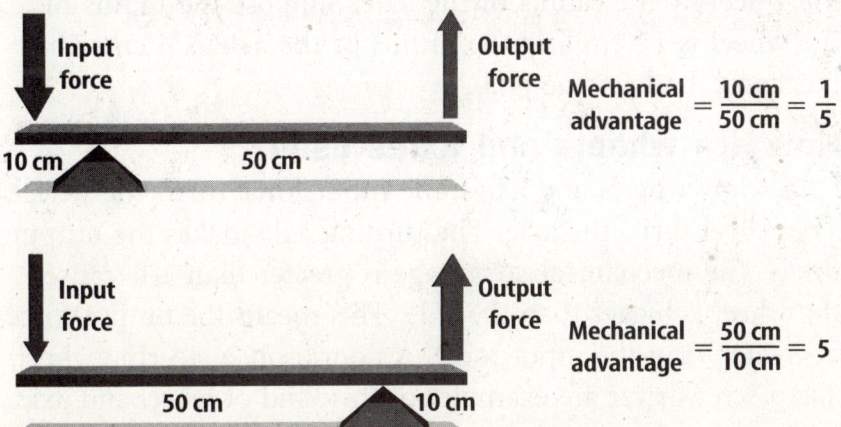

The fulcrum of a lever can be in different positions. When the fulcrum is closer to the output force than the input force, the mechanical advantage is greater than 1. Scissors, a wheelbarrow, and a baseball bat are all levers.

✔ **Reading Check**

3. **Explain** What holds a screw into an object?

_____

_____

_____

Picture This

4. **Calculate** Use the figure. Suppose the distance from the input force to the fulcrum is 60 cm. The distance from the fulcrum to the output force is 20 cm. What is the mechanical advantage? Show your work.

_____

_____

_____

Wheel and Axle

Imagine a doorknob the size of a pencil. Would it be easy to turn? No, it would be hard to turn. A doorknob is a simple machine that makes opening a door easier. A doorknob is a wheel and axle. A **wheel and axle** is made up of two round objects of different sizes that are attached so they turn together. The larger object is the wheel and the smaller object is the axle. The faucet handle shown in the figure is a wheel and axle.

<u>Picture This</u>

5. Analyze Does the wheel or the axle of the faucet make the output force?

_____

_____

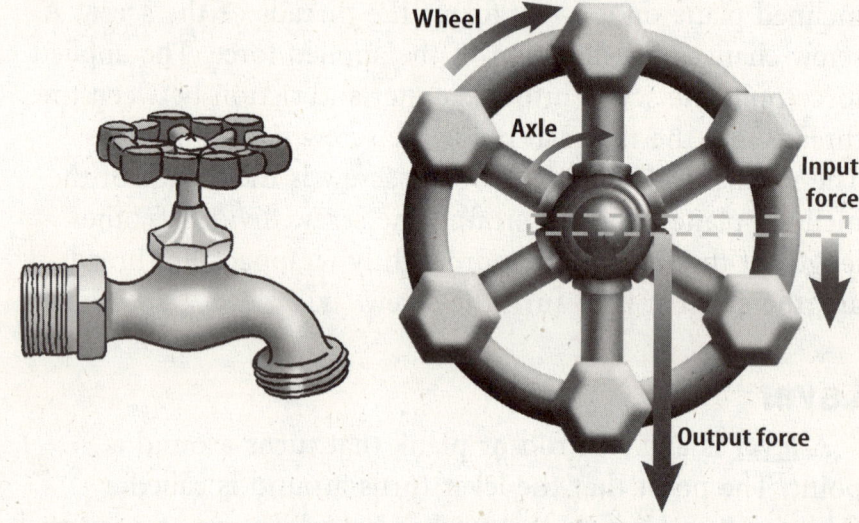

Wheel

Axle

Input force

Output force

Applying Math

6. Calculate A wheel has a radius of 15 cm. The axle has a radius of 3 cm. What is the mechanical advantage of the wheel and axle? Show your work.

_____

_____

How do you find the MA of a wheel and axle?

The MA (mechanical advantage) of a wheel and axle is usually greater than 1. You find it by dividing the radius of the wheel by the radius of the axle. Suppose the radius of the wheel is 12 cm and the radius of the axle is 4 cm. The mechanical advantage is 3.

How are wheels and axles used?

In some wheels and axles, the input force turns the wheel. The wheel turns the axle. The turning axle makes the output force. The mechanical advantage is greater than 1 because the wheel is bigger than the axle. This means the output force is greater than the input force. A doorknob, a steering wheel, and a screwdriver are examples of this kind of wheel and axle.

In other wheels and axles, the input force turns the axle. The axle turns the wheel. The wheel makes the output force. The mechanical advantage is less than 1. The output force is less than the input force. A fan and a Ferris wheel are examples of this kind of wheel and axle.

Pulley

To raise a sail, a sailor pulls down on a rope. The rope uses a simple machine called a pulley to change the direction of the force needed. A **pulley** is a grooved wheel with a rope or cable wrapped over it.

What is a fixed pulley?

Think about the pulley on a sail. The pulley is attached to something above your head. This kind of pulley is called a fixed pulley. When you pull down on the rope, the sail is pulled up. Look at the figure of the fixed pulley below. A fixed pulley does not change the force you apply. It also does not change the distance over which you apply a force. A fixed pulley does change the direction in which you apply your force. The mechanical advantage of a fixed pulley is 1. ☑

What is a movable pulley?

You can also attach a pulley to the object you need to lift. This is called a movable pulley. A movable pulley lets you apply a smaller force to lift the object. The mechanical advantage of a movable pulley is always 2. The middle pulley in the figure below is a movable pulley.

You often will see movable and fixed pulleys used together. This is called a pulley system. The mechanical advantage of a pulley system is equal to the number of sections of rope pulling up on the object. The mechanical advantage for the pulley system in the figure is 3.

☑ **Reading Check**

7. Describe How does a pulley change the force you apply?

_____

_____

_____

Picture This

8. Explain How does a movable pulley change the input force?

_____

_____

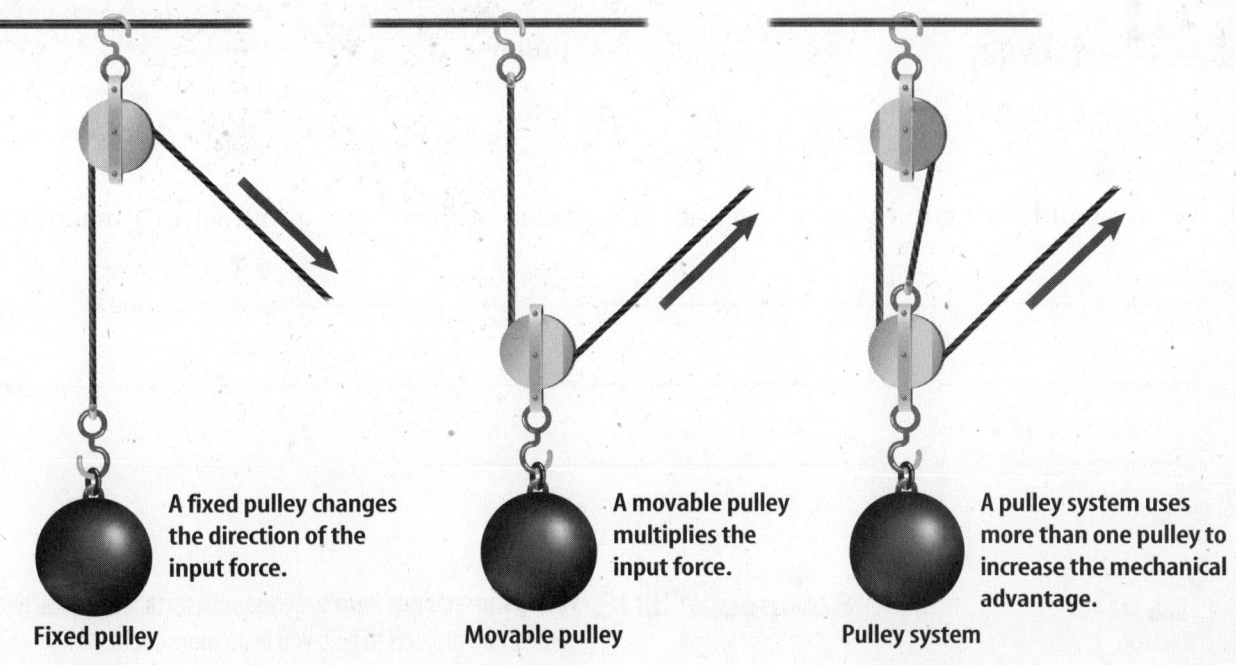

A fixed pulley changes the direction of the input force.

Fixed pulley

A movable pulley multiplies the input force.

Movable pulley

A pulley system uses more than one pulley to increase the mechanical advantage.

Pulley system

● After You Read

Mini Glossary

compound machine: a machine made up of more than one simple machine

inclined plane: a flat, sloped surface

lever: any stiff rod or plank that turns around a point

pulley: a grooved wheel with a rope or cable wrapped over it

screw: an inclined plane wrapped around a cylinder or post

simple machine: a machine that does work with only one movement

wedge: an inclined plane that moves

wheel and axle: two round objects of different sizes that are attached so they turn together

1. Review the terms and their definitions in the Mini Glossary. Write a sentence describing how a screw and a wedge are related.

_____

_____

2. Match each simple machines with the correct example. Write the letter of each simple machine in Column 2 on the line in front of the example it is or uses in Column 1.

Column 1	Column 2
_____ 1. tooth	a. inclined plane
_____ 2. doorknob	b. wheel and axle
_____ 3. threads of a lightbulb	c. wedge
_____ 4. wheelbarrow	d. screw
_____ 5. ramp	e. pulley
_____ 6. flagpole rope	f. lever

3. What would be a good way to teach an elementary science class about simple machines?

_____

_____

_____

End of Section

Science Online Visit **ips.msscience.com** to access your textbook, interactive games, and projects to help you learn more about simple machines.

Thermal Energy

section ❶ Temperature and Thermal Energy

● Before You Read

Do you consider the temperature when you choose the clothes you wear each day? List two examples of when it's important to know the temperature of something.

What You'll Learn
■ how temperature and kinetic energy are related
■ three scales used to measure temperature
■ what thermal energy is

● Read to Learn

What is temperature?

Think of a hot summer day. You jump into a swimming pool to cool off. At first, the water feels cold on your skin. But the water might feel warm to your friend who has been swimming for awhile. How do you know if something is hot or cold? Your sense of touch tells you if something feels hot, warm, or cold.

Why do some things feel hot and others feel cold?

How hot or cold something feels depends on its temperature. To understand temperature, think of water in a glass. Water is made of molecules. Water molecules are always moving. When something moves, it has energy of motion, or kinetic energy.

The molecules in the glass of water move at different speeds. Some move quickly and some move slowly. The fastest molecules have the most kinetic energy. **Temperature** is a measure of the average kinetic energy of the molecules. The more kinetic energy the molecules have, the higher the temperature. Molecules have more kinetic energy when they are moving faster. So the higher the temperature, the faster the molecules are moving.

Study Coach

Make Flash Cards As you read this section, think of questions your teacher might ask on a test. Write each question on the front of a flash card. Write each answer on the back of each card.

FOLDABLES

Ⓐ Find Main Ideas Make a Foldable like the one below. As you read this section, write the main ideas about temperature and thermal energy.

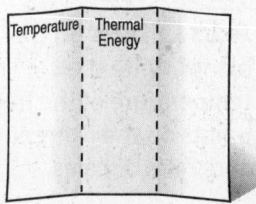

What are some effects of temperature?

Sometimes in hot weather, a sidewalk will crack. When the temperature of an object increases, its molecules move faster and farther apart. This causes the object to get larger, or to expand. Concrete in a sidewalk, like almost all substances, expands as its temperature increases. When an object cools, its molecules slow down and move closer together. The object gets smaller, or contracts.

Most materials expand when they are heated and contract when they are cooled. How much a material expands or contracts depends on the material and the change in temperature. Liquids usually expand more than solids. And, the greater the change in temperature, the more an object expands or contracts.

Measuring Temperature

Recall that an object's temperature depends on the average kinetic energy of all its molecules. How do you measure the kinetic energy of all those tiny molecules? One way to measure temperature is to use a thermometer. ☑

One type of thermometer is made of a glass tube with a liquid inside. When the temperature of the liquid increases, the liquid expands and moves higher inside the tube. How high the liquid moves depends on the temperature.

What temperature scales are used?

A thermometer has to have a scale on it to measure temperature change. Two temperature scales are the Fahrenheit scale and the Celsius scale. The thermometer in the figure has both scales. Water freezes at 32°F and boils at 212°F on the Fahrenheit scale. There are 180 equal degrees between the freezing and boiling point of water on the Fahrenheit scale. On the Celsius scale, water freezes at 0°C and boils at 100°C. There are 100 Celsius degrees between the boiling and freezing point of water on the Celsius scale. Celsius degrees are bigger than Fahrenheit degrees.

Copyright © Glencoe/McGraw-Hill, a division of The McGraw-Hill Companies, Inc.

✔ Reading Check

1. **Describe** What does the temperature of an object depend on?

_____

_____

Picture This

2. **Add Labels** In the blanks, write the temperature of the freezing points on the Fahrenheit and Celsius scales.

Freezing point of water

Freezing point of water

_____ °F

_____ °C

How is one scale changed to the other?

Use the following equations to convert, or change, °F to °C and °C to °F.

To convert temperature in °F to °C: $°C = \frac{5}{9}(°F - 32)$

To convert temperature in °C to °F: $°F = \frac{9}{5}(°C) + 32$

For example, to change 68°F to degrees Celsius, first subtract 32, multiply by 5, then divide by 9. So 68°F = 20°C.

Are there other temperature scales?

Another temperature scale is the Kelvin scale. On the Kelvin scale, 0 K is the lowest temperature possible. 0 K is also called absolute zero. It equals −273°C. To change from Celsius degrees to Kelvin degrees, add 273 to the Celsius temperature.

$$K = °C + 273$$

Thermal Energy

Recall that molecules in motion have kinetic energy. Molecules also have potential energy. Potential energy is energy that can be changed into kinetic energy. The sum of the kinetic energy and potential energy of all the molecules in an object is its **thermal energy**.

What is potential energy?

A ball being held above the ground has potential energy. When the ball is dropped, the potential energy changes to kinetic energy because the ball is now moving.

Molecules in a material attract each other. The molecules have potential energy because of this attraction. The potential energy of the molecules changes as the molecules get closer together or farther apart.

How are temperature and thermal energy different?

Suppose you have two glasses filled with the same amounts of milk. The milk in both glasses is at the same temperature. If you pour both glasses of milk into a pitcher, the temperature of the milk doesn't change. But the thermal energy of the milk does change. The milk in the pitcher has more thermal energy than the smaller amounts of milk in either glass did. That's because there are more molecules of milk in the pitcher than there were in either glass.

Applying Math

3. **Use Formulas** Convert 80°F to degrees Celsius. Show all your work.

_____

_____

_____

Applying Math

4. **Calculate** Convert 30°C to Kelvin degrees. Show all your work.

_____

_____

Think it Over

5. **Interpret** When you hold a yo-yo in your hand, what kind of energy does the yo-yo have?

_____

● After You Read
Mini Glossary

temperature: a measure of the average kinetic energy of an object's molecules

thermal energy: the sum of the kinetic energy and potential energy of all the molecules in an object

1. Review the terms and definitions in the Mini Glossary. Write a sentence using the term *thermal energy.*

_____

_____

_____

2. Write the letter of the term in **Column 2** that matches the example in **Column 1**.

Column 1	Column 2
_____ 1. water boils at 212°F	a. kinetic energy
_____ 2. the type of energy a ball has when you hold it above the ground	b. Fahrenheit scale
	c. temperature
_____ 3. water freezes at 0°C	d. potential energy
_____ 4. energy of motion	e. Kelvin scale
_____ 5. measure of the average kinetic energy of an object's particles	f. Celsius scale
_____ 6. water freezes at 273 K	

3. You were asked to think of questions your teacher might ask on a test then write each question on the front of a flash card. How could you use the flash cards to help you study for a test?

_____

_____

_____

End of Section

 **Science** ●nline Visit **ips.msscience.com** to access your textbook, interactive games, and projects to help you learn more about temperature and thermal energy.

chapter 15 Thermal Energy

section ② Heat

● Before You Read

Write down two things you do to make yourself feel warmer.

_____

_____

What You'll Learn

- compare thermal energy and heat
- three ways heat moves
- what insulators and conductors are

● Read to Learn

Heat and Thermal Energy

It's cold, turn up the heat. Heat the oven to 375°F. A heat wave has hit the Midwest. You've often heard the word *heat*, but what exactly is it? <u>Heat</u> is the transfer of thermal energy from one object to another when the objects are at different temperatures. Thermal energy is transferred when two objects are in contact with each other. More heat is transferred when the difference in temperature between the objects is large. Less heat is transferred when the temperature difference is small.

For example, no thermal energy moves between two pots of boiling water that are touching. The water in both pots is the same temperature. Suppose a pot of boiling water touches a pot of cold water. Thermal energy is transferred from the hot pot to the cold pot. The hot water loses thermal energy and the cold water gains it. Thermal energy will transfer until both objects are the same temperature.

How does thermal energy move?

Thermal energy always is transferred from warmer objects to cooler objects. It never transfers from a cooler object to a warmer one. The warmer object loses thermal energy and becomes cooler. The cooler object gains thermal energy and becomes warmer. Thermal energy can be transferred in three ways—by conduction, radiation, or convection.

Mark the Text

Identify Main Ideas As you read this section, highlight the main ideas about conduction, radiation, and convection.

FOLDABLES

A Find Main Ideas Add the label "Heat" to the right column of Foldable A as shown below. As you read this section, write down the main ideas about heat.

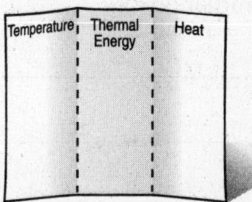

Conduction

When you eat something hot, conduction occurs. As the hot food touches your mouth, heat moves from the food to your mouth. **Conduction** is the movement of thermal energy between objects that are touching.

When you hold an ice cube in your hand, conduction is occurring. The ice cube starts to melt and your hand starts to feel cold. The fast-moving molecules in your warm hand bump into the slow-moving molecules in the cold ice. When the faster-moving molecules touch the slower-moving molecules, energy passes from molecule to molecule. As a result, thermal energy moves from your warm hand to the cold ice. The slow-moving molecules in the ice start moving faster. With more energy, the ice warms and its temperature rises. The ice begins to melt. The fast-moving molecules in your hand move more slowly. They lose thermal energy and your hand becomes cooler.

When does conduction work best?

Conduction works best in solids and liquids. That's because the molecules and atoms are closer together in a solid or a liquid than in a gas. The molecules and atoms have to move only a short distance before they bump into each other and transfer energy to another molecule or atom. So, thermal energy is transferred by conduction faster in liquids and solids than in gases.

Radiation

On a clear day, you walk outside and feel the warmth of the Sun. How does the Sun heat Earth? The Sun transfers thermal energy to Earth, but not by conduction. The Sun and Earth do not touch. Instead, the Sun transfers thermal energy to Earth by radiation. **Radiation** is the transfer of energy by electromagnetic waves. Electromagnetic waves can carry energy through empty space, like the space between Earth and the Sun. These waves can also carry energy through solids, liquids, and gases. ☑

The Sun is not the only object that transfers thermal energy by radiation. Sit next to a fire in a fireplace. You feel heat transferred by radiation from the fire to your skin. All objects give off electromagnetic radiation. Warm objects give off more radiation than cool objects.

Think it Over

1. **Describe** Give an example of heat being transferred by conduction.

_____

_____

_____

Reading Check

2. **Explain** What transfers thermal energy when objects are heated by radiation?

_____

_____

Convection

When you heat a pot of water, thermal energy transfers by conduction from the stove to the pot. It can be transferred in another way, too. As gas and liquid molecules move, they carry energy with them. **Convection** is the transfer of thermal energy through the movement of molecules from one part of a material to another.

How is thermal energy transferred by convection?

Thermal energy is transferred by convection as a pot of water is heated. First, thermal energy is transferred to the water molecules near the bottom of the pot. These water molecules begin to move faster as their thermal energy increases. The faster the molecules move, the farther apart they get. Now the molecules in the warm water are farther apart than the molecules in the cooler water near the top of the pot. So, the warm water is less dense than the cool water. The warm, less dense water rises to the top of the pot. The cool, more dense water moves down to the bottom of the pot. As the cool water is heated, it rises to the top. This repeats until all the water in the pot is the same temperature.

What is natural convection?

Natural convection takes place when a cool, dense fluid, pushes away a warm, less dense fluid. Think of the shore of a lake. The water is cooler than the land during the day. The warm land heats the air above it by conduction. As the air gets hotter, its particles move faster and farther away from each other. The hot air is less dense and it rises. The cooler, denser air from above the lake moves toward the land. You feel this movement of cool air as wind. The figure shows that as cool air moves over the land, it pushes the warm, less dense air up. The land heats the cool air and the cycle repeats.

Think it Over

3. **Explain** Why is a warm fluid less dense than a cool fluid?

_____

_____

_____

Picture This

4. **Describe** Using a highlighter, make one long stroke that follows the arrows in the figure to show the path of the air flow. Describe the movement of the air.

_____

_____

_____

Warm air

Cool air

Thermal Conductors	Thermal Insulators

What is forced convection?

Forced convection takes place when an outside force pushes a fluid to make the fluid move and transfer thermal energy. A fan is an example of an outside force. Computers use fans to keep their electronic parts from getting too hot. The fan blows cool air onto the hot parts.

Thermal energy from the computer parts is transferred to the air around them by conduction. Warm air is pushed up away from the hot parts and cool air moves in. The hot parts transfer thermal energy to the cool air around them.

Thermal Conductors

Why are most cooking pans made of metal? Why does a metal spoon in a bowl of hot soup feel warm? Metal is a good conductor of heat. A **conductor** is any material that transfers thermal energy easily. Some materials are good conductors because of the types of atoms or chemical compounds they contain.

Remember that an atom has a nucleus surrounded by one or more electrons. In some materials, like metals, these electrons are not held tightly in place. They can move around freely. These electrons can transfer thermal energy by bumping into other atoms. Metals such as gold and copper are the best conductors of thermal energy.

Thermal Insulators

When you cook, you want the pan to conduct thermal energy from the hot burner to your food. But you do not want thermal energy to move easily to the pan's handle. An insulator is a material that thermal energy does not flow through easily. Most cooking pans have handles made of insulators.

Liquids and gases are usually better insulators than solids are. Air is a good insulator. Many insulating materials have spaces filled with air. The air prevents thermal energy from moving by conduction. Metals and other good conductors of thermal energy are poor insulators. Air and other good insulators are poor thermal energy conductors.

Houses and other buildings contain insulating materials. These materials reduce the thermal energy conduction between the inside and outside. Insulating windows are made of two layers of glass. There is a layer of air or other gas in between the layers of glass. This layer reduces thermal energy conduction. It keeps thermal energy from going outside in winter and from coming inside in summer.

💡 **Think it Over**

5. **Determine** Would a container made of a conductor or of an insulator be better for keeping hot food from getting cold?

Heat Absorption

On a hot day, you can walk barefoot across the lawn. But the pavement of the street is too hot to walk on. Why is the pavement hotter than the grass? The change in temperature of an object as it is heated depends on the material it is made of.

What is specific heat?

The **specific heat** of a material is the amount of thermal energy needed to raise 1 kilogram of that material by 1°C. More heat is needed to change the temperature of a material with a high specific heat than a material with a low specific heat. ☑

For example, the sand on a beach has a lower specific heat than water in a lake. On a hot summer day, the sand feels warmer than the water. Both are warmed by radiation from the Sun. But, the sand heats up faster than the water because it has a lower specific heat than the water. At night, the sand feels cool and the water feels warmer. They both lose thermal energy to the cooler night air. However, the temperature of the sand decreases faster than the temperature of the water.

Thermal Pollution

Some power plants and factories use water to cool hot equipment. The cooling water becomes hot. This hot water may be released into lakes, rivers, or the ocean. The hot water increases the temperature of the nearby water. **Thermal pollution** is the increase in the temperature of a body of water caused by warmer water being added to it. Rainwater falling on warm roads and parking lots can also cause thermal pollution.

Why is thermal pollution harmful?

Warmer water has less dissolved oxygen than cooler water. Warm water causes fish and other animals to use more oxygen. Some animals may die because there is not enough oxygen in the water. Also, in warmer water, parasites and diseases are a bigger problem for many organisms. Factories and power plants can reduce thermal pollution by cooling hot water in cooling towers before it's released.

✔ **Reading Check**

6. **Explain** Which kind of material needs more thermal energy to raise its temperature, one with a high specific heat or one with a low specific heat?

_____

💡 **Think it Over**

7. **Identify** Name one harmful result of thermal pollution.

_____

_____

_____

● After You Read

Mini Glossary

conduction: the movement of heat between objects that are touching

conductor: any material that transfers heat easily

convection: the transfer of thermal energy through the movement of molecules from one part of a material to another

heat: thermal energy that moves from one object to another when the objects are at different temperatures

radiation: the transfer of energy by electromagnetic waves

specific heat: the amount of heat needed to raise 1 kilogram of a material by 1°C

thermal pollution: the increase in the temperature of a body of water caused by adding warmer water

1. Review the terms and definitions in the Mini Glossary. Chose one of the terms that describes how heat can be moved and use it in a sentence.

_____

_____

2. Fill in the blanks on the web diagram below with examples of the different methods of heat transfer.

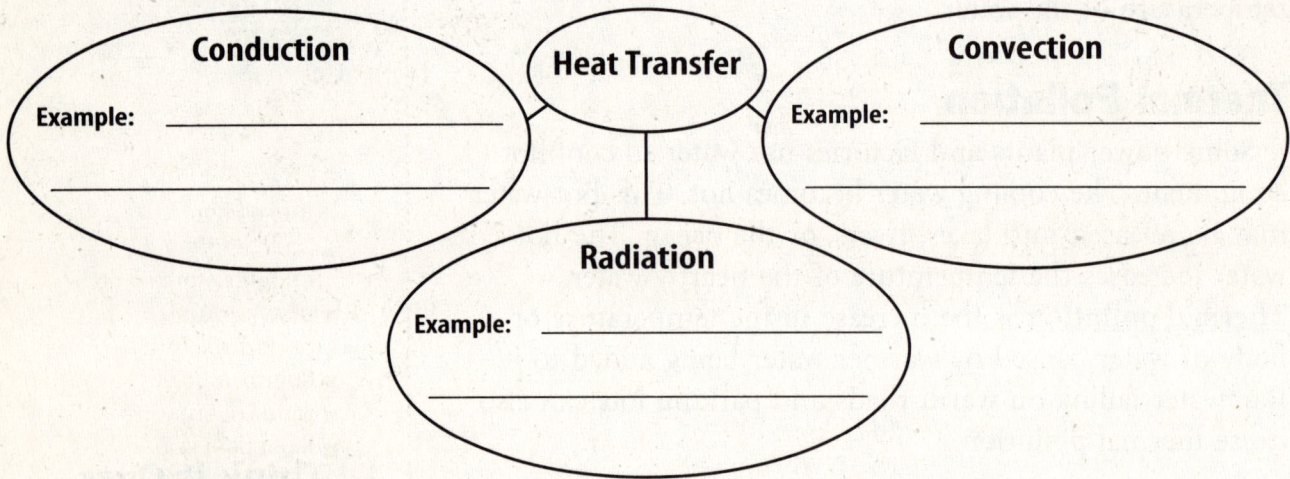

3. You were asked to highlight the main ideas about conduction, radiation, and convection. What would be another way to learn about these three methods of heat movement?

_____

_____

 **Science** online Visit **ips.msscience.com** to access your textbook, interactive games, and projects to help you learn more about heat.

Thermal Energy

section ❸ Engines and Refrigerators

● Before You Read

List at least two machines that have engines. Describe what each engine does.

_____

_____

_____

What You'll Learn

■ what heat engines do
■ forms of energy
■ energy is never created or destroyed
■ how internal combustion engines work
■ how refrigerators move heat

● Read to Learn

Heat Engines

Cars, motorcycles, and trucks use heat engines. A **heat engine** is a machine that changes thermal energy into mechanical energy. Mechanical energy is the sum of an object's kinetic energy and potential energy. The heat engine in a car converts thermal energy into mechanical energy when it makes the car move faster. This causes the car's kinetic energy to increase.

What are some other forms of energy?

Thermal energy and mechanical energy are not the only forms of energy. Chemical energy is stored in chemical bonds between atoms. Radiant energy is carried by electromagnetic waves. Nuclear energy is stored in the nuclei of atoms. Electrical energy is carried by electric charges as they move in an electric circuit. Devices such as heat engines convert one form of energy into other useful forms. ☑

Can energy be created or destroyed?

When energy changes from one form to another, the total amount of energy stays the same. The law of conservation of energy states that energy only can be changed from one form to another. Energy cannot be created or destroyed.

Mark the Text

Identify Main Ideas As you read this section, underline the answer to each question that is asked in a heading.

✔ Reading Check

1. **List** four types of energy.

_____

_____

_____

_____

What is the most common heat engine?

If you have ever ridden in a car, plane, bus, boat, or truck, you have used a type of heat engine. Most vehicles use a heat engine called an internal combustion engine. An **internal combustion engine** is an engine that burns fuel in a combustion chamber inside the engine.

How does an internal combustion engine work?

The engines in most cars have four or more combustion chambers, or cylinders. Usually the more cylinders an engine has, the more power it can produce. Inside each cylinder is a piston that moves up and down. A mixture of fuel and air is sent into the combustion chamber. A spark from a spark plug causes the fuel mixture to burn. The explosive force of the reaction pushes the piston down. As the piston moves up and down, it turns a rod called a crankshaft. The crankshaft then turns the wheels of the car. An internal combustion engine converts thermal energy to mechanical energy. The process is called a four-stroke cycle. ✓

Are all internal combustion engines the same?

There are different types of internal combustion engines. In a diesel engine, the air in a cylinder is pushed together, or compressed, until the pressure is very high. Instead of a spark plug, the high pressure ignites the fuel, or starts the fuel burning.

Many lawn mowers use a two-stroke engine. During the first stroke, fuel and air move into the chamber and are compressed. During the second stroke, the fuel mixtures burn and push the piston down.

Refrigerators

Recall that thermal energy will flow only from something warm to something that is cooler. So how can a refrigerator be cooler inside than the air in the kitchen? Refrigerators keep food cold by moving heat. A refrigerator absorbs heat from the food and other materials inside the refrigerator. Then, it moves the heat to the outside of the refrigerator. There the heat is transferred to the air around the refrigerator.

Inside a refrigerator, a liquid material called a coolant moves through pipes. The coolant absorbs heat from inside and carries the heat outside the refrigerator through the pipes.

✔ Reading Check

2. **Explain** What type of energy does an internal combustion engine convert to mechanical energy?

FOLDABLES

C Organize Information
Make a Foldable like the one below. Use the Foldable to help you organize information about refrigerators, air conditioners, and heat pumps.

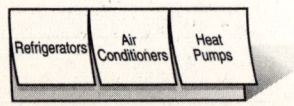

How does a coolant absorb heat?

The figure below shows how a refrigerator works. Liquid coolant is forced through a pipe. The liquid coolant passes through an expansion valve where it changes from a liquid into a gas. When it changes into a gas, it becomes cold. The cold gas moves through the pipes inside the refrigerator. Because the coolant gas is so cold, it absorbs heat from the inside of the refrigerator and gets warmer.

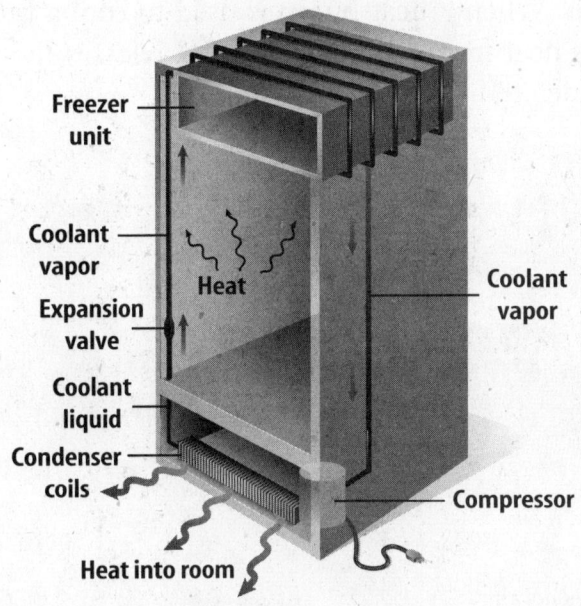

Freezer unit

Coolant vapor

Expansion valve

Coolant liquid

Condenser coils

Heat

Coolant vapor

Compressor

Heat into room

Picture This

3. Interpret Figures
Highlight the part of the refrigerator where the coolant changes from a liquid to a gas.

How does warm coolant get rid of heat?

Even though the coolant has absorbed heat, it is still cooler than the air outside the refrigerator. The heat in the coolant cannot be transferred to the outside air. The warm coolant gas passes through a compressor. The gas is compressed and gets warmer. Next, the gas passes through the condenser coils. Inside the coils, the coolant releases its heat to the cooler air outside the refrigerator. The coolant gas cools and changes back into a liquid. The liquid coolant is pumped through the expansion valve. The cycle is repeated.

Think it Over

4. Draw Conclusions
What must happen to the coolant before it can transfer the heat it has absorbed to the air?

_____

_____

_____

How do air conditioners work?

You've probably seen air-conditioning units outside of many houses. Most air conditioners cool the same way that a refrigerator does. Just like a refrigerator, the coolant inside the pipes of an air conditioner absorbs heat. The coolant passes through a compressor and becomes warmer. The warm coolant travels through pipes outside the house where the heat from the coolant moves to the outside air.

How do heat pumps work?

Some buildings use heat pumps to heat and to cool. Just like an air conditioner or a refrigerator, a heat pump moves heat from one place to another.

The figure shows how a heat pump heats the air inside a building. First, the coolant absorbs heat from the outside air through the coils outside the building. Then, the coolant is warmed as it passes through the compressor. The warm coolant releases its heat inside the building through the inside coils. When a heat pump is used to cool a building, it removes heat from the air inside and releases the heat outside. ☑

☑ Reading Check

5. **Explain** When a heat pump is used for cooling, where does it release the heat?

_____

_____

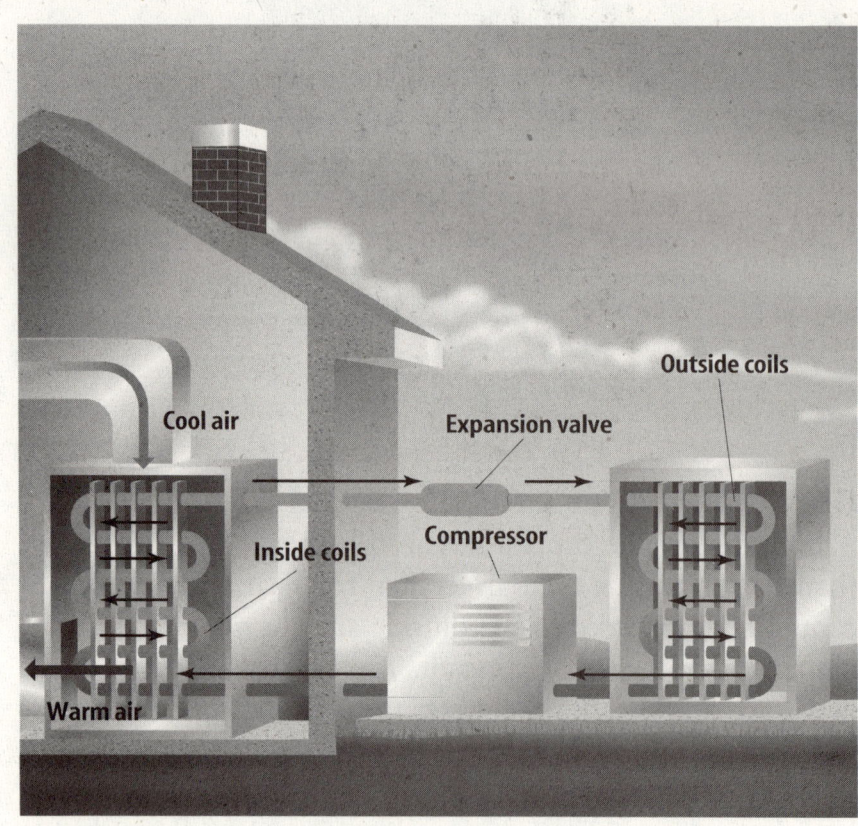

Cool air

Inside coils

Expansion valve

Compressor

Outside coils

Warm air

Picture This

6. **Identify** Highlight the path of the heat when a heat pump is used for heating.

● After You Read

Mini Glossary

heat engine: a machine that changes thermal energy into mechanical energy

internal combustion engine: an engine that burns fuel in a combustion chamber inside the engine

1. Review the terms and definitions in the Mini Glossary. Write one sentence that uses both terms.

   _____

   _____

2. Complete the flow chart below to help you organize what you learned about how an internal combustion engine works.

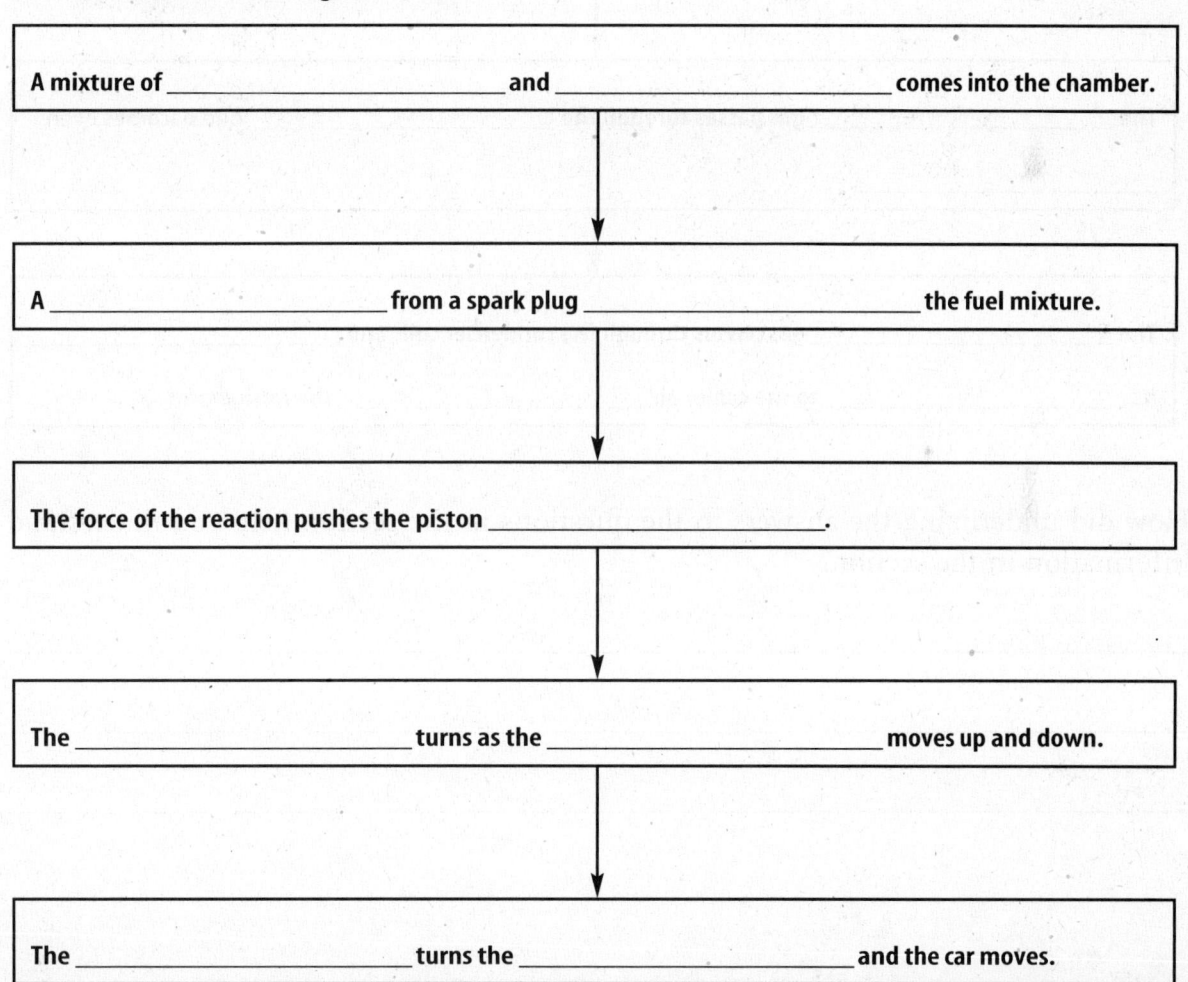

A mixture of _____ and _____ comes into the chamber.

A _____ from a spark plug _____ the fuel mixture.

The force of the reaction pushes the piston _____.

The _____ turns as the _____ moves up and down.

The _____ turns the _____ and the car moves.

3. Use the flow chart below to list the steps that explain how a refrigerator cools.

The liquid coolant becomes _____ as it changes to a _____.

↓

The _____ gas moves through _____ inside the refrigerator.

↓

The _____ gas absorbs _____ from the inside of the refrigerator and becomes _____.

↓

The _____ gas passes through the _____ and becomes even _____.

↓

The _____ gas passes through the condenser coils and _____ its _____ to the cooler air _____ the refrigerator.

4. How did underlining the answers to the questions in the section help you understand the information in the section?

_____

_____

_____

Science Online Visit ips.msscience.com to access your textbook, interactive games, and projects to help you learn more about engines and refrigerators.

Waves

section ❶ What are waves?

● Before You Read

Describe what comes to mind when you think of waves.

_____

_____

What You'll Learn

■ how waves, energy, and matter are related
■ the difference between transverse waves and compressional waves

● Read to Learn

What is a wave?

Imagine that you are floating on an air mattress in a swimming pool and someone jumps into the pool near you. You and your air mattress bob up and down after the splash. What happened? Energy from the person jumping in made your air mattress move. But the person did not touch your air mattress. The energy from the person jumping in moved through the water in waves. **Waves** are regular disturbances that carry energy without carrying matter. The waves disturbed, or changed the motion of, your air mattress.

What do waves do?

Water waves carry energy. Sound waves also carry energy. Have you ever felt a clap of thunder? If so, you felt the energy in a sound wave. You also move energy when you throw a ball. But, there is a difference between a moving ball and a wave. A ball is made of matter. When you throw a ball, you move matter as well as energy. A wave moves only energy.

A Model for Waves

How can a wave move energy without moving matter? Imagine several people standing in a line. Each person passes a ball to the next person. The ball moved, but the people did not. Think of the ball as the energy in a wave and the people as the molecules that move the energy.

Study Coach

Create a Quiz As you read this section, write quiz questions based on what you have learned. After you write the quiz questions, answer them.

FOLDABLES

Ⓐ Identify Make the following Foldable from a sheet of notebook paper to help you organize information about waves.

What are waves?

Mechanical Waves

In the model of the wave, the ball (energy) could not be moved if the people (molecules) were not there. The same thing happens when a rock is thrown into a pond. Waves form where the rock hits the water. The molecules in the water bump into each other and pass the energy in the waves. The energy of a water wave cannot be moved or transferred if there are no water molecules.

Waves that use matter to move or transfer energy are **mechanical waves.** Water waves are mechanical waves. The matter that a mechanical wave travels through is called a medium. In a water wave, the medium is water. Solids, liquids, and gases are also mediums. For example, sound waves can travel through air, water, solids, and other gases. Without one of these mediums, there would be no sound waves. There is no air in outer space, so sound waves cannot travel in space.

What are transverse waves?

One kind of mechanical wave is a transverse wave. Transverse means to pass through, across, or over. In a **transverse wave,** the energy of the wave makes the medium move up and down or back and forth at right angles to the direction the wave moves. Think of a long rope stretched out on the ground. If you shake one end of the rope up and down, you make a wave that seems to slide along the rope, like the wave shown in the figure.

Picture This

2. **Draw and Label** In the figure, draw a circle around each crest in the wave. Then, use a different color of pen or pencil to draw a square around each trough.

Transverse Wave

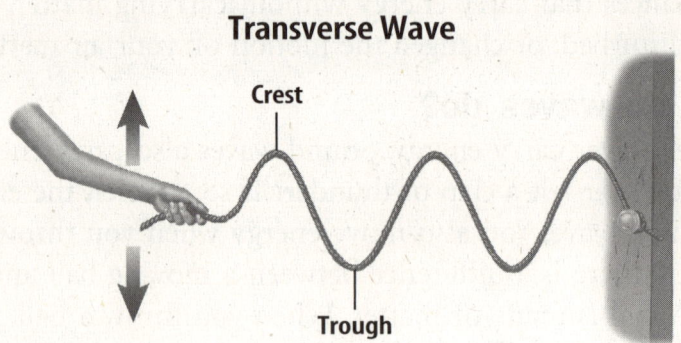

Crest

Trough

It might seem that the rope is moving away from you, but only the wave is moving away from your hand. The energy of the wave travels through the rope. But the matter in the rope does not move. Look at the figure. You can see that the wave has peaks and valleys that are spaced apart at even and regular distances. The high points of transverse waves are called crests. The low points are called troughs.

What are compressional waves?

Mechanical waves can be either transverse or compressional. Compress means to press or squeeze together. In a **compressional wave,** matter in the medium moves forward and backward along the same direction that the wave travels.

An example of a compressional wave made with a coiled spring is shown in the figure. A string is tied to the spring to show how the wave moves. Some coils on one end are compressed and then let go. As the wave begins, the coils near the end are close together. The other coils are far apart. The wave travels along the spring.

Compressional Wave

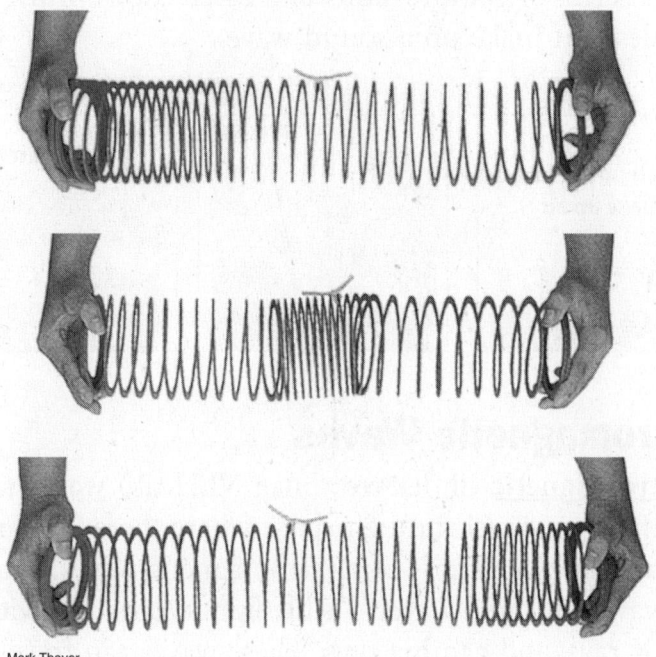

Mark Thayer

The coils and string move only as the wave passes them. Then, they go back to where they were. Compressional waves carry only energy forward along the spring. The spring is the medium the wave moves through, but the spring does not move along with the wave.

Sound Waves Sound waves are compressional waves. How do you make sound waves when you talk or sing? Hold your fingers against your throat while you hum. You can feel your vocal cords vibrating, or moving back and forth very quickly. You can also feel vibrations when you touch a stereo speaker while it is playing. All waves are made by something that is vibrating. ☑

Picture This

3. **Describe** Look at the figures. Describe the coils of the spring when the wave passes through them. Are they close together or far apart?

_____

☑ Reading Check

4. **Identify** What kind of waves are sound waves?

_____

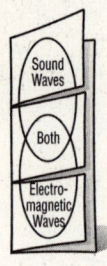

Sound Waves

Both

Electro-magnetic Waves

Picture This

5. Identify Look at the figure. What do the dots above the drum represent?

_____

_____

6. Classify What is radiant energy?

_____

_____

Making Sound Waves

A vibrating object causes the air molecules around it to vibrate. Look at the figure. When the drum is hit, the drumhead vibrates up and down. When the drumhead moves up, the air molecules next to it are pushed closer, or compressed, together. The group of compressed molecules is called a compression. The compression moves away from the drumhead.

When the drumhead moves down, the air molecules near it have more room and can spread apart. This group of molecules is a rarefaction. Rarefaction means something that has become less dense. The rarefaction also moves away from the drumhead. As the drumhead vibrates up and down, it makes a series of compressions and rarefactions in the air molecules that make up a sound wave.

Molecules that make up air

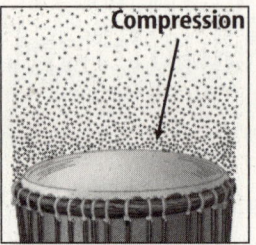
Compression

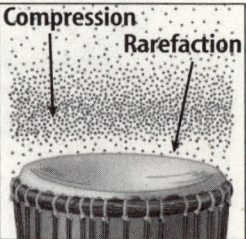
Compression
Rarefaction

Electromagnetic Waves

Electromagnetic (ih lek troh mag NEH tik) **waves** are waves that can travel through space where there is no matter. There are different kinds of electromagnetic waves, such as radio waves, infrared waves, visible light waves, ultraviolet waves, X rays, and gamma rays. These waves can travel in matter or in space. For example, radio waves from TV and radio stations travel through air. They can be reflected from a satellite in space. Then, they travel through air and the walls of your house to your TV or radio.

How does the Sun emit light and heat?

The Sun emits electromagnetic waves that travel through space and reach Earth. The energy carried by electromagnetic waves is called radiant energy. Almost 92 percent of the radiant energy that reaches Earth from the Sun is carried by infrared and visible light waves. Infrared waves make you feel warm. Visible light waves make it possible for you to see. Some of the Sun's radiant energy is carried by ultraviolet waves. These are the waves that can cause sunburn. ☑

After You Read

Mini Glossary

compressional wave: a type of mechanical wave in which matter in the medium moves forward and backward along the same direction that the wave travels

electromagnetic waves: waves that can travel through space where there is no matter

mechanical waves: waves that use matter to move energy

transverse wave: a type of mechanical wave in which the energy of the wave makes the medium move up and down or back and forth at right angles to the direction the wave travels

waves: regular disturbances that carry energy without carrying matter

1. Read the key terms and definitions in the Mini Glossary above. Write a sentence using the term *mechanical wave* on the lines below.

   _____

   _____

2. Use the Venn diagram to compare and contrast transverse and compressional waves. Arrange the characteristics of the waves according to whether they are true for transverse waves, compressional waves, or both.

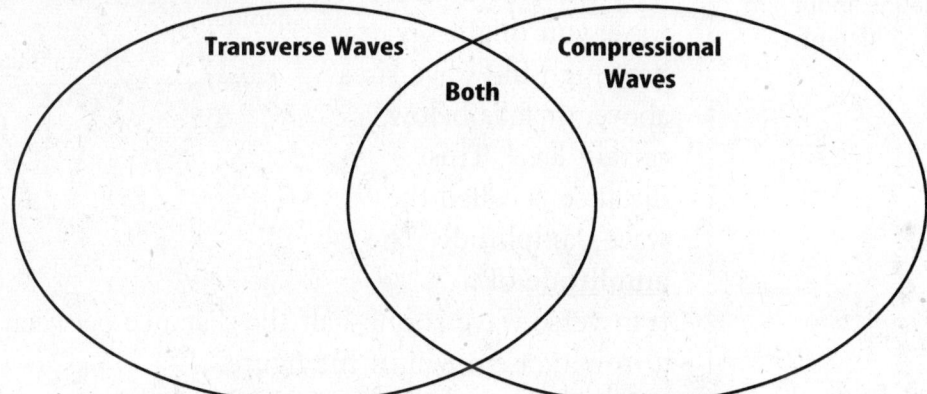

Transverse Waves   Both   Compressional Waves

3. How did the examples of the rope and the spring toy help you understand the difference between transverse and compressional waves?

   _____

   _____

   _____

Science nline Visit **ips.msscience.com** to access your textbook, interactive games, and projects to help you learn more about waves.

End of Section

Reading Essentials **255**

Waves

section ❷ Wave Properties

What You'll Learn

- about the frequency and the wavelength of a wave
- why waves travel at different speeds

● Before You Read

Think about waves in an ocean and waves in a pond. How would you describe each kind of wave?

_____

_____

Mark the Text

Underline Terms As you read this section, underline each property of a wave. Then, highlight information about each property in a different color.

● Read to Learn

Amplitude

To describe a water wave, you might say how high the wave rises above, or falls below, a certain level. This distance is called the wave's amplitude. The **amplitude** of a

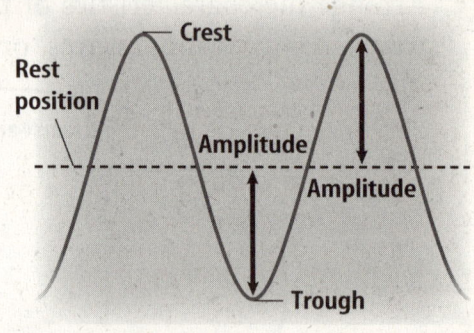

transverse wave is one-half the distance between a crest and a trough, as shown in the figure.

In a compressional wave, the amplitude depends on how close together the particles of the medium are. The amplitude is greater when the particles of the medium are squeezed closer together in each compression and spread farther apart in each rarefaction.

How are amplitude and energy related?

A wave's amplitude is related to the energy that the wave carries. For example, electromagnetic waves of bright light carry more energy and have greater amplitudes than electromagnetic waves of dim light. Loud sound waves carry more energy and have greater amplitudes than soft sound waves. A very loud sound can carry enough energy to damage your hearing.

FOLDABLES

C Organize Information Make the following Foldable to help you organize information about the different properties of waves.

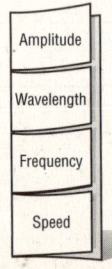

Amplitude

Wavelength

Frequency

Speed

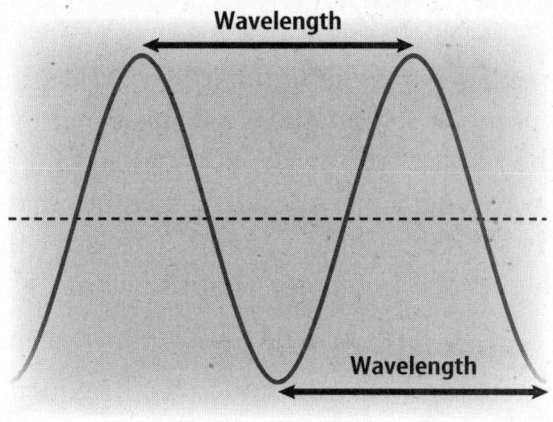

Transverse Wave

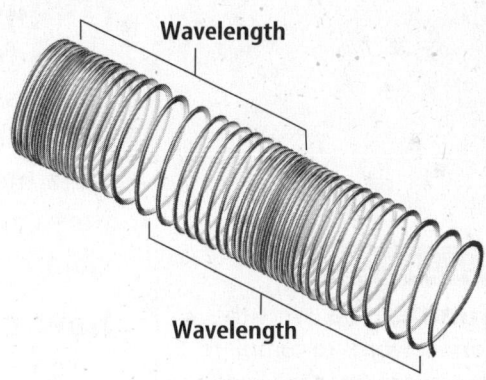

Compressional Wave

Wavelength

You also can describe a wave by its wavelength. Look at the figure above. For a transverse wave, **wavelength** is the distance from the top of one crest to the top of the next crest, or from the bottom of one trough to the bottom of the next trough. For a compressional wave, the wavelength is the distance between the center of one compression and the center of the next compression, or from the center of one rarefaction to the center of the next rarefaction.

The wavelengths of electromagnetic waves can vary from extremely short to longer than a kilometer. X rays and gamma rays have wavelengths that are smaller than the diameter of an atom.

This range of wavelengths is called the electromagnetic spectrum. The figure at the right shows the names given to different parts of the electromagnetic spectrum. Visible light, or light you can see, is only a small part of the electromagnetic spectrum. The wavelength of visible light gives light its color. For example, red light waves have longer wavelengths than green light waves.

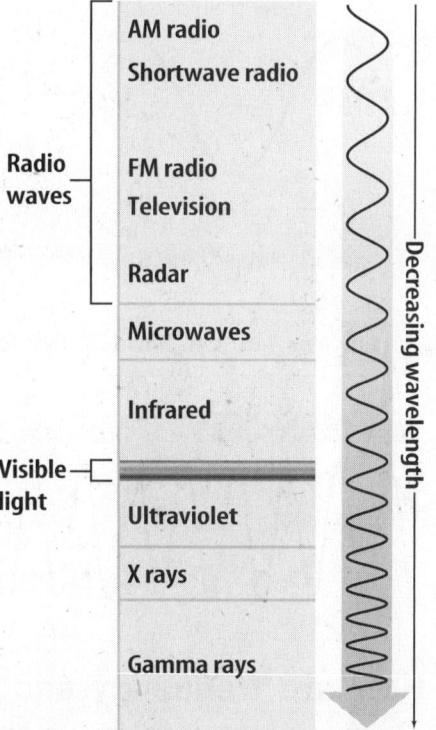

Picture This

1. **Describe** Look at the figure of the transverse wave. Compare the wavelengths between two crests to the wavelength between two troughs. Describe what you find.

_____

_____

Picture This

2. **Use Graphs** Which of the following has the greatest wavelength?
   a. microwaves
   b. X rays
   c. AM radio waves
   d. FM radio waves

Frequency

The **frequency** of a wave is the number of wavelengths that pass a given point in 1 s. Frequency is measured in hertz (Hz). Hertz are the number of wavelengths per second. So, 1 Hz means one wavelength per second. Remember that waves are made by something that vibrates. The faster the vibration is, the higher the frequency is of the wave. ☑

How can you model frequency?

You can use a model to help you understand frequency. If two waves travel with the same speed, their frequency and wavelength are related. Look at the figure below. Imagine people on two moving sidewalks next to each other. One sidewalk has four people on it. They are spaced 4 m apart. The other sidewalk has 16 people on it. They are spaced 1 m apart.

Imagine both sidewalks are moving at the same speed. The sidewalks move toward a pillar. On which sidewalk will more people go past the pillar? The sidewalk with 16 people on it has a shorter distance between people. Four people on this sidewalk will pass the pillar for every one person on the other sidewalk.

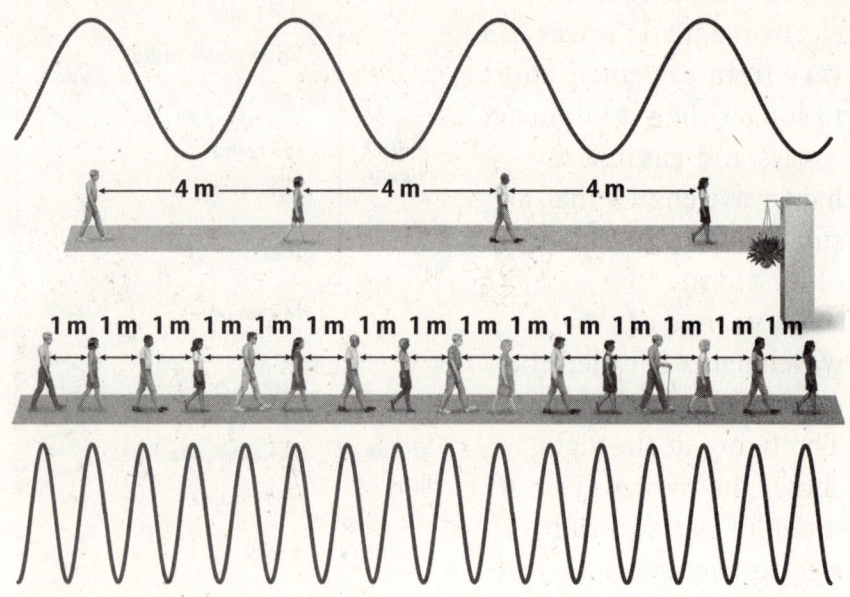

How are frequency and wavelength related?

Suppose that each person on the sidewalks represents the crest of a transverse wave. The movement of the people on the first sidewalk is like a wave with a 4 m wavelength. For the second sidewalk, the wavelength would be 1 m.

✔ Reading Check

3. **Summarize** Write the correct words to complete the sentence on the lines below.

Waves that vibrate fast have ____a.____ frequencies. Waves that vibrate slowly have ____b.____ frequencies.

a._____

b._____

Picture This

4. **Use Models** On the bottom sidewalk, circle groups of four people each. Then draw a line from each group of four people to one person on the top sidewalk.

Applying Math

5. **Calculate** If three people on the top sidewalk pass the pillar, how many people on the bottom sidewalk will have passed the pillar?

_____

The sidewalk with the longer, 4 m, wavelength carries a person past the pillar less frequently. Longer wavelengths have lower frequencies. On the second sidewalk, people pass the pillar more frequently. There, the wavelength is shorter—only 1 m. Shorter wavelengths have higher frequencies. This is true for all waves that travel at the same speed. As the frequency of a wave increases, its wavelength decreases.

What makes different colors and pitches?

The color of a light wave depends on the wavelength or the frequency of the light wave. For example, blue light has a higher frequency and shorter wavelength than red light.

Pitch is how high or how low a sound seems to be. Either the wavelength or the frequency determines the pitch of a sound wave. The pitch and frequency increase from note to note when you sing a musical scale. High-sounding pitches have higher frequencies. As the frequency of sound waves increases, their wavelengths decrease. Lower pitches have lower frequencies. As the frequency of a sound wave decreases, their wavelengths increase. ☑

Wave Speed

You have probably watched a thunderstorm on a hot summer day. You see lightning flash between a dark cloud and the ground. If the thunderstorm is far away, it takes many seconds before you will hear the sound of the thunder that goes with the lightning. This happens because light travels much faster in air than sound does. Light travels through air at about 300 million m/s. Sound travels through air at about 340 m/s. You can calculate the speed of any wave using this equation. The Greek letter lambda, λ, represents wavelength.

Wave Speed Equation

wave speed (m/s) = **frequency** (Hz) × **wavelength** (m)
$$v = f\lambda$$

Mechanical waves, such as sound, and electromagnetic waves, such as light, change speed when they travel in different mediums. Mechanical waves usually travel fastest in solids and slowest in gases. Electromagnetic waves travel fastest in gases and slowest in solids. For example, the speed of light is about 30 percent faster in air than in water.

✔ Reading Check

6. **Summarize** What determines color and pitch? Circle your answer.
   a. wavelength
   b. frequency
   c. wavelength and frequency
   d. wavelength or frequency

Applying Math

7. **Use an Equation** What is the speed in m/s of a wave with a frequency of 50 Hz and wavelength of 2 m? Show your work.

_____

_____

_____

● After You Read

Mini Glossary

amplitude: transverse wave—one-half the distance between a crest and a trough; compressional wave—how close together the particles of the medium are

frequency: the number of wavelengths that pass a given point in 1 s

wavelength: transverse wave—the distance from the top of one crest to the top of the next crest, or from the bottom of one trough to the bottom of the next trough; compressional wave—the distance between the center of one compression and the center of the next compression, or from the center of one rarefaction to the center of the next rarefaction

1. Review the terms and their definitions in the Mini Glossary. Explain in your own words how wavelength and frequency are related.

_____

_____

2. Label the parts of the transverse wave in the diagram below.

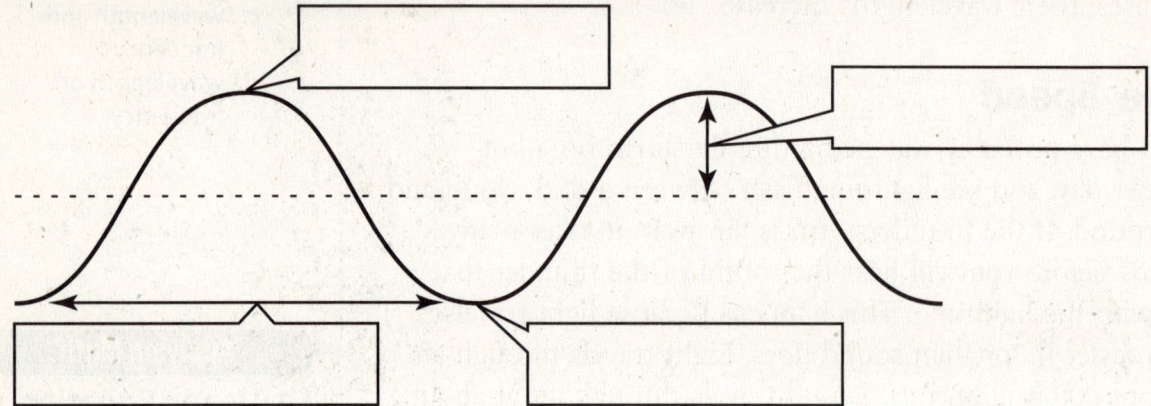

3. You were asked to underline properties of waves and highlight information about them. How did this help you understand and learn about properties of waves?

_____

_____

_____

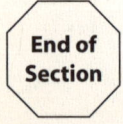
End of Section

Science Online Visit **ips.msscience.com** to access your textbook, interactive games, and projects to help you learn more about properties of waves.

Waves

section ❸ Wave Behavior

● Before You Read

Have you ever shouted and heard an echo? On the lines below, write about what you think causes an echo.

_____

_____

● Read to Learn

Reflection

You can see yourself in a mirror because waves of light are reflected. Reflect means to throw back. **Reflection** happens when a wave hits an object or surface and bounces off. Light waves from the Sun or a lightbulb bounce off of your face. The light waves hit the mirror and reflect back to your eyes. So you see your reflection in the mirror.

You can see your reflection in the smooth surface of a pond, too. But, if the water has ripples or waves, it is harder to see your reflection. You cannot see a sharp image when light reflects from an uneven surface like ripples on the water. This is because the reflected light goes in many different directions.

Refraction

A wave changes direction when it reflects from a surface. Waves can also change direction in another way. Have you ever tried to grab a sinking object in a swimming pool, but missed it? You were probably sure you grabbed right where it was. But, the light waves from the object changed direction when they moved from the water to the air. The bending of a wave as it moves from one medium to another is **refraction**.

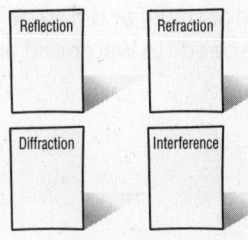

How are refraction and wave speed related?

Remember that the speed of a wave can be different in different materials. For example, light waves travel faster in air than in water. Refraction happens when the speed of a wave changes as it moves from one medium to another.

Picture This

1. **Display** In the water of the first figure, draw an arrow from the light ray to the normal that shows how the light ray bends toward the normal.

   In the air of the second figure, draw an arrow from the normal to the light ray to show how the light ray bends away from the normal.

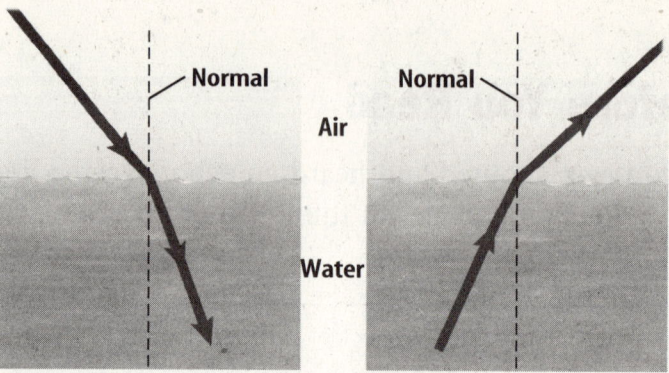

Wave Speed The figures above show how a light wave bends when it passes from air to water and water to air. A line that is perpendicular to the water's surface is called the normal. A light ray slows down and bends toward the normal when it passes from air into water. A light ray speeds up and bends away from the normal when it passes from water into air. If the speed of the wave changes a lot between mediums, the direction of the wave will change a lot too.

Refraction The figure below shows refraction of a fish in a fishbowl. Refraction makes the fish appear to be closer to the surface. It also appears farther away from you than it really is. Light rays reflected from the fish are bent away from the normal as they pass from water to air. Your brain assumes that light rays always travel in straight lines. So, the light rays seem to be coming from a fish that is closer to the surface.

Picture This

2. **Use an Illustration** In the figure, trace the line that shows how the light would travel if light rays did not travel at different speeds in water and air.

Refraction

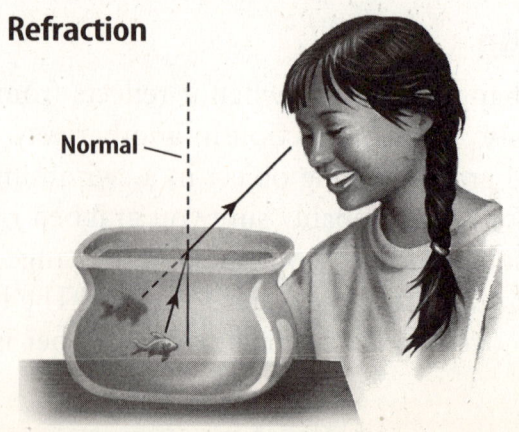

How does refraction make color?

Recall that different wavelengths make different colors. You can separate the colors in sunlight using a prism. A prism is an object or medium used to break light into its different wavelengths. Light is refracted twice when it passes through a prism—once it when it enters and once when it leaves. Since each color has a different wavelength, each color is refracted by a different amount. The colors of light are separated when they leave the prism. Violet light has the shortest wavelength. It is refracted, or bent, the most. Red light has the longest wavelength. It is refracted the least.

How are the colors of a rainbow made?

Each raindrop is a tiny prism. Light rays refract when they enter and again when they leave a raindrop. The colors refract at different angles because they have different wavelengths. The wavelengths separate into all the colors you can see. The colors you see in a rainbow are in order of decreasing wavelength: red, orange, yellow, green, blue, indigo, and violet.

Diffraction

Why can you hear music from the band room when you are down the hall? Sound waves bend as they pass through an open doorway. This is why you can hear the music. This bending is caused by diffraction. **Diffraction** is the bending of waves around a barrier. ☑

Light waves can diffract, too. But, they cannot diffract as much as sound waves. You can hear the band playing music when you are down the hall, but you cannot see the musicians until you actually look inside the band room door.

How are diffraction and wavelength related?

The wavelengths of light are much shorter than the opening of the band room door. This is why the light waves do not diffract as much as the sound waves do when they pass through the door. Light waves have wavelengths that are very short—between about 400 and 700 billionths of a meter. The doorway is about 1 m wide. The wavelengths of sound waves you can hear can be as long as 10 m. Sound waves are much closer in measurement to the opening of the door. A wave diffracts more when its wavelength is similar to the size of the barrier or opening.

💡 **Think it Over**

3. **Explain** why the color violet is refracted the most.

_____

_____

✔ **Reading Check**

4. **Define** What is diffraction?

_____

_____

_____

💡 **Think it Over**

5. **Communicate** A garage door is 3 m wide. Which sound waves will diffract most easily when they pass through the door—ones with a wavelength of 2 m or ones with a wavelength of 0.2 m?

_____

Can water waves diffract?

Imagine water waves in the ocean. What happens when the waves hit a barrier like an island? They go around the island. If the wavelength of the water waves is close to the size and spacing between the islands, the water waves diffract around the islands and keep moving. If the islands are bigger than the wavelength of the water waves, the water waves diffract less.

What happens when waves meet?

Suppose you throw two pebbles into a still pond. Waves spread out from where each pebble hits the water. When two waves meet, will they hit each other and change direction? No, they pass right through each other and keep moving. ✔

How do waves interfere with each other?

What happens when two waves overlap? The two waves add together, or combine, and make a new wave. The ability of two waves to combine and make a new wave when they overlap is **interference**. There are two kinds of interference—constructive and destructive as shown in the figure.

Constructive Interference In constructive interference, the crest of one wave overlaps the crest of another wave. They form a larger wave with greater amplitude. Then the original waves pass through each other and keep traveling as they were before.

Constructive Interference

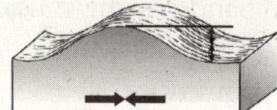
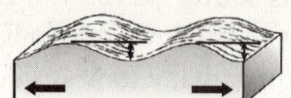

Destructive Interference In destructive interference, the crest of one wave overlaps the trough of another. The amplitudes of the waves combine to make a wave with a smaller amplitude. If the waves have equal amplitudes, they will cancel each other out while the waves overlap. Then the original waves pass through each other and keep traveling as they were before.

Destructive Interference

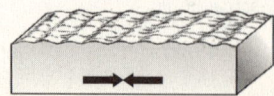
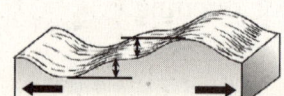

✔ **Reading Check**

6. **Infer** What happens when two waves meet?

_____

_____

Picture This

7. **Conclude** Look at the figure of destructive interference. When can two waves cancel each other out?

How are particles and waves different?

Diffraction When light travels through a small opening, it spreads out in all directions on the other side of the opening. What would happen if particles were sent through the small opening? They would not spread out. They would keep going in a straight line. Diffraction, or spreading, happens only with waves.

Interference Interference does not happen with particles, either. When waves meet, they interfere and then keep going. If particles meet, either they hit each other and scatter, or miss each other. Interference and diffraction both are properties of waves but not particles. ☑

How can noise be reduced?

A lawn mower and a chain saw make loud noises. These loud noises can damage hearing.

Ear Protectors That Absorb Noise Loud sounds have waves with larger amplitudes than softer sounds. Loud sound waves carry more energy than softer sound waves. You have cells in your ears that vibrate and send signals to your brain. Energy from loud sound waves can damage these cells and can cause you to lose your hearing. Ear protectors can help prevent loss of hearing. The protectors absorb, or take in, some of the energy from sound waves. The ear is protected because less sound energy reaches it.

Ear Protectors That Interfere With Noise Pilots of small planes have a similar problem. The airplane's engine makes a lot of noise. But, pilots cannot wear ear protectors to shut out all of the engine's noise. If they did, they would not be able to hear instructions from air-traffic controllers.

Instead, pilots wear special ear protectors. These ear protectors have electronic circuits. The circuits detect noise from the airplane. Then they make sound frequencies that destructively interfere with the noise. Remember that destructive interference makes a smaller wave. The frequencies interfere only with the engine's noise. Pilots can still hear the air-traffic controllers. So, destructive interference can be helpful.

✔ **Reading Check**

8. **Determine** What two properties do waves have that particles do not have?

_____

_____

💡 **Think it Over**

9. **Explain** How do the ear protectors some pilots wear work?

_____

_____

_____

● After You Read

Mini Glossary

diffraction: the bending of waves around a barrier

interference: the ability of two waves to combine and make a new wave when they overlap

reflection: occurs when a wave hits an object or surface and bounces off

refraction: the bending of a wave as it moves from one medium to another

1. Review the terms and their definitions in the Mini Glossary. Write one or two sentences describing how refraction can make a rainbow.

_____

_____

_____

2. In the graphic organizer below, name the four different wave properties. Give an example of each.

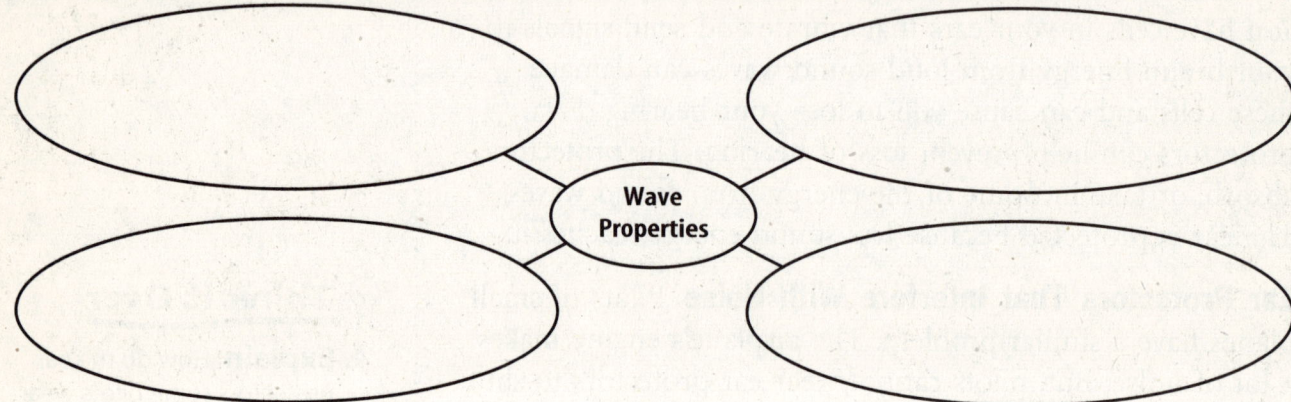

3. You were asked to highlight each question head and the answer to each question as you read this section. Name another strategy that would help you learn the properties of wave.

_____

_____

_____

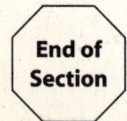

End of Section

Science Online Visit ips.msscience.com to access your textbook, interactive games, and projects to help you learn more about wave behavior.

Sound

section ❶ What is sound?

● Before You Read

Write down two sounds that you like to hear and two sounds that you do not like to hear. Why are some sounds pleasant and others not?

_____

_____

_____

_____

What You'll Learn
- what the characteristics of sound waves are
- how sound travels
- what the Doppler effect is

● Read to Learn

Sound and Vibration

Think of all the sounds you have heard today. You may have heard the sound of your alarm clock, people talking, or locker doors slamming. All sounds have one thing in common. Every sound is made by something that vibrates.

What are sound waves?

How does an object that is vibrating make sound? When you speak, vocal cords in your throat vibrate. The vibrations make sound waves that travel through the air to other people's ears. Their brain understands the sound waves and they hear your voice.

A wave carries energy from one place to another. A wave does not move matter from one place to another. An object that is vibrating in air makes a sound wave. The vibrating object causes air molecules to move back and forth. These air molecules bump into other air molecules nearby and make them move back and forth, too. This happens again and again. In this way, energy is transferred from one place to another as a sound wave.

Mark the Text

Identify Characteristics
As you read this section, highlight the sentences that describe characteristics of sound waves.

What type of wave is a sound wave?

A sound wave is a compressional wave. Have you ever played with a coiled spring toy? When you hold both ends of the spring and someone squeezes together the coils at one end, a wave moves along the spring. You can see the coils of the spring move together and then apart as the compressional wave moves along the spring. The coils move back and forth as the compressional wave moves past them but the toy stays in the same place. ☑

In a compressional wave, the material the wave passes through moves back and forth along the same direction that the wave moves. In the toy, the coils of the spring move back and forth along the same direction the wave is moving as energy is transferred. In a sound wave, air molecules move back and forth along the same direction the sound wave is moving.

How are sound waves made?

Look at the tuning fork on the left in the figure below. When the end moves outward into the air, it pushes the air molecules together. This makes an area where the air molecules are closer together, or more dense. This area of higher density is called a compression.

When the end of the tuning fork moves inward, the air molecules next to it spread farther apart. This area of lower density is called a rarefaction. You can see a rarefaction on the right in the figure below.

As the tuning fork vibrates, it forms a series of compressions and rarefactions. The compressions and rarefactions move away from the tuning fork as the molecules bump into other molecules. Energy is transferred from one molecule to the next as the compressions and rarefactions move away from the tuning fork.

☑ **Reading Check**

1. **Describe** What type of wave is a sound wave?

Picture This

2. **Compare and Contrast** In the figures, how do the molecules in the rarefaction differ from the molecules in the compression?

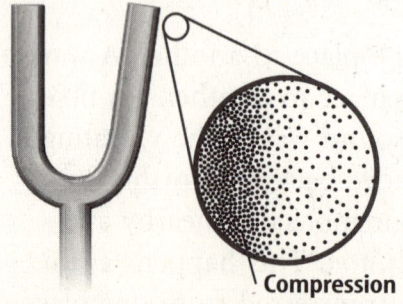

Compression

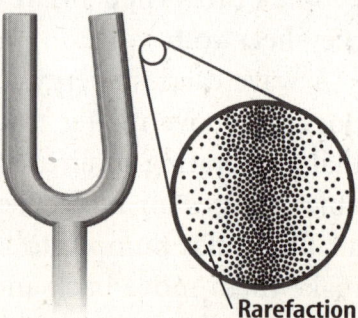

Rarefaction

Sound Wave

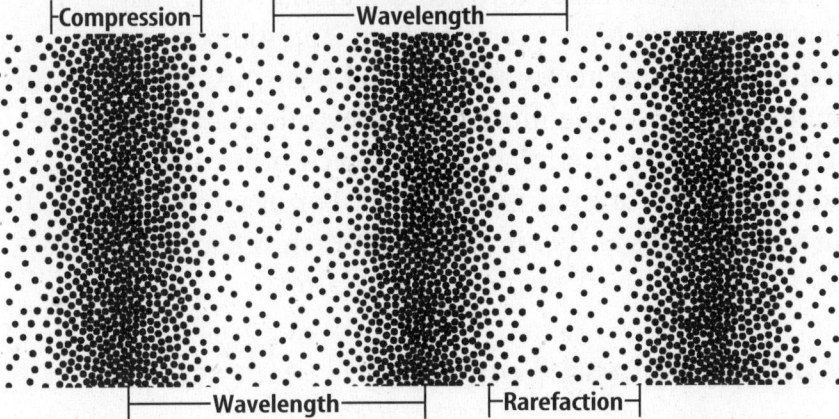

├─Compression─┤ ├─────Wavelength─────┤

├─────Wavelength─────┤ ├─Rarefaction─┤

Picture This

3. Determine How many compressions can be seen in the wave in the figure?

_____

What is a wavelength?

Like other waves, a sound wave can be described by its wavelength and frequency. Look at the figure above. It shows the wavelength of a compressional wave. The wavelength is the distance from one compression to the next or from one rarefaction to the next.

What is frequency?

The frequency of a sound wave is the number of compressions or rarefactions that pass by a certain point in one second. The faster an object vibrates, the higher the frequency of the sound wave it forms.

The Speed of Sound

Sound waves can travel through other materials in the same way they travel through air. But, sound waves travel at different speeds through different materials.

As a sound wave travels through a material, molecules in that material bump into each other. Molecules in a solid are closer together than they are in a liquid or a gas. That means they do not have to travel far before they bump into nearby molecules. Sound travels fastest in solids because the molecules are closest together. Sound usually travels slowest through gases because the molecules are farthest apart. For example, sound travels through air at 343 m/s. Sound travels through water at 1,483 m/s, and through glass at 5,640 m/s.

FOLDABLES

Ⓐ Organize Information
Make the following Foldable to help you organize information about sound. Give examples under each tab.

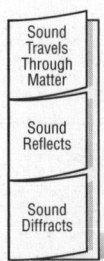

Sound Travels Through Matter

Sound Reflects

Sound Diffracts

Does temperature change the speed of sound?

Sound travels faster through a material when that material is at a higher temperature. The molecules of a material move faster as the material heats up. The faster the molecules move, the more often they bump into each other. The more times the molecules hit each other, the faster sound travels through the material. For example, the speed of sound in air at 0°C is 331 m/s. When the air warms to 20°C, sound travels through it at 343 m/s.

Amplitude and Loudness

What's the difference between loud sounds and quiet sounds? Play a radio loudly, then play it quietly. You will hear the same instruments and voices, but something is different. The difference is that loud sound waves usually carry more energy than soft sound waves do. **Loudness** is a person's understanding of how much energy a sound wave carries.

How are amplitude and energy related?

The amount of energy a wave carries depends on its amplitude. The amplitude of a sound wave shows how spread out the molecules are in the compressions and rarefactions of the wave.

Compare the two sound waves in the figure below. In the sound wave on the right, the molecules are closer together in the compressions and farther apart in the rarefactions than in the wave on the left. This wave has a higher amplitude. The vibrating object that made the wave on the right transferred more energy to move the particles closer together or spread them farther apart. Sound waves with greater amplitude carry more energy and sound louder. Sound waves with smaller amplitude carry less energy and sound quieter.

Applying Math

4. **Calculate** How much faster does sound travel through air at 20°C than through air at 0°C?

_____

_____

Picture This

5. **Infer** Circle the sound wave that will sound louder.

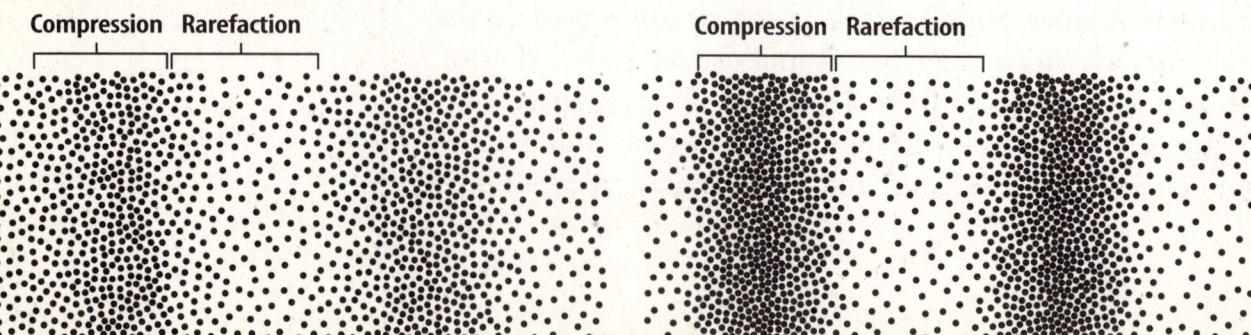

Compression Rarefaction   Compression Rarefaction

How is the energy of a sound wave measured?

How loud a sound seems is different for each person. But, the energy carried by sound waves can be measured by a scale called the decibel (dB) scale. The figure to the right shows the decibel scale. An increase of 10 dB means the energy carried by a sound has increased ten times. But, an increase of 20 dB means that the sound carries 100 times more energy.

Hearing Damage Hearing damage begins to happen at sound levels of about 85 dB. The amount of damage depends on the frequencies of the sound and how long a person is exposed to the sound. Some music concerts produce sound levels as high as 120 dB. The energy carried by these sound waves is about 30 billion times greater than the energy carried by sound waves made by whispering.

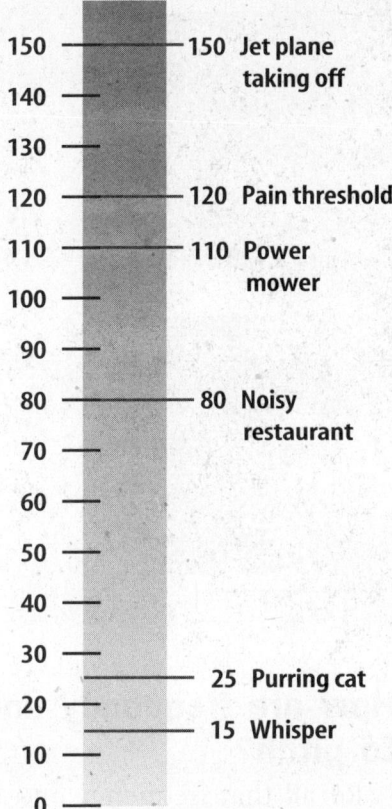

Decibel Scale

- 150 — 150 Jet plane taking off
- 140
- 130
- 120 — 120 Pain threshold
- 110 — 110 Power mower
- 100
- 90
- 80 — 80 Noisy restaurant
- 70
- 60
- 50
- 40
- 30
- 25 Purring cat
- 20 — 15 Whisper
- 10
- 0

Frequency and Pitch

The **pitch** of a sound is how high or how low it sounds. A flute makes a high-pitched sound. A tuba makes a low-pitched sound. The pitch of a sound depends on the frequency of the sound. The higher the pitch is, the higher the frequency is. For example, a sound wave with a frequency of 440 hertz (Hz) has a higher pitch than a sound wave with a frequency of 220 Hz.

What frequencies can be heard?

You can hear sound waves with frequencies between about 20 Hz and 20,000 Hz. Some animals can hear sounds with even higher or lower frequencies. Dogs can hear frequencies up to almost 50,000 Hz. Dolphins and bats can hear frequencies as high as 150,000 Hz.

Picture This

6. Use Diagrams How many decibels is the sound level of a purring cat?

_____

_____

Think it Over

7. Think Critically Explain why going to some music concerts could damage your hearing.

_____

_____

_____

_____

Think it Over

8. Infer Which sound wave has a higher pitch, a wave with a frequency of 100 Hz or a wave with a frequency of 500 Hz?

_____

_____

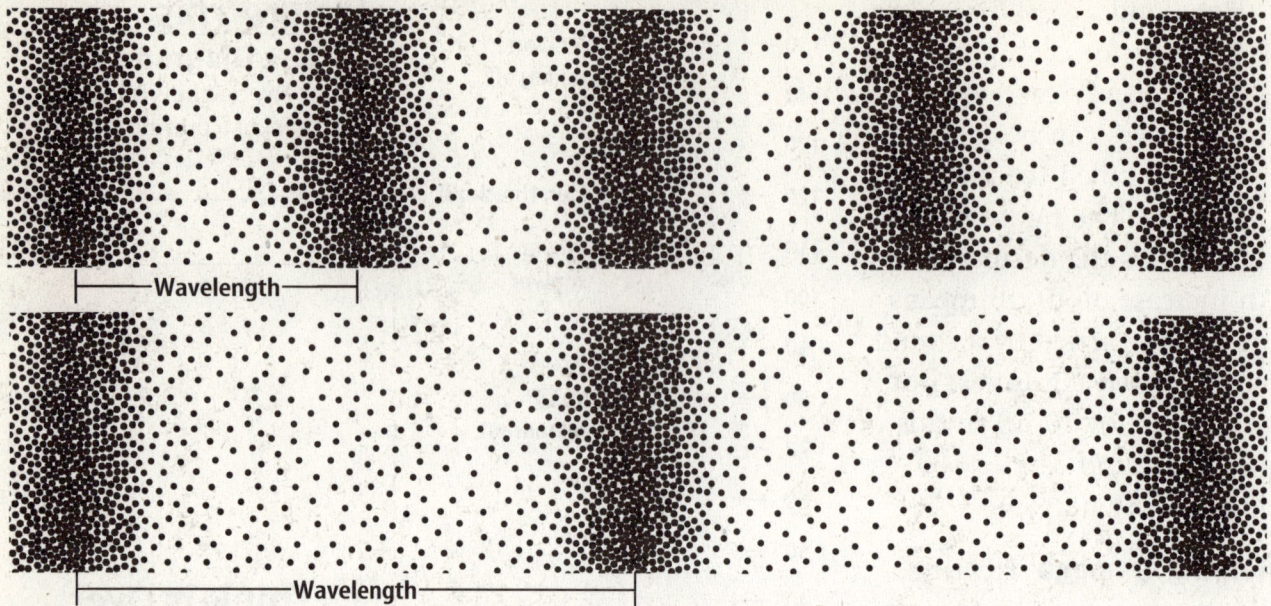

Picture This

9. Interpret Scientific Diagrams Circle the sound wave that has the higher pitch.

How are frequency and wavelength related to pitch?

Recall that frequency and wavelength are related. If two sound waves travel at the same speed, the wave with the shorter wavelength has the higher frequency. Look at the sound waves in the figure. The wavelength of the upper wave is shorter. So, more compressions and rarefactions will go past a certain point every second than for the wave at the bottom of the figure. This means the upper sound wave has a higher frequency than the lower sound wave. It also means that the upper wave is higher in pitch. Sound waves with a higher pitch have shorter wavelengths than those with a lower pitch.

Why do some human voices sound higher than others?

When you make a sound, you breathe out past your vocal cords and your vocal cords vibrate. Not everyone's vocal cords are the same length and thickness. Shorter, thinner vocal cords vibrate at higher frequencies than longer or thicker ones. Children have shorter, thinner vocal cords because their vocal cords are still growing. So, most children's voices sound higher than adults' voices. Muscles in the throat can stretch the vocal cords tighter. People can change the pitch of their voices by controlling these muscles. ✔

✔ Reading Check

10. Identify Who has shorter vocal cords, a child or an adult?

_____

Echoes

Sound waves reflect off hard surfaces just like a water wave bounces off rocks at the beach. An **echo** is a reflected sound wave. The amount of time it takes an echo to return to where the sound wave was first made depends on how far away the reflecting surface is.

Sonar systems use reflected sound waves to find the location and shape of objects under water. A pulse of sound is sent toward the ocean floor. The wave reflects off the ocean floor and back to a receiver. By measuring the length of time it took the echo to return, the distance to the ocean floor can be measured. Sonar can be used to map the ocean floor, locate submarines, schools of fish, and other objects under water. ☑

✔ **Reading Check**

11. Explain What does sonar measure to find the depth of the ocean floor?

_____

_____

_____

_____

What is echolocation?

Some animals use echoes to tell their location and to hunt. This is called echolocation. Bats give off high-pitched squeaks, then listen for the echoes. The type of echo a bat hears tells it exactly where an insect is. Dolphins also use echolocation to help navigate or find their way and to locate objects in the ocean. People who have vision problems might use echolocation to estimate the size and shape of a room.

The Doppler Effect

Have you listened to an ambulance siren as the ambulance sped toward you, then passed you? The pitch gets higher as the ambulance moves toward you. It gets lower as the ambulance moves away from you. The **Doppler effect** is the change in frequency of a sound wave when the source of a sound moves compared to the listener.

The Doppler effect occurs if the source of the sound is moving or if the listener is moving. Suppose you drive past a factory as its whistle blows. As you move toward the factory, the whistle will sound higher pitched. As you move closer, you meet each sound wave a little earlier than you would if you were not moving. You hear more wavelengths per second, so the whistle sounds higher in pitch. As you move away from the factory, each sound wave takes a little longer to reach you. You hear fewer wavelengths per second. So, the whistle sounds lower in pitch.

💡 **Think it Over**

12. Determine Which sound would have a higher pitch, a sound moving toward you or a sound moving away from you?

_____

_____

Radio Waves Radar guns can measure the speed of cars and baseball pitches by using the Doppler effect. A radar gun sends out a radio wave instead of a sound wave. The radio wave reflects off a moving object. The frequency of the radio wave changes depending on the speed of the moving object and whether the object is moving toward the radar gun or away from it. The radar gun uses the change in the frequency of the reflected wave to find the speed of the object. ☑

Diffraction of Sound Waves

A sound wave diffracts when it bends around an object in its path or when it spreads out after passing through a small opening. The amount a wave diffracts depends on whether the wavelength is bigger or smaller than the object or opening. A wave barely diffracts if its wavelength is much smaller than the object or opening. As the size of the wavelength becomes closer to the size of the object or opening, the amount of diffraction increases.

You can hear the diffraction of sound waves if you go to the band room while the band is practicing. If you stand in the open doorway, you will hear the band normally. If you stand around the corner from the band room, you will hear the tubas and other low-pitched instruments better than the high-pitched instruments.

The low-pitched instruments make sound waves with longer wavelengths than the high-pitched instruments. The long wavelengths are closer to the size of the door opening than the shorter wavelengths made by the high-pitched instruments. The longer wavelengths diffract more. So, you hear them even when you are not standing in the doorway. In the same way, the lower frequencies in the human voice allows you to hear someone talking even when the person is around the corner.

Using Sound Waves

Sound waves can be used to treat some health problems. A process called ultrasound uses sound waves with high frequencies. Ultrasound can be used to make an image of the inside of the body. Ultrasound can be used to see how a fetus is developing and to study the heart. Ultrasound along with the Doppler effect can be used to determine if a heart is working properly.

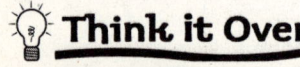

✔ Reading Check

13. Describe What does a radar gun use to find the speed of an object?

_____

_____

💡 **Think it Over**

14. Evaluate Which instrument are you more likely to hear if you are standing around the corner from the band room, a trombone or a flute?

_____

_____

● After You Read

Mini Glossary

Doppler effect: the change in frequency of a sound wave when the source of a sound moves compared to the listener

echo: a reflected sound wave

loudness: a person's understanding of how much energy a sound wave carries

pitch: how high or how low a sound is

1. Review the terms and their definitions in the Mini Glossary. Write one or two sentences that describe how animals use echoes.

   _____

   _____

   _____

2. Place each characteristic of sound in the correct space on the concept map below.

   wavelength   frequency   loudness   amplitude   pitch

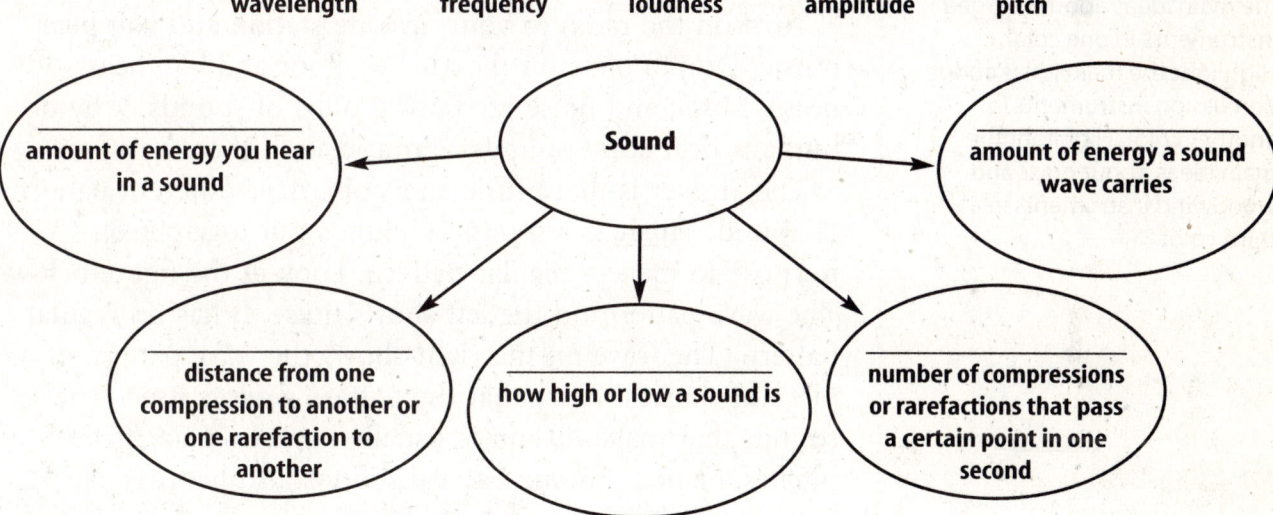

3. You were asked to highlight the sentences that describe characteristics of sound waves. How did this help you learn the content of the section?

   _____

   _____

Science Online Visit **ips.msscience.com** to access your textbook, interactive games, and projects to help you learn more about sound.

End of Section

Sound

section ➋ Music

What You'll Learn

- the difference between music and noise
- how instruments produce music
- how you hear

Mark the Text

Locate Information As you read this section, highlight the main ideas about stringed instruments in one color. Highlight the main ideas about percussion instruments in another color. Highlight the main ideas about brass and woodwind instruments in a third color.

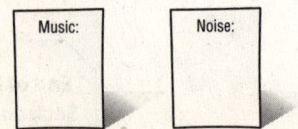

B Compare and Contrast Make the following Foldable to help you understand the differences between music and noise. Include examples of each.

Music: Noise:

● Before You Read

What is your favorite song? What do you like about it?

_____

_____

● Read to Learn

What is music?

Turn on the radio to your favorite station and you hear music. Drop a plate on the kitchen floor and you hear noise. Music and noise are both groups of sounds. Why do humans hear some sounds as music and others as noise?

The answer is that music and noise have different patterns of sound. **Music** is a group of sounds put together on purpose to make a regular pattern. Look at the figure below. The wave pattern on the left shows noise. It has no regular pattern. The wave on the right shows the wave pattern of a piece of music. Notice that the pattern repeats itself. The sounds that make up music usually have a regular pattern of pitches, or notes. Some natural sounds also have regular patterns. That is why sounds like rain on a roof or birds singing may sound musical to some people.

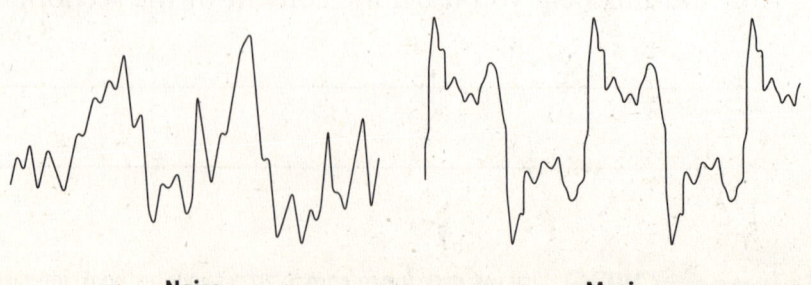

Noise   **Music**

How is music made?

Music is made by vibrations. Your vocal cords vibrate when you sing. When you play a guitar, the strings vibrate. When you beat a drum, the drumhead vibrates. ☑

Tap a bell with a hard object and the bell makes a sound. Tap another bell that has a different size or shape and you will hear a different sound. The bells sound different because each bell vibrates at different frequencies.

The frequencies at which a bell vibrates depend on the bell's shape and the material it is made from. The certain frequencies at which an object vibrates are called its **natural frequencies**.

How are natural frequencies used in music?

When an object is struck or plucked, it vibrates at one or more natural frequencies. The natural frequencies of any object depend on its size, shape, and the material it is made from. Musical instruments use the natural frequencies of strings, drumheads, and air inside pipes to produce different musical notes. ☑

What is resonance?

Sometimes sound waves can make an object vibrate. When a tuning fork is struck, it vibrates at its natural frequency and produces a sound wave. The sound wave has the same frequency as the natural frequency of the tuning fork.

Suppose you have two tuning forks that have the same natural frequency. You strike one tuning fork. The sound waves it makes strike the other tuning fork. These sound waves cause the tuning fork that was not struck to absorb energy and vibrate. This is an example of resonance. **Resonance** happens when an object is made to vibrate at its natural frequencies by absorbing energy from a sound wave or another object vibrating at these frequencies.

How do musical instruments use resonance?

Musical instruments use resonance to make their sounds louder. If a vibrating tuning fork is placed against a table, the sound waves made by the tuning fork may make the table resonate, or vibrate at the same frequency. The vibrations of the tuning fork and the table combine. This makes the sound louder.

☑ Reading Check

1. **Explain** how music is made.

_____

_____

☑ Reading Check

2. **List** the three things the natural frequencies of an object depend on.

_____

_____

_____

Overtones

The same note played on different instruments sounds different even though it has the same pitch. Why? A tuning fork produces a single frequency called a pure tone. Musical instruments do not produce pure tones. Most objects vibrate at more than one natural frequency, so they produce sound waves of more than one frequency.

If you play one note on a guitar, the pitch you hear is the lowest frequency produced by the vibrating string. The **fundamental frequency** is the lowest frequency produced by a vibrating object. The string also vibrates at higher frequencies. **Overtones** are the frequencies higher than an instrument's fundamental frequency. The figure below shows that overtones are multiples of the fundamental frequency. The different overtones produced by each musical instrument make them sound different from each other.

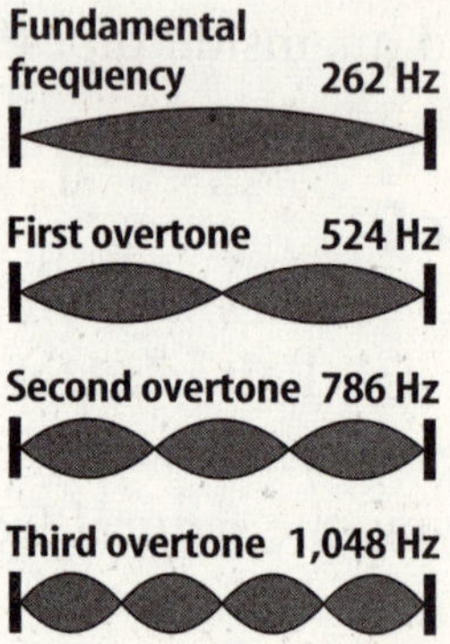

Fundamental
frequency   262 Hz

First overtone   524 Hz

Second overtone   786 Hz

Third overtone   1,048 Hz

Musical Scales

A musical instrument produces musical sounds. The sounds often are part of a musical scale that is a series of notes with certain frequencies. The figure below shows a sequence of notes from a musical scale. Notice that the frequency of the eighth note in the scale is twice that of the first note. The frequency doubles every eight notes.

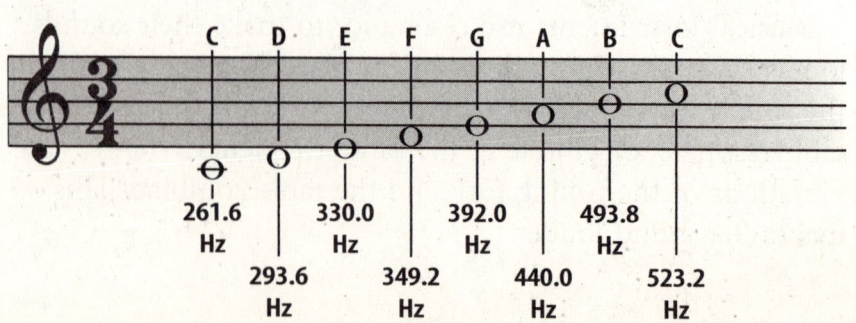

C 261.6 Hz   D 293.6 Hz   E 330.0 Hz   F 349.2 Hz   G 392.0 Hz   A 440.0 Hz   B 493.8 Hz   C 523.2 Hz

Copyright © Glencoe/McGraw-Hill, a division of The McGraw-Hill Companies, Inc.

Think it Over

3. **Think Critically** Why does a note played on a piano sound different than the same note played on a guitar?

_____

_____

_____

Applying Math

4. **Calculate** the frequency of the next C note in the sequence shown in the figure.

_____

_____

Stringed Instruments

A piano, a violin, and a guitar are all stringed instruments. Stringed instruments produce music by making strings vibrate. Piano strings are struck. A bow is slid across violin strings. Guitar strings are plucked.

Strings for these instruments often are made of wire. The pitch of a note depends on the length of the string, the thickness of the string, and how tight the string is. Shorter, narrower, or tighter strings produce higher pitches. For example, a thinner guitar string produces a higher pitch than a thicker string.

How do stringed instruments use resonance?

A vibrating string usually produces a soft sound. To make the sound louder, most stringed instruments have a hollow space that contains air. This space is called a resonator. The resonator absorbs energy from the vibrating strings. Then, it begins to vibrate at its natural frequencies. ☑

For example, the body of a guitar is a resonator. It makes the vibrating strings sound louder. When a guitar is played, the strings vibrate. The vibrating strings make the guitar's body and the air inside it resonate. The vibrating guitar strings sound louder, just as the vibrating tuning fork that was placed against a table sounded louder.

Percussion

You have to strike a percussion instrument to make a sound. Drums are percussion instruments. When you strike a drumhead, it vibrates. The vibrating drumhead is attached to a hollow chamber filled with air. The chamber resonates and makes the sound louder.

Can you change the pitch of a drum?

Some drums have a fixed pitch, but some can be tuned to play different notes. If the drumhead is tightened, the natural frequency of the drumhead is increased. The pitches produced by the drum get higher. A steel drum plays different notes in the scale when you hit different areas in the drum.

A xylophone is another percussion instrument. It is made of wood or metal bars of different lengths. You strike these bars when you play a xylophone. The longer the bar is, the lower the note it produces when it is struck.

✔ Reading Check

5. **Explain** What makes the sound of stringed instruments louder?

_____

FOLDABLES™

C Classify Make the following Foldable to organize musical instruments into groups. Give examples of instruments and how they make music.

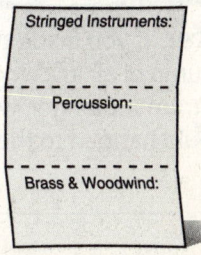

Stringed Instruments:

Percussion:

Brass & Woodwind:

Brass and Woodwinds

Brass and woodwind instruments, like those in the figure below, are made of one or more pipes of different lengths. The pipes can be straight or twisted. The pipes contain columns of air. The columns of air have different natural frequencies. Music is made when the air in an instrument vibrates at different frequencies. ☑

There are different ways to make the air column vibrate. To play a brass instrument like a trumpet, a musician vibrates the lips and blows into the mouthpiece. The air column vibrates and a note sounds on the trumpet.

Some woodwind instruments, like clarinets, saxophones, and oboes, use one or two reeds to make sounds. The reeds vibrate when the musician blows on them. This makes the air column vibrate and you hear a note. Flutes also are woodwind instruments. To play the flute, a musician blows across a small opening in the instrument to make the air column vibrate.

How do you change pitch in woodwinds?

To change a note played on a woodwind, a musician changes the length of the vibrating air column. If the length of the vibrating air column is made shorter, the pitch of the sound goes higher. Musicians change the length of the vibrating column of air by closing and opening finger holes along the instrument.

How is pitch changed in brass instruments?

Musicians playing a brass instrument can blow harder to change the pitch of a sound. Blowing harder makes the air column resonate at a higher natural frequency. This makes the pitch higher. Another way to change the pitch is by pressing valves that change the length of the tube.

✔ Reading Check

6. Identify What vibrates inside the pipes of a brass instrument?

_____

_____

💡 Think it Over

7. Infer If you made the column of air in a woodwind instrument longer, what would happen to the pitch?

_____

_____

Beats

Interference happens when two waves overlap and combine to make a new wave. The new wave can have a different frequency, wavelength, and amplitude than the two original waves.

Suppose two notes close in frequency are played at the same time. The two notes interfere to make a new sound. The new sound gets louder and softer several times each second. If you were listening to the sound, you would hear a series of beats as the sound got louder and softer. The number of beats you would hear each second is called the beat frequency. The beat frequency is equal to the difference in the frequencies of the two notes.

Suppose the frequency of the first note is 329 Hz and the frequency of the second note is 332 Hz. The beat frequency is the difference in these two frequencies, or 332 Hz − 329 Hz. The beat frequency is 3 Hz. That means you would hear the sound get louder and softer three times each second. In other words, you would hear three beats each second.

What are beats used for?

In music, beats are used to help tune instruments. To tune a piano, a piano tuner might hit a tuning fork that vibrates at a certain frequency. The piano tuner then hits the key on the piano that makes a note of the same frequency. Beats are heard if the pitch of the note from the piano is different from the pitch of the tuning fork. When the piano string is tuned correctly, the beats disappear.

Reverberation

You have learned that sound reflects off hard surfaces. If you stand in an empty gym and speak in a loud voice, the sound of your voice will be reflected back and forth several times. The sound of your voice will bounce off the floor, walls, and ceiling. It will reverberate. Repeated echoes of sound are called **reverberation**. ☑

In the gym, reverberation makes the sound of your voice stay for awhile before the sound dies out. Some reverberation makes voices and music sound lively. Not enough reverberation makes sounds flat and lifeless. But sometimes, reverberation can produce a confusing mess of noise. This happens when too many sounds stay for too long.

Applying Math

8. **Calculate** What is the beat frequency between a note with a frequency of 293 Hz and a note with a frequency of 295 Hz?

✔ **Reading Check**

9. **Describe** What do you hear when you hear repeated echoes of sound?

How is reverberation controlled?

Concert halls and theaters are designed to produce just the right amount of reverberation. Sometimes the walls, floors, and ceiling are covered with soft materials. This reduces echoes. Sometimes special panels are put on walls or hung from the ceiling. These panels are designed to reflect sound toward the audience.

The Ear

You hear sounds with your ears. Your ear is an organ that can hear sounds of many different frequencies. It can hear sounds from about 20 Hz to 20,000 Hz. Your ear can also hear very loud sounds and very soft sounds. The softest sounds you can hear have about one trillionth the amount of energy as the loudest sounds you can hear. The figure below shows a human ear. It has three parts—the outer ear, the middle ear, and the inner ear.

Picture This
10. Identify Highlight the ear canal in the figure.

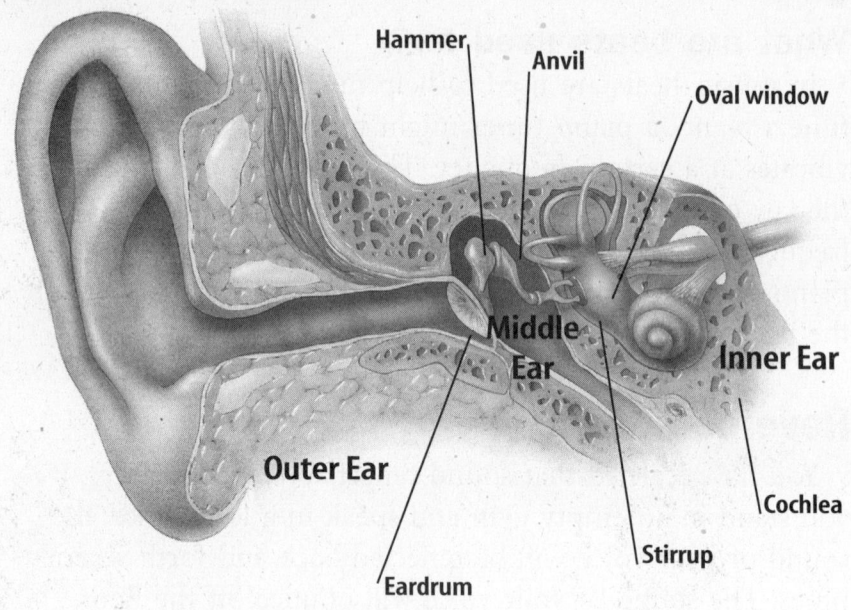

FOLDABLES™

D Make Drawings Make the following Foldable to show how the ear works. Draw the parts of the ear under the tabs. Label each part.

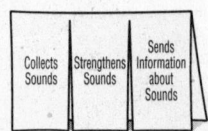

What does the outer ear do?

Your outer ear collects sound waves. Then, it directs the sound waves into the ear canal. Look at the figure again. Notice that the outer ear looks something like a funnel. This shape helps it collect the sound waves.

How do animals collect sounds?

Some animals use their ears to find food or to keep away from danger. These animals often have large outer ears. A barn owl uses its ears to hunt for food at night. Its outer ears are not made of flesh. The feathers on the owl's face help direct the sound to its ears. Sea mammals hear very well even though their outer ears are small holes.

What does the middle ear do?

When sound waves reach the middle ear, they make the eardrum vibrate. The **eardrum** is a thin skin-like structure, or membrane, that stretches across the ear canal like a drumhead. When the eardrum vibrates, it makes the three small bones next to it vibrate. These bones are connected to each other. They are called the hammer, the anvil, and the stirrup. These bones make the vibrations stronger. ☑

What does the inner ear do?

The stirrup vibrates another membrane. This membrane is called the oval window. The inner ear begins at the oval window. The inner ear is filled with fluid. The fluid vibrates when the vibrations from the middle ear reach the inner ear. When the fluid vibrates, it makes special hair-tipped cells inside the cochlea vibrate.

Different sounds vibrate these cells in different ways. The cells send signals with information about the sound's frequency and strength. The cells also send information about how long a sound lasts. The signals travel to the brain along the auditory nerve. They go to the part of the brain that is responsible for hearing.

Hearing Loss

Some diseases can damage your ears. Loud sounds also can damage your ears. If you listen to loud sounds for long periods of time, the sounds can damage the hair cells in the cochlea. These hair cells may die if they are damaged. You cannot grow new hair cells. When the hair cells die, some of your hearing is lost. ☑

As people get older, they can lose their hearing. Some hair cells and nerves in the ear stop working properly. About 30 percent of people older than 65 have some hearing loss due to aging.

☑ **Reading Check**

11. **Identify** What vibrates when sound waves reach the middle ear?

_____

☑ **Reading Check**

12. **Explain** why listening to loud sounds for a long time can damage your hearing.

_____

_____

_____

_____

● After You Read

Mini Glossary

eardrum: a very thin layer of skin that stretches across the ear canal like a drumhead

fundamental frequency: the lowest frequency made by a vibrating object

music: a group of sounds put together on purpose to make a regular pattern

natural frequencies: the certain frequencies at which an object vibrates

overtones: the frequencies higher than an instrument's fundamental frequency

resonance: happens when an object is made to vibrate at its natural frequencies by absorbing energy from a sound wave or another object vibrating at these same frequencies

reverberation: repeated echoes of sound

1. Review the terms and their definitions in the Mini Glossary. Use the terms *music* and *eardrum* in a sentence together.

_____

_____

2. Write the letter of the statement in Column 2 that best matches the term in Column 1.

Column 1	Column 2
_____ 1. music	**a.** used by musical instruments to make their sounds louder
_____ 2. musical scale	**b.** part of the ear that collects sounds
_____ 3. outer ear	**c.** percussion instrument
_____ 4. resonance	**d.** organized sound
_____ 5. xylophone	**e.** a series of notes
_____ 6. hair cells	**f.** send signals about a sound to the brain

3. How could you show a younger student what reverberation is?

_____

_____

_____

End of Section

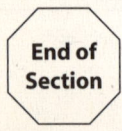 **Science** nline Visit **ips.msscience.com** to access your textbook, interactive games, and projects to help you learn more about music.

 **Electromagnetic Waves**

section ❶ The Nature of Electromagnetic Waves

● Before You Read

What do you think of when you hear the word *wave*?

_____

_____

_____

What You'll Learn

- how electromagnetic waves are made
- what electromagnetic waves are like

● Read to Learn

Waves in Space

When you are at the beach, you may enjoy swimming or surfing in the ocean waves. Did you know that you are also enjoying another type of wave when you are at the beach? You can feel the warmth of the Sun on your skin. You can see its brightness with your eyes. The energy from the Sun that you can feel and see comes to you in the form of waves. These waves are a lot like those that bring you TV and cell phone signals. These are the same type of waves that a dentist uses to take X rays.

How is energy moved by waves?

A wave moves or transfers energy from one place to another without transferring matter. How do waves transfer energy? Waves, like water or sound waves, transfer energy by making particles of matter move. Energy passes from particle to particle when the particles bump into each other. Waves that use matter to transfer energy are called mechanical waves.

The space between the Sun and Earth is almost empty. How can the Sun's energy reach Earth if there is no matter to transfer the energy? A different type of wave called an electromagnetic wave is what carries energy from the Sun to Earth. An **electromagnetic wave** is a wave of charged particles that can travel through empty space or through matter.

Mark the Text

Underline New Ideas
Look for words or sentences that you do not understand. Underline them. When you finish reading, ask a classmate or your teacher to help you understand the things you underlined.

Forces and Fields

Electromagnetic waves are made of two parts—an electric field and a magnetic field. These fields are force fields. A force field is what lets one object put forces on another object without the objects touching. Earth is surrounded by a force field called the gravitational field. This field exerts or puts the force of gravity on all objects that have mass. ☑

How does Earth's force field work? When you throw a ball into the air, it always falls back to Earth. That is because the force of gravity pulls down on the ball. Gravity pulls on the ball when it is still in your hand. It pulls on the ball when it is flying through the air. Earth's gravitational field even goes out into space. Earth's gravitational field is what keeps the Moon orbiting Earth.

What are magnetic fields?

Have you ever played with magnets? You may know they come together, or attract each other, without touching. They also push each other apart, or repel, without touching. Two magnets put forces on each other without touching because they are surrounded by a force field called a magnetic field. Remember how a gravitational field exerts a force on anything with mass? A magnetic field exerts a force on another magnet or magnetic materials. Magnetic fields cause other magnets to line up along the direction of the magnetic field.

What are electric fields?

Remember that atoms contain protons, neutrons, and electrons. Protons and electrons have a property called electric charge. Protons have a positive electric charge. Electrons have a negative electric charge.

A particle with electric charge is surrounded by an electric field, just like a magnet is surrounded by a magnetic field. The figure shows a charged particle. The arrows show the force put on the particle by the electric field.

☑ **Reading Check**

1. **Explain** What does a force field do?

_____

_____

_____

Picture This

2. **Identify** Which of these could be the charged particle in the figure?
   a. a neutron
   b. a wave
   c. a field
   d. an electron

Electric field

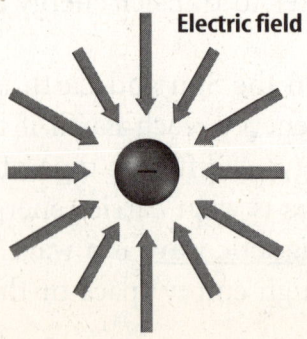

Making Electromagnetic Waves

An electromagnetic wave is made of electric and magnetic fields. How is this kind of wave produced? Think about a wave on a rope. You can make a wave on a rope by shaking one end of the rope up and down. An electromagnetic wave is produced when charged particles, such as electrons, move back and forth or vibrate.

You know that a charged particle is surrounded by an electric field. But, a charged particle that is moving also is surrounded by a magnetic field. For example, electrons flow in a wire that carries an electric current. Because of this, the wire is surrounded by a magnetic field as shown in the figure below. So, a moving charged particle is surrounded by an electric field and a magnetic field.

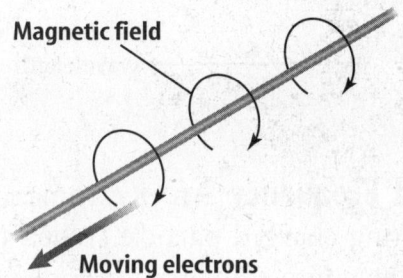

Magnetic field

Moving electrons

How are electromagnetic waves produced?

When you shake a rope up and down, you make a wave that moves away from your hand. When a charged particle vibrates up and down, it makes changing electric and magnetic fields that move away from the vibrating charge in many directions. These changing fields form an electromagnetic wave. The figure below shows how electric and magnetic fields change as they move along one direction.

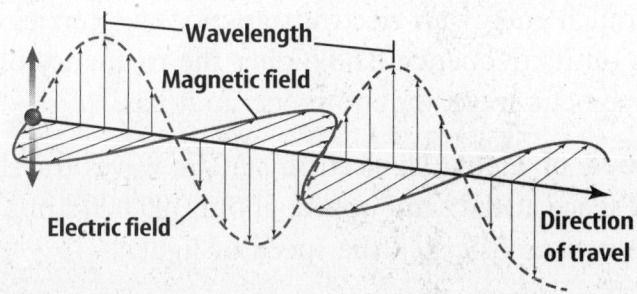

Wavelength

Magnetic field

Electric field

Direction of travel

Picture This

3. Identify Highlight the direction the electrons are moving in the figure. Use another color to highlight the magnetic field.

Picture This

4. Interpret Data What fields produce electromagnetic waves?

_____

_____

_____

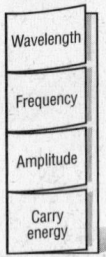

Wavelength

Frequency

Amplitude

Carry energy

Picture This

5. **Label** a crest and a trough on the wave in the figure.

✔ **Reading Check**

6. **Describe** What causes a charged particle to move when it is struck by an electromagnetic wave?

_____

_____

Properties of Electromagnetic Waves

Like all waves, an electromagnetic wave has a wavelength and a frequency. You can make a wave on a rope when you move your hand up and down while holding the rope. Look at the figure. Frequency is how many times you move the rope up and back down in 1 s. Wavelength is the distance from one crest to the next or from one trough to the next.

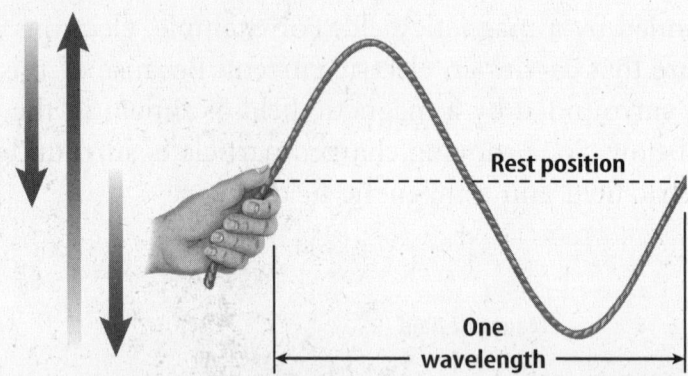

Rest position

One wavelength

Wavelength and Frequency An electromagnetic wave is made by a vibrating charged particle. When the charged particle makes one complete vibration, one wavelength is made. Look at the figure on the previous page. The frequency of an electromagnetic wave is the number of wavelengths that pass by a point in 1 s. This is the same as the number of times the charged particle makes one complete vibration in 1 s.

Energy The energy carried by an electromagnetic wave is **radiant energy**. What happens if an electromagnetic wave hits a charged particle? The electric field part of the wave exerts a force on the particle and causes it to move. Some of the radiant energy carried by the wave is transferred into the energy of motion of the particle. ☑

How much energy an electromagnetic wave carries depends on its frequency. The higher the frequency of an electromagnetic wave, the more energy it has.

The Speed of Light All electromagnetic waves travel through space at the same speed, about 300,000 km/s. This speed sometimes is called the speed of light.

● After You Read

Mini Glossary

electromagnetic wave: a wave of moving charged particles that can travel through empty space or through matter

radiant energy: the energy carried by an electromagnetic wave

1. Review the terms and their definitions in the Mini Glossary. Use the terms *radiant energy* and *electromagnetic wave* to describe why you can feel the warmth of the Sun.

   _____

   _____

2. Fill in the blanks in the concept map below to describe electromagnetic waves.

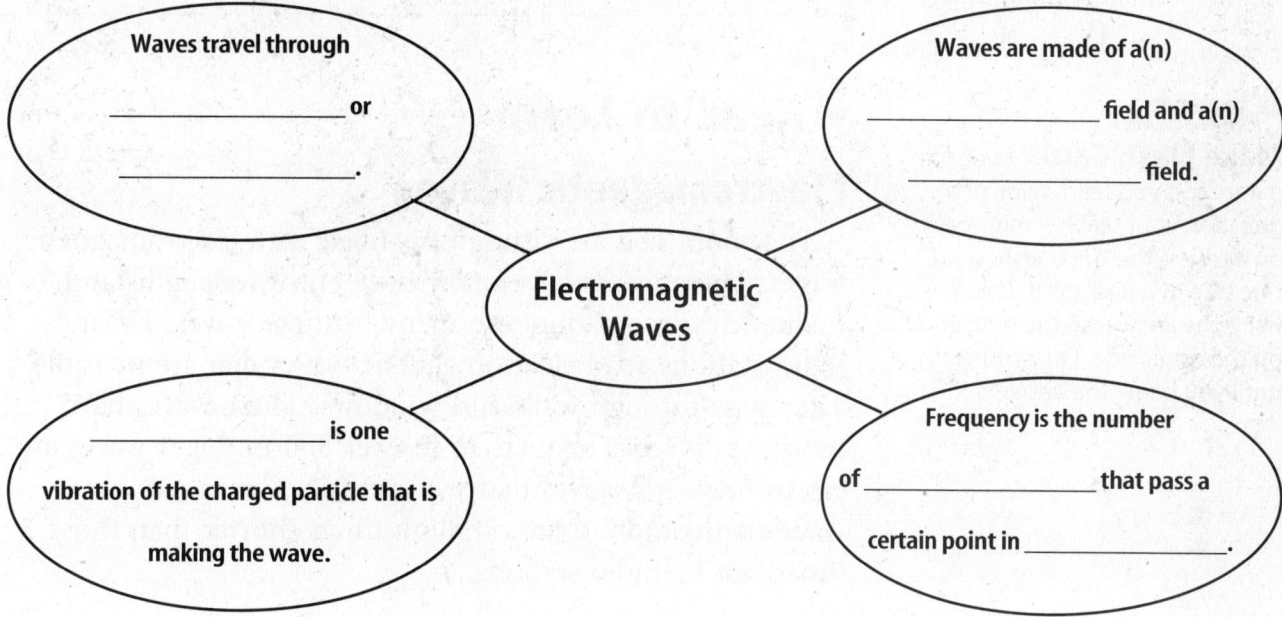

Waves travel through

_____ or

_____.

Waves are made of a(n)

_____ field and a(n)

_____ field.

Electromagnetic Waves

_____ is one vibration of the charged particle that is making the wave.

Frequency is the number

of _____ that pass a

certain point in _____.

3. At the beginning of the section, you were asked to underline words or sentences that you did not understand. How did this help you learn the material?

   _____

   _____

   _____

Science Online Visit **ips.msscience.com** to access your textbook, interactive games, and projects to help you learn more about the nature of electromagnetic waves.

End of Section

chapter 18 Electromagnetic Waves

section 2 The Electromagnetic Spectrum

What You'll Learn

- the differences among kinds of electromagnetic waves
- uses for different kinds of electromagnetic waves

Before You Read

What happens when you stay in the sun too long without sunscreen? Why do you think this happens?

____

____

Study Coach

Make Flash Cards For each paragraph you read, think of a question your teacher might ask on a test. Write the question on one side of a flash card. Then, write the answer to the question on the other side. Quiz yourself until you know the answers.

Read to Learn

Electromagnetic Waves

The room you are sitting in is filled with electromagnetic waves. These waves have many different wavelengths and frequencies. You cannot see many of these waves. TV and radio stations send electromagnetic waves that are invisible. They pass through walls and windows. These waves have wavelengths from about 1 m to over 500 m. Light waves are electromagnetic waves that you can see. They have wavelengths more than a million times shorter than those broadcast by radio stations.

How are electromagnetic waves grouped?

The **electromagnetic spectrum** is the wide range of electromagnetic waves with different wavelengths and frequencies. The electromagnetic spectrum is divided into different parts. Look at the figure at the top of the next page. It shows the electromagnetic spectrum and the names of the different waves that make up different parts of the spectrum. Even though electromagnetic waves have different names, they all travel at the same speed in empty space—the speed of light. Remember that for waves that travel at the same speed, the frequency increases as the wavelength decreases. So, as the frequency of electromagnetic waves increases, their wavelength decreases.

Radio waves | Microwaves | Infrared waves | Visible light | Ultraviolet waves | X rays | Gamma rays

Increasing frequency, decreasing wavelength

Radio Waves

Electromagnetic waves with wavelengths longer than about 0.3 m are called radio waves. **Radio waves** have the lowest frequencies of all the electromagnetic waves. They also carry the least energy. Television signals and AM and FM radio signals are types of radio waves. Like all electro-magnetic waves, radio waves are produced by moving charged particles. One way to make radio waves is to make electrons vibrate in a piece of metal. This piece of metal is called an antenna. The moving electrons in an antenna cause radio waves to move outward from the antenna. By changing how fast the electrons in the antenna vibrate, radio waves of different frequencies can be made.

How are radio waves received?

Radio waves can cause electrons in another antenna to vibrate. Vibrating electrons in a receiving antenna form an alternating electric current. This current can be used to produce a picture on a TV screen and sound from a loudspeaker. Changing the frequency of the waves changes the alternating current in the receiving antenna. This produces the different pictures and sounds you hear on your TV. ☑

What are microwaves?

Radio waves with wavelengths from about 0.3 m to 0.001 m are called microwaves. Microwaves have a higher frequency and shorter wavelength than the waves used in your home radio. Cellular and portable phones use microwaves.

Microwave ovens use microwaves to heat food. The microwaves cause water molecules in food to vibrate. As the molecules vibrate faster, the food becomes warmer.

Picture This

1. **Interpret** What type of electromagnetic waves have the shortest wavelength?

_____

☑ Reading Check

2. **Describe** What happens when radio waves cause electrons in an antenna to vibrate?

_____

_____

_____

What is radar?

Some animals, like bats, use echolocation. In echolocation, sound waves are sent out and bounce off objects. The reflected sound waves help determine the size and location of objects. Radar uses electromagnetic waves to detect objects in this way. Radar stands for RAdio Detecting And Ranging. Radar was first used in World War II to find enemy aircraft.

A radar station sends out radio waves that bounce off objects such as aircraft. Electronic equipment receives the reflected signals. The equipment measures the time it takes for the waves to return. Because the speed of the radio waves is known, the distance to the airplane can be found from the measured time. Electromagnetic waves travel so fast that this process takes less than one second. ☑

Infrared Waves

Have you ever stood close to a fireplace to warm up? It does not take long to feel the heat from the glowing fire. You can also warm up by standing next to a hot object that is not glowing. The heat you feel is from electromagnetic waves called infrared waves. **Infrared waves** are electromagnetic waves with wavelengths between about one thousandth and 0.7 millionths of a meter.

All objects send out electromagnetic waves. In any material, the atoms and molecules are always moving. The electrons in the atoms and molecules also vibrate and send out electromagnetic waves. Most electromagnetic waves given off by an object at room temperature are infrared waves. They have wavelengths of about 0.000 01 m, or one hundred thousandth of a meter.

Infrared detectors can detect infrared waves from objects that are warmer or cooler than their surroundings. Forests are usually cooler than their surroundings. Infrared detectors on satellites can be used to map forests, water, rock, and soil. Some types of night vision devices use infrared detectors. They make it possible for objects to be seen in nearly total darkness.

How do animals use infrared waves?

Some animals can detect infrared rays. Snakes called pit vipers have a pit between the nostril and the eye that detects infrared waves. These pits help them hunt at night by detecting the infrared waves their prey gives off.

✔ **Reading Check**

3. **Identify** What is known about radio waves that helps measure distance?

_____

_____

Visible Light

Have you ever wondered why some hot objects glow? As an object gets warmer, its atoms and electrons vibrate faster and faster. The vibrating electrons produce electromagnetic waves with higher frequencies and shorter wavelengths. The object might glow if the temperature is high enough. Some of the electromagnetic waves the object gives off, or emits, can be seen. Electromagnetic waves you can see with your eyes are called **visible light**. ☑

Visible light has wavelengths between about 0.7 and 0.4 millionths of a meter. The different colors you see are electromagnetic waves with different wavelengths. Red light has the longest wavelength. Blue light has the shortest wavelength. Most objects you see do not give off visible light. They reflect visible light from another source, such as the Sun or a lightbulb.

Ultraviolet Radiation

Ultraviolet radiation is higher in frequency than visible light and has even shorter wavelengths—between 0.4 millionths and about 10 billionths of a meter. Ultraviolet radiation also carries more energy. The radiant energy carried by an ultraviolet wave can damage or kill the molecules that make up living cells.

The figure below shows the intensity, or strength, of the electromagnetic waves from the sun. Most of the Sun's waves are infrared waves and visible light. But, too much exposure to the Sun's ultraviolet waves can cause sunburn. Exposure to these waves over a long period of time can cause the skin to age and might cause skin cancer. To protect your skin from ultraviolet waves, wear sunscreen and do not stay in the Sun too long.

☑ Reading Check

5. **Describe** Why do some hot objects glow?

_____

_____

_____

Electromagnetic Waves from the Sun

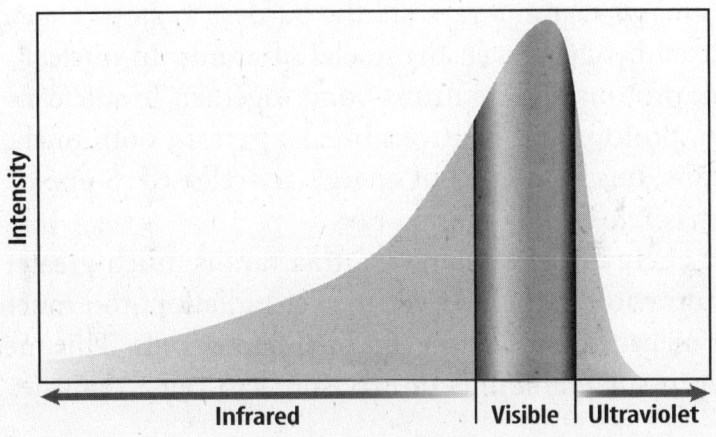

Infrared   Visible   Ultraviolet

Picture This

6. **Identify** Most of the electromagnetic waves emitted by the Sun are what type?

_____

Is ultraviolet radiation helpful?

Your body uses ultraviolet radiation from the Sun to produce vitamin D. A few minutes of sunlight each day is enough to produce all the vitamin D your body needs. Human skin tans as protection against too much ultraviolet radiation. But, a tan can be a sign that the skin has received too much ultraviolet radiation.

Because ultraviolet radiation can kill cells, it is used to disinfect objects such as surgical instruments and goggles.

What is the ozone layer?

Ozone is a molecule that contains three oxygen atoms. It is formed high in Earth's atmosphere. The ozone layer absorbs most of the ultraviolet radiation from the Sun before it reaches Earth's surface.

Chemical compounds called CFCs can react with ozone molecules and break them apart. CFCs are used in some air conditioners and refrigerators. There is evidence that CFCs in the atmosphere reduce the ozone over Antarctica at certain times of the year. This reduction is known as the ozone hole. To prevent this, CFCs are being used less.

The atmosphere also absorbs other types of electromagnetic radiation. It absorbs higher energy waves like X rays and gamma rays. Radio waves, light waves, and some infrared waves can pass through the atmosphere.

X Rays and Gamma Rays

Ultraviolet rays can go through, or penetrate, the top layer of your skin. **X rays** have an even higher frequency than ultraviolet rays and enough energy to go right through skin and muscle. A shield made out of a dense metal such as lead is needed to stop X rays.

Gamma rays have the highest frequency and carry the most energy. Gamma rays are the hardest to stop. They are produced by changes in the nuclei of atoms. In nuclear fusion, protons and neutrons bond together. In nuclear fission, protons and neutrons break apart. In both of these reactions, huge amounts of energy are released. Some of this energy is released as gamma rays.

The energy of X rays and gamma rays is much greater than ultraviolet rays. Like ultraviolet radiation, too much X-ray or gamma radiation can hurt or kill cells. This means that even small amounts of exposure can cause damage.

Think it Over

7. Draw Conclusions Why is the ozone layer important to life on Earth?

_____

_____

_____

Think it Over

8. Explain Why do gamma rays have more energy than other forms of electromagnetic energy?

_____

_____

How are X rays used?

Have you ever had an X ray? If so, you know that doctors can use X rays to see inside your body. X rays can pass through the less dense tissues of skin and muscle. X rays strike a film and leave shadow images of bones and denser tissues. Doctors use X-ray images to find injuries and diseases such as broken bones and cancer. A CT scanner uses X rays to produce images of the human body as if it had been sliced like a loaf of bread.

Like ultraviolet waves, X rays can be harmful to cells and tissue. One or two X rays are not harmful. However, a large number of X rays could be harmful to your body. X-ray machine operators usually stand behind shields to protect themselves. A patient who gets an X ray usually wears a lead apron or shield to protect the body parts that are not receiving the X ray.

How are gamma rays used?

Even though gamma rays are dangerous to living organisms, they have some helpful uses, just like X rays do. A beam of gamma rays can be used to kill cancerous tumors. Gamma rays can also kill disease-causing bacteria in food. More than 1,000 Americans die each year from *Salmonella* bacteria in poultry and *E. coli* bacteria in meat. Gamma radiation has been used since 1963 to kill bacteria in food. However, this method is not often used in the food industry.

Astronomy with Different Wavelengths

Astronomers use more than visible light to study objects in space. Some objects that produce no visible light can be found through X rays, infrared waves, or radio waves that they give off. Scientists use instruments that can detect many types of electromagnetic radiation to study these objects.

Recall that Earth's atmosphere blocks X rays, gamma rays, most ultraviolet rays, and some infrared rays. To study these types of radiation from space, scientists have to get above Earth's atmosphere. Satellites have been built for this. Three of these sattelites are the Extreme Ultraviolet Explorer (EUVE), the Chandra X-Ray Observatory, and the Infrared Space Observatory (ISO). ✔

Think it Over

9. **Draw Conclusions**
Why do doctors use X rays only when necessary?

_____

_____

_____

Reading Check

10. **Describe** How are astronomers able to study X rays and gamma rays given off by objects in space?

_____

_____

_____

● After You Read

Mini Glossary

electromagnetic spectrum: the wide range of electro-magnetic waves with different wavelengths and frequencies

gamma rays: electromagnetic waves that have the highest frequency and carry the most energy

infrared waves: electromagnetic waves with wavelengths between about one thousandth and 0.7 millionths of a meter

radio waves: electromagnetic waves that have the lowest frequencies and carry the least energy

ultraviolet radiation: electromagnetic waves that are higher in frequency than visible light and have even shorter wavelengths—between 0.4 millionths and about 10 billionths of a meter

visible light: electromagnetic waves that you can see with your eyes

X rays: electromagnetic waves that have an even higher frequency than ultraviolet rays and enough energy to go right through skin and muscle

1. Review the terms and their definitions in the Mini Glossary. What are the two types of electromagnetic radiation that humans can sense?

_____

2. In the table below, place the six types of electromagnetic radiation you learned about in this section in order from longest wavelength to shortest wavelength.

Types of Electromagnetic Radiation

3. Which type of harmful electromagnetic radiation are you exposed to most often? How can you protect yourself from too much exposure?

_____

_____

End of Section

Science Online Visit **ips.msscience.com** to access your textbook, interactive games, and projects to help you learn more about the electromagnetic spectrum.

Electromagnetic Waves

section ❸ Using Electromagnetic Waves

● Before You Read

When was the last time you listened to the radio or watched TV? How often do you do these things each week?

_____

_____

● Read to Learn

Telecommunications

You use electromagnetic waves each time you talk on the telephone, listen to the radio, watch TV, or do research on the Internet. Today you can talk to someone far away or send an email almost instantly. Communicating with electromagnetic waves is called telecommunications.

Using Radio Waves

Radio waves are used to send and receive information over long distances. Using radio waves to communicate has several advantages. Radio waves pass through walls and windows easily. They are not harmful to people like X rays and ultraviolet waves are. So, most telecommunication devices like radios, TVs, and telephones use radio waves to send and receive sounds and images.

How does radio transmission work?

Radio and TV stations are given a frequency at which they broadcast radio waves. **Carrier waves** are the radio waves broadcast by a station at its assigned frequency. To receive the station's signals, you tune your radio or TV to the frequency of the station's carrier waves. The amplitude or frequency of the carrier wave is changed to send information.

What You'll Learn

■ the different ways of using electromagnetic waves to communicate

■ the differences between AM and FM radio signals

Study Coach

Identify the Main Point
As you read, write down the main point under each heading in the text.

FOLDABLES™

Ⓑ Organize Information
Make the following Foldable to organize information about using electromagnetic waves.

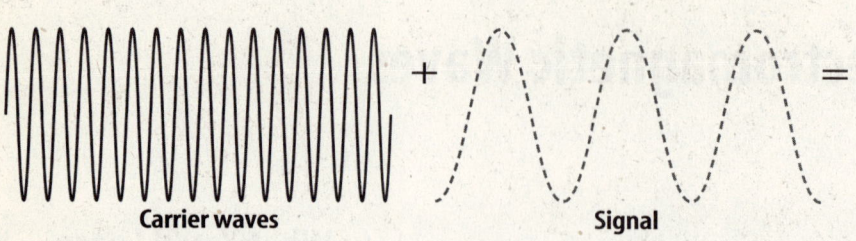

Carrier waves + Signal =

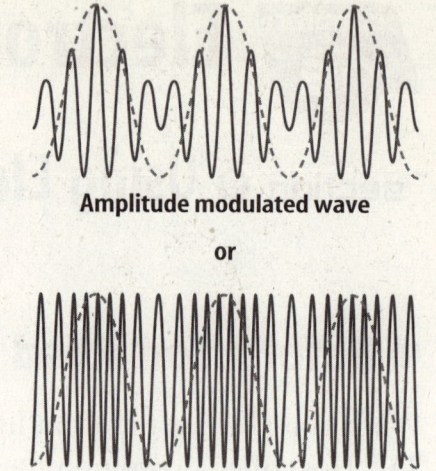

Amplitude modulated wave

or

Frequency modulated wave

1. Explain Look at the figure above. How does the carrier wave change in an AM wave?

_____

_____

_____

✔ Reading Check

2. Describe What part of the carrier wave is changed in FM transmission?

_____

What is amplitude modulation?

The letters *AM* in AM radio stand for amplitude modulation. In amplitude modulation, the amplitude of the carrier wave is changed to send information. The original sound is changed into an electrical signal. The electrical signal is used to vary the amplitude of the carrier wave. You can see an example in the figure above. The frequency of the carrier wave does not change—only the amplitude changes. In the receiver, the varying amplitude of the carrier waves produces an electric signal. The radio's loudspeaker uses the signal to produce the original sound.

What is frequency modulation?

FM stands for frequency modulation. FM radio works much the same way as AM radio. The difference is the frequency of the carrier wave is changed instead of the amplitude. You can see this in the figure at the top of the page. The FM receiver uses the varying frequency of the carrier wave to produce an electric signal. The radio's speaker converts the electric signal into sound waves. ☑

Telephones

Telephones change a sound wave into an electric signal. The signal travels through a wire to the telephone switching system. The electric signals may be sent through wires or changed to radio or microwave signals and sent through the air. The electric signal may also be changed to a light wave to be sent through fiber-optic cables. At the receiving end, the signal is changed back to sound waves.

How do wireless phones and pagers work?

Many phones do not use wires to send signals. Some use radio waves. Cordless phones change the electrical signal from the microphone of a telephone into a radio signal. The signal is sent to the base station of the telephone. The phone changes the electrical signals into sound waves. If you are receiving a call on a cordless telephone, the base station sends electrical signals to the phone. The phone changes the signals into sound waves. Cellular phones work the same way. But they work over distances of many kilometers. The base station uses a large antenna. The antenna communicates with the cell phone and with other base stations.

Pagers also use base stations. When you dial a pager, the signal is sent to a base station. The base station sends an electromagnetic signal to the pager. The pager receives the signal and beeps or vibrates to indicate that someone has called. You can also send information to a pager if you have a touch-tone phone. The pager receives and displays information such as your phone number.

Communications Satellites

How can radio signals be sent to the other side of the world? Radio signals cannot travel directly through Earth. Instead, radio signals are sent to satellites. The satellites can communicate with other satellites or with ground stations. Some satellites are designed to move at the same speed as Earth, so they are above the same place on Earth at all times. This is called geosynchronous (jee oh SIHN kroh nus) orbit.

The Global Positioning System

The **Global Positioning System,** or GPS, is a system of ground stations, satellites, and receivers that is used to locate objects on Earth. A GPS receiver measures the time it takes radio waves to travel from several satellites to the receiver. The receiver uses this information to find its latitude, longitude, and elevation. Handheld GPS receivers can give a location within a few hundred meters. Some GPS units are used to measure the movements of Earth's crust. They can measure the movement to within a few centimeters. ☑

Copyright © Glencoe/McGraw-Hill, a division of The McGraw-Hill Companies, Inc.

Think it Over

3. **Explain** When you are using a wireless phone, you are really using a radio. Explain.

_____

_____

_____

Reading Check

4. **Identify** What is the purpose of the Global Positioning System?
   a. to locate satellites in space
   b. to locate objects on Earth
   c. to locate radios on Earth
   d. to communicate with other parts of Earth

● After You Read

Mini Glossary

carrier waves: the radio waves broadcast by a station at its assigned frequency

Global Positioning System (GPS): a system of ground stations, satellites, and receivers that is used to locate objects on Earth

1. Review the terms and their definitions in the Mini Glossary. Write a sentence explaining how the carrier wave is changed in frequency modulation.

   _____

   _____

   _____

2. In the space provided, give three examples of telecommunication devices that use radio waves. Then, describe how the device uses radio waves.

Example	Use of Radio Waves

3. Which form of telecommunication do you use most often? Explain.

   _____

   _____

   _____

End of Section

Science Online Visit ips.msscience.com to access your textbook, interactive games, and projects to help you learn more about using electromagnetic waves.

chapter 19

Light, Mirrors, and Lenses

section 1 Properties of Light

Before You Read

When someone says the word *light,* what do you think of?

____

____

What You'll Learn

- about light waves
- how light behaves
- why objects have color

Read to Learn

What is light?

What happens when you drop a rock on the smooth surface of a pond? The rock makes a wave. You can see the wave spread out in a circle. A wave is a disturbance that carries energy through matter or space. The matter in this case is water. The energy comes from the rock hitting the water. As the ripples spread out, they carry some of the energy.

Light is another type of wave that carries energy. A light source is something that gives off light. Light bulbs and the Sun are light sources. Light sources give off light waves into space like the ripples in a pond. While ripples from a rock spread out only on the surface of a pond, light waves spread out in all directions from a light source.

Sometimes it is easier to think of light as a ray. A **light ray** is a narrow beam of light that travels in a straight line. A light source gives off many light rays that travel away from the source in all directions. ☑

How does light travel through space?

Waves on a pond need a material to travel through. The material through which a wave travels is a **medium.** Light is an electromagnetic wave. Electromagnetic waves can travel through empty space. They do not need a medium in which to travel. They also can travel through materials such as air, water, and glass.

Mark the Text

Highlight Highlight the words and phrases that you do not understand. When you finish reading, ask another student or your teacher to help you understand the things you highlighted.

Reading Check

1. **Describe** How does a light ray travel?

____

____

Light and Matter

What can you see in a dark room with no windows or lights? You cannot see anything until you turn on the lights or open the door to let in light from outside the room. Most objects around you are not light sources. For these objects to be seen, light from a light source bounces off the objects and into your eyes. The figure below shows how this works. The process of light hitting an object and bouncing off is called reflection. You can read the words on this page because of reflection. Light from a source reflects from the page and into your eyes. But, not all light rays reflected from the page go to your eyes. Light rays reflect in many directions.

Picture This

2. **Identify** Circle the light source in the figure.

What are opaque, translucent, and transparent objects?

Three things can happen to light waves that strike an object. Some of the waves are absorbed by the object, some are reflected by it, and some might pass through it. What happens to light when it strikes an object depends on what material the object is made of.

Materials that let no light pass through them are opaque (oh PAYK). You cannot see other objects through opaque materials. Materials that let almost all light pass through them are transparent. You can see other objects clearly through transparent materials. Examples of transparent materials are clear glass and clear plastic. Materials that let only some light pass through them are translucent (trans LEW sent). You can see objects through translucent materials, but not clearly. Waxed paper and frosted glass are translucent materials. ✔

✔ **Reading Check**

3. **Describe** Which of the following best describes an object that you can see clearly through? Circle your answer.

   a. opaque
   b. transparent
   c. translucent
   d. opaque and translucent

Color

Light from the Sun looks white. But, it is really a mixture of colors. Each color of light is a light wave with a different wavelength. Red light waves have the longest wavelengths. Violet light waves have the shortest wavelengths. When white light passes through a prism, the light separates into different colors. When light waves from all the colors enter your eye at the same time, your brain sees the mixture as white.

Why do objects have color?

Why does grass look green and a rose look red? When white light hits an object that is not transparent, the object absorbs some of the light waves. The other light waves are reflected. If an object reflects red light waves and absorbs all the other waves, it looks red. If an object reflects only blue waves, it looks blue. An object that looks white reflects all of the light waves that strike it. Objects that look black absorb all light waves. ☑

What are the primary light colors?

White light is made up of red, orange, yellow, green, blue and violet light. But, there are many more colors. Humans can see thousands of colors, including brown, pink, and purple.

You can make almost any color light by mixing different amounts of red, green, and blue light. Red, green, and blue are known as primary light colors. Mix all three primary light colors together and you get white light. Mix red and green light to get yellow light. You see yellow because that is how your brain interprets the mixture of red and green. To your brain, a mixture of red and green light waves looks the same as yellow light made by a prism.

What are primary pigment colors?

Materials that are used to change the color of objects, like paint, are called pigments. Mixing pigments together forms colors in a different way than mixing colored light does.

The color of the pigment you see is the color of the light waves that are reflected from it. But, the primary pigment colors are not red, blue, and green. They are yellow, magenta, and cyan (SI an). The primary colors for light and pigments are different, but related. Yellow pigment absorbs blue light and reflects red and green light. Magenta pigment absorbs green light and reflects red and blue light. Cyan pigment absorbs red light and reflects blue and green light.

✔ Reading Check

4. Explain What happens to light waves that strike an object that looks white?

_____

_____

FOLDABLES

Ⓐ Organize Information
Use a half-sheet of notebook paper to help you organize information about color.

Investigating Color

● After You Read

Mini Glossary

light ray: a narrow beam of light that travels in a
straight line

medium: the material through which a wave travels

1. Review the terms and their definitions in the Mini Glossary. Explain the difference between light waves and waves caused by a rock tossed into a pond.

_____

_____

_____

2. Fill in the graphic organizer below to compare and contrast what happens when light hits different types of materials.

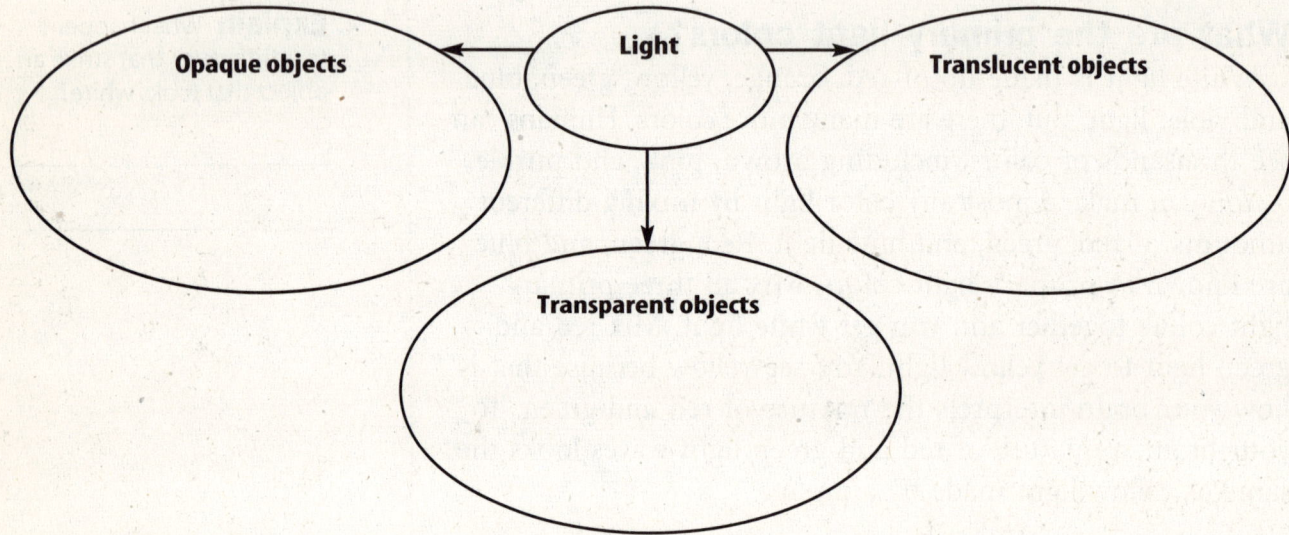

3. At the beginning of the section, you were asked to highlight things that you did not understand. How did this help you learn the material?

_____

_____

_____

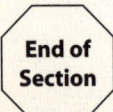

End of
Section

Science●**nline** Visit **ips.msscience.com** to access your textbook, interactive games, and projects to help you learn more about the properties of light.

 **Light, Mirrors, and Lenses**

section ❷ **Reflection and Mirrors**

● Before You Read

On the lines below, describe at least one way that you use a mirror.

_____

_____

What You'll Learn

■ how light is reflected from rough and smooth surfaces.
■ how mirrors work
■ about concave and convex mirrors

● Read to Learn

The Law of Reflection

Have you ever seen your reflection on the surface of a pond? You can see yourself because light reflects from the surface to your eyes. The figure below shows a light ray reflecting from a mirror. The normal is an imaginary line that is perpendicular to the surface where the light ray strikes. The angle between the incoming light ray and the normal is called the angle of incidence. The angle between the reflected light ray and the normal is called the angle of reflection. The **law of reflection** states that the angle of incidence is equal to the angle of reflection.

Study Coach

Outline As you read, make an outline of this section. Use the headings of the section to make your outline. Write down the main points or ideas under each heading.

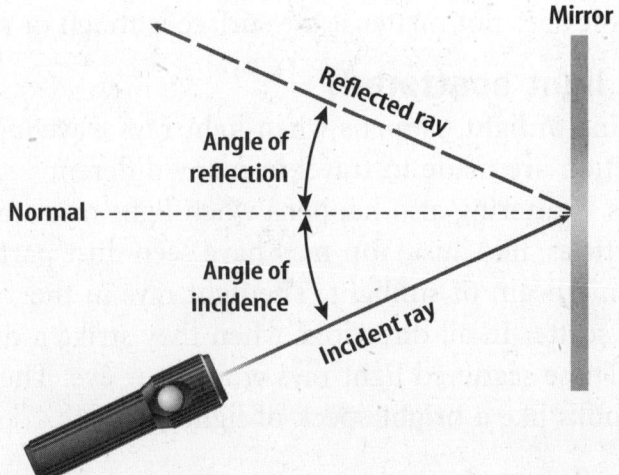

Picture This

1. **Determine** Look at the angles marked in the figure. If the angle of reflection measures 20 degrees, what is the measure of the angle of incidence?

_____

Reflection from Surfaces

Why can you see your reflection in some surfaces but not others? Why does a piece of metal make a better mirror than a piece of paper? Reflection depends on the smoothness of the surfaces.

What are the types of reflection?

The surface of a piece of paper may seem smooth. But it is not as smooth as the surface of a mirror. If you look at the surface of the paper under a microscope, you would see how rough it is. Look at the first figure below. A rough surface causes light rays to reflect in many directions. This uneven reflection of light waves from a rough surface is called diffuse reflection.

Picture This

2. **Describe** the pattern of light rays and reflections in the second figure.

_____

_____

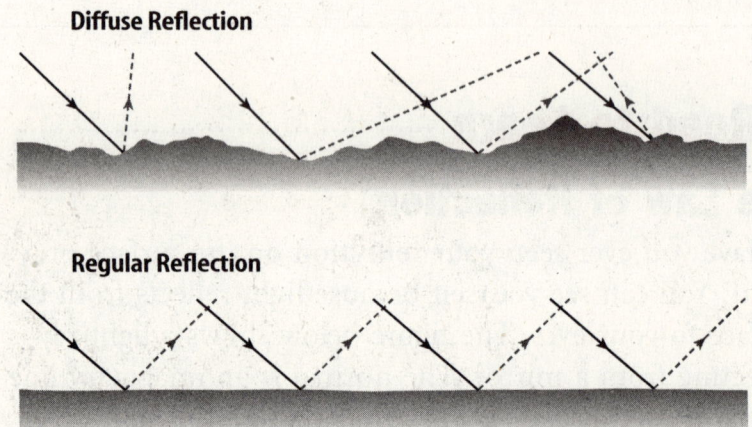

Diffuse Reflection

Regular Reflection

Smooth surfaces, like the one in the second figure, reflect light rays in a more regular way. Reflection from mirrors or other very smooth surfaces is called regular reflection. Light rays that are regularly reflected make the image you see in a mirror. Every light ray that strikes a surface obeys the law of reflection. It does not matter if the surface is rough or smooth.

How is light scattered?

Scattering of light happens when light rays traveling in one direction are made to travel in many different directions. Scattering also happens when light rays strike small particles, like dust. You may have seen dust particles floating in a beam of sunlight. The light rays in the sunbeam scatter in all directions when they strike a dust particle. These scattered light rays enter your eye. The dust particle looks like a bright speck of light to you.

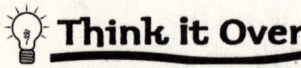 **Think it Over**

3. **Explain** What kind of surface causes light waves to scatter?

_____

_____

What is a plane mirror?

A plane mirror has a flat reflecting surface. Your image in a plane mirror looks like it would in a photograph. You and your image in a plane mirror are facing in opposite directions. Your left and right sides switch places on your mirror image. Your image also seems to be coming from behind the mirror.

The figure below shows a person looking into a plane mirror. Light rays from a light source strike each part of the person. These light rays bounce off the person. Some of these light rays strike the mirror and are reflected off the mirror. The figure shows the path of some of these light rays that have been reflected off the person and reflected back to the person's eye by the mirror.

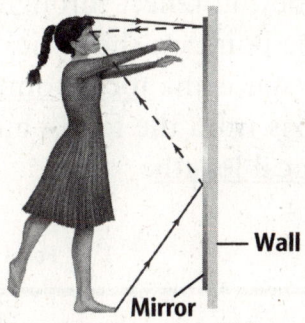

Wall

Mirror

Picture This

4. Highlight Trace the path of the light rays.

How is an image formed in a plane mirror?

The image in a plane mirror seems to be behind the mirror because of the way your brain understands the light rays that enter your eyes. When the light rays bounce off the mirror's surface, your brain thinks they followed the path shown by the dashed lines in the figure below. Your brain thinks light rays travel in straight lines without changing direction. This makes reflected light rays look like they are coming from behind the mirror. The image also looks like it is the same distance behind the mirror as the person is in front of the mirror.

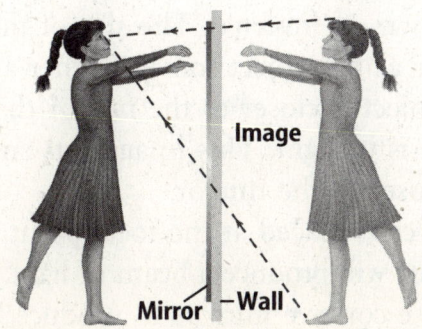

Image

Mirror   Wall

Picture This

5. Explain The dashed line shows where the light waves seem to be coming from. Where are they really coming from?

_____

_____

6. Contrast Explain the difference between concave and convex mirrors.

_____

_____

_____

Concave and Convex Mirrors

Some mirrors are not flat. A concave mirror has a surface that is curved inward, like the bowl of a spoon. Concave mirrors cause reflected light rays to come together, or converge.

A convex mirror has a surface that curves outward, like the back of a spoon. Convex mirrors cause reflected light rays to spread out, or diverge. Concave and convex mirrors form images that are different from the images that are formed by plane mirrors. To help you remember which is concave and which is convex, think of a concave mirror as caved in. ✔

How do concave mirrors form images?

The figure below shows a concave mirror. The optical axis is a straight line drawn perpendicular to the center of the mirror. When light rays travel parallel to the optical axis and strike the mirror, they all reflect through the same point on the optical axis. The point on the optical axis that reflected light rays pass through is the **focal point**. The distance along the optical axis from the focal point to the center of the mirror is the **focal length**.

Picture This

7. Identify Use a highlighter to mark the focal length in the figure.

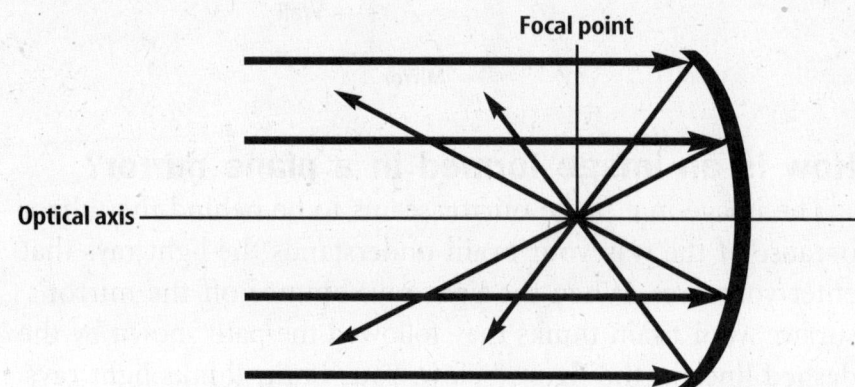

Focal point

Optical axis

How does the focal point affect an image?

The image formed by a concave mirror depends on where the object is compared to the focal point. If the object is farther from the mirror than the focal point, the image looks upside down, or inverted. The size of this image becomes smaller as the object moves farther away from the mirror. If the object is closer to the mirror than the focal point the image is upright. This image gets smaller as the object moves closer to the mirror.

If a light source is placed at the focal point of a concave mirror, the mirror will produce a beam of light. Flashlights and car headlights use concave mirrors to produce beams of light.

ⓑ Compare and Contrast Use two quarter-sheets of notebook paper to help you compare and contrast concave and convex mirrors.

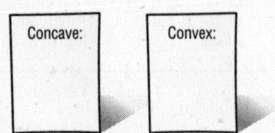

Concave: Convex:

How do convex mirrors form images?

A convex mirror has a reflecting surface that curves outward. Convex mirrors cause light rays to spread out, or diverge, as shown in the figure below.

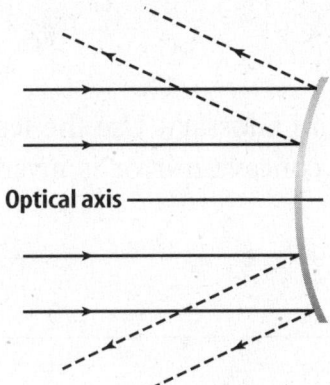

Optical axis

Picture This
8. Explain Why does a convex mirror not have a focal point?

_____

_____

_____

Images formed by convex mirrors are upright. They also seem to be behind the mirror like images in plane mirrors. But, convex mirrors are different from plane mirrors. As shown in the figure below, the image in a convex mirror is always smaller than the actual object. It does not matter how far away the object is.

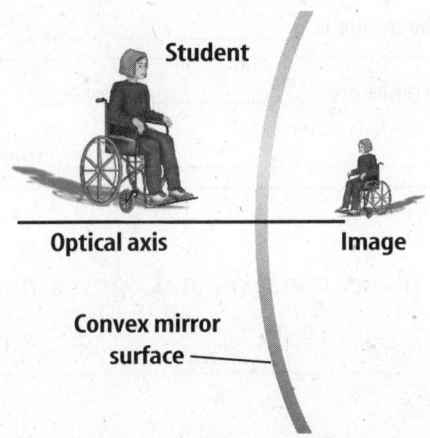

Student

Optical axis · Image

Convex mirror surface

Picture This
9. Describe How does the size of the image formed by a convex mirror compare to the size of the actual object?

_____

_____

_____

How are convex mirrors used?

Since you can see a larger area reflected in a convex mirror than you can in other mirrors convex mirrors often are used as security mirrors in stores. Outside rearview mirrors on cars and trucks also are made from convex mirrors.

● After You Read

Mini Glossary

focal length: the distance along the optical axis from the focal point to the center of the mirror

focal point: the point on the optical axis that reflected light rays pass through

law of reflection: says that the angle of incidence is equal to the angle of reflection

1. Review the terms and their definitions in the Mini Glossary. Use the term *focal point* in a sentence or two to explain when the image in a concave mirror is inverted.

   _____

   _____

2. Fill in the table below about the images formed by different types of mirrors.

Type of Mirror	What does the mirror look like?	What kind of image is formed?
Plane		Image is upright.
Concave		If the object is past the focal point, the image is _____. If the object is between the focal point and the mirror, the image is _____.
Convex	Curved outward	Images are _____ and _____ than the object.

3. What objects in your house are examples of plane, concave, and convex mirrors?

   _____

   _____

   _____

   _____

 Visit **ips.msscience.com** to access your textbook, interactive games, and projects to help you learn more about reflection and mirrors.

Light, Mirrors, and Lenses

section ❸ Refraction and Lenses

● Before You Read

Why do some people wear eyeglasses? What do eyeglasses do?

_____

_____

What You'll Learn

■ why light waves change direction
■ how concave and convex lenses form images

● Read to Learn

Bending of Light Rays

Have you ever looked at a glass of water with a straw in it? Did the water make the straw look like it was bent? Sometimes a penny in the bottom of a cup cannot be seen from the side until water is added. These things happen because light rays bend as they pass from one material to another. What causes light rays to change direction?

What is the speed of light and how does it change?

The speed of light in empty space is about 300 million meters per second (m/s). Light passing through air, water, or glass travels more slowly. This is because atoms that make up these materials slow the light waves down. The table below shows the speed of light in some different materials.

The Speed of Light Through Some Materials	
Material	**Speed of Light**
Air	About 300 million m/s
Water	About 227 million m/s
Glass	About 197 million m/s
Diamond	About 125 million m/s

Study Coach

Identifying the Main Point As you read each paragraph, find the main point or main idea and write it down. When you finish reading, make sure you understand each main idea that you have written.

Applying Math

1. **Calculate** About how much faster is the speed of light through air than through glass?

_____

_____

_____

The Refraction of Light Waves

Suppose you are looking at a straw in a glass of water. Light waves from the part of the straw that is underwater travel through water, glass, and air before they reach your eyes. The speed of light is different in each of these materials.

What happens to a light wave when it travels from one material into another? If the light wave is traveling at an angle to the boundary between the two materials, it changes direction, or bends. This bending happens because the speed of the light wave changes when it passes from one material into the other. The bending of light waves due to a change in speed is called refraction. The greater the change in speed, the more the light wave bends, or refracts. ☑

How does light bend?

Why does a change in speed cause a light wave to bend? The figure below shows what happens to the wheels of a car as they move from pavement to mud at an angle. The wheel that enters the mud first slows down a little. The other wheel is still on pavement and is still turning at the original speed. The difference in speed between the two wheels on the axle causes the axle to turn a little bit. This makes the car turn a little bit, too.

Now imagine a light wave traveling at an angle from air into water. Like the car wheels, the first part of the wave to enter the water slows down. The rest of the wave travels at the original speed until it enters the water. The light wave bends as long as one part is traveling faster than the rest of the wave.

✓ Reading Check

2. Explain What is refraction?

_____

_____

_____

<u>Picture This</u>

3. Identify Look at the figure. Find the axle with one wheel in the mud and the other on the pavement. Circle the wheel that is traveling slower.

Convex and Concave Lenses

Do you like to take pictures? Have you ever looked at something far away through binoculars or a telescope? Maybe you have looked at a small insect with a magnifying glass. You can do all of these things because of lenses. A **lens** is a transparent object with at least one curved side that causes light to bend. How much the light bends depends on how curved the sides of the lens are. The more curved the sides of a lens are, the more light will be bent after it enters the lens. ☑

How do convex lenses bend light?

A **convex lens** is thicker in the center than it is at the edges. Look at the figure below. In a convex lens, light rays traveling parallel to the optical axis are bent so they pass through the focal point. The more curved the lens is, the closer the focal point is to the lens. This makes the focal length shorter. Convex lenses are sometimes called converging lenses because they cause light waves to meet, or converge.

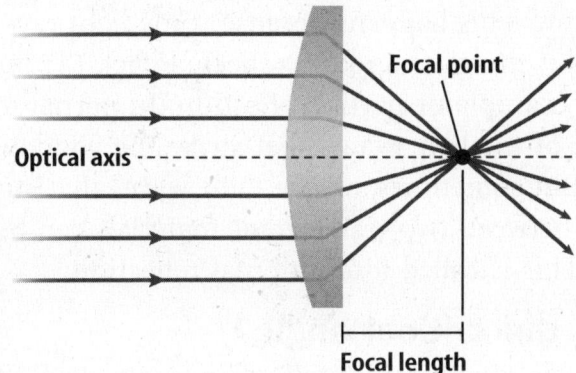

Focal point

Optical axis

Focal length

How do magnifying glasses work?

The image formed by a convex lens is like the image formed by a concave mirror. For both, the type of image depends on how far the object is from the mirror or lens. If the object is farther than two focal lengths from the lens, the image seen through the lens is upside down, or inverted. The image also is smaller than the actual object.

If the object is closer to the lens than one focal length, the image is right-side up and larger than the object. A magnifying glass is a convex lens. As long as the magnifying glass is less than one focal length from the object, you can make the image appear larger by moving the magnifying glass away from the object.

Picture This

5. Describe What happens to the light rays after they pass through the lens?

_____

_____

How do concave lenses bend light?

A <u>concave lens</u> is thicker at the edges than in the middle. A concave lens causes light waves to spread out, or diverge. They are not brought to a focus. The figure below shows how light waves that travel parallel to the optical axis are bent after passing through a concave lens. The image formed by a concave lens is like the image formed by a convex mirror. The image is upright and smaller than the actual object.

Picture This

6. **Draw Conclusions**
   Which type of mirror forms an image like the image formed by a concave lens?

_____

_____

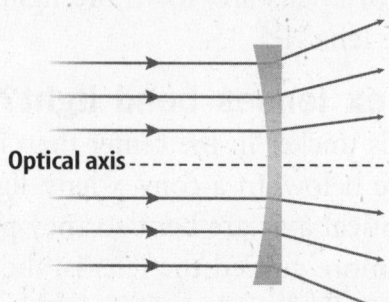

Optical axis

Total Internal Reflection

Sometimes you can see your reflection when you look at a glass window. This happens because some light rays reflected from you hit the glass and are reflected back to your eyes. This is an example of partial reflection. In partial reflection, only some of the light waves that strike the window are reflected. But sometimes, all the light waves that strike the boundary between two transparent materials can be reflected. This is called total internal reflection.

What is the critical angle?

To see how total internal reflection works, look at the figure at the top of the next page. Light travels faster in air than in water, so the refracted beam bends away from the normal. As the angle between the incident beam and the normal increases, the refracted beam bends closer to the air-water boundary. At the same time, more of the light is reflected, and less passes into the air.

Total internal reflection occurs when a light beam in water strikes the boundary at the angle with the normal called the critical angle. All the light waves are reflected when an angle is equal to or greater than the critical angle at the air-water boundary. It is as if there were a mirror at the boundary. The size of the critical angle depends on the two materials. The critical angle for an air-water boundary is about 48 degrees. ☑

 **Reading Check**

7. **Identify** What happens to a light beam that strikes the air-water boundary at an angle greater than the critical angle?

_____

_____

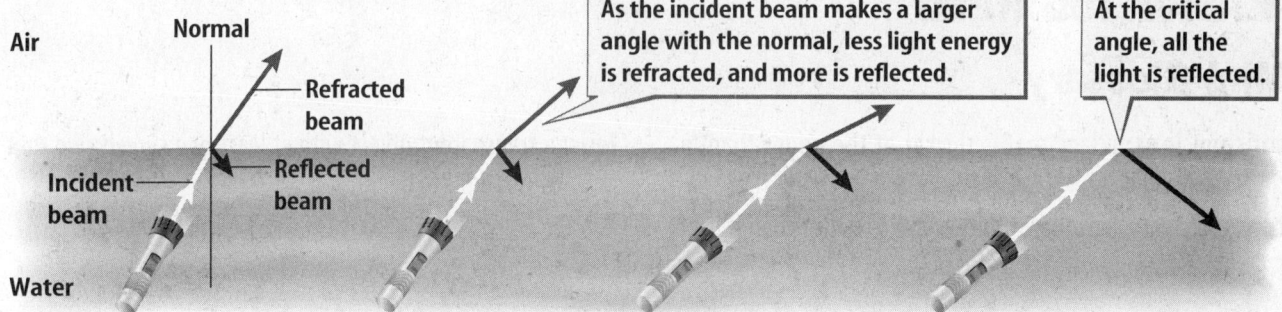

Air

Normal

Refracted beam

Incident beam

Reflected beam

Water

As the incident beam makes a larger angle with the normal, less light energy is refracted, and more is reflected.

At the critical angle, all the light is reflected.

What are optical fibers?

Optical fibers are thin, flexible, transparent fibers. Think of an optical fiber as a light pipe. A light beam can travel for many kilometers through an optical fiber. It loses almost no energy. Even if the fiber is bent, light that goes in one end of the fiber comes out the other end.

Light moves through optical fibers because of total internal reflection. A thin fiber of glass or plastic is covered with a material called cladding. Light travels faster in the cladding than in the fiber. When light strikes the boundary between the fiber and the cladding, total internal reflection happens. The beam of light bounces from boundary to boundary as it travels down the fiber, as shown in the figure below.

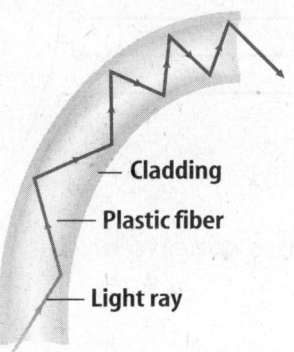

— Cladding

— Plastic fiber

— Light ray

How are optical fibers used?

Optical fibers are used in the communications industry. Television programs, computer information, and telephone conversations can be changed into light signals. These signals are sent from place to place through optical fibers. Signals cannot leak from one fiber to another because of total internal reflection. This means the signal stays clear. One optical fiber the thickness of a human hair can carry thousands of phone conversations.

Picture This

8. **Describe** What happens to the light beam at the critical angle?

_____

_____

Picture This

9. **Identify** Use a highlighter to highlight the path light takes through the optical fiber in the figure.

● After You Read

Mini Glossary

concave lens: a lens that is thicker at the edges than in the middle

convex lens: a lens that is thicker in the center than it is at the edges

lens: a transparent object with at least one curved side that causes light to bend

1. Review the terms and their definitions in the Mini Glossary. Write a sentence about how concave lenses bend light.

_____

_____

2. On the figure below, continue each light wave to show what happens to the light waves after they pass through the lenses.

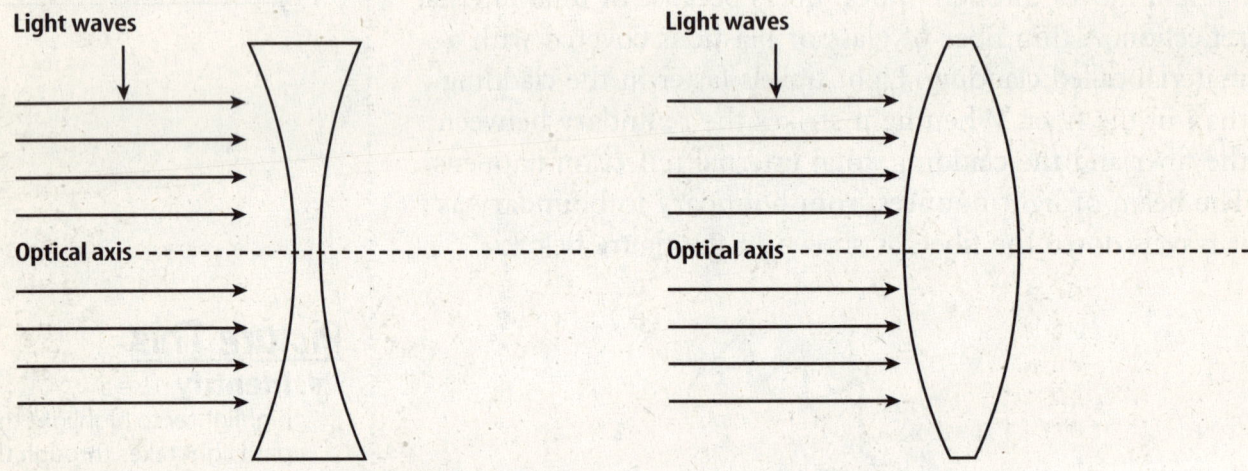

3. Look carefully at a pair of eyeglasses. Are the lenses concave or convex? How do you know?

_____

_____

_____

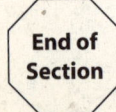

End of Section

Science ⬭nline Visit **ips.msscience.com** to access your textbook, interactive games, and projects to help you learn more about refraction and lenses.

Light, Mirrors, and Lenses

section ❹ Using Mirrors and Lenses

● Before You Read

List some ways you use mirrors. List some ways you use lenses.

_____

_____

● Read to Learn

Microscopes

For almost 500 years, lenses have been used to see very small objects. Early microscopes were simple. They had only one lens and magnified less than 100 times. Today, compound microscopes use more than one lens and can magnify objects up to 2,500 times.

The figure shows how a compound microscope forms an image. An object is placed close to a convex lens called the objective lens. This lens produces a larger image inside the microscope. The light rays from the image pass through a second convex lens called the eyepiece lens. The eyepiece lens makes the image larger, or magnifies it. When two lenses are used, a much larger image is formed than with just one lens.

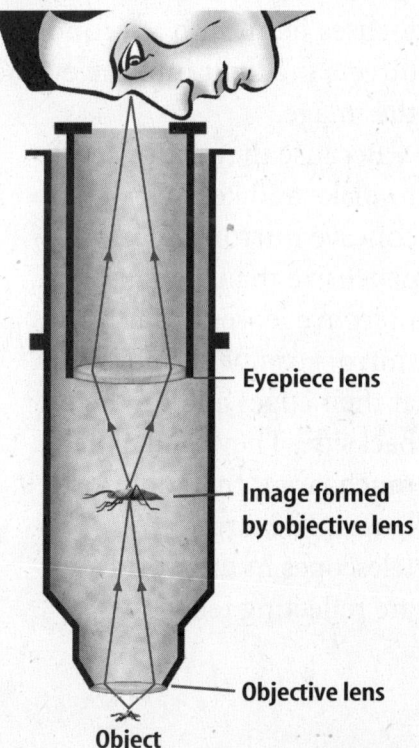

Eyepiece lens

Image formed by objective lens

Objective lens

Object

FOLDABLES

⊙ Organize Information
Make the following Foldable to help you organize information about microscopes, telescopes, and cameras.

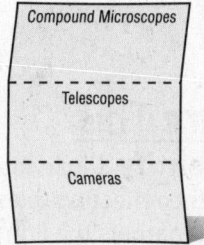

Compound Microscopes

Telescopes

Cameras

Telescopes

Telescopes are used to look at objects that are far away. The first telescopes were made about the same time as the first microscopes. Much of what we know about the solar system and the distant universe has come from images gathered by telescopes.

What is a refracting telescope?

The simplest **refracting telescope** uses two convex lenses to form an image of a distant object. Just like a compound microscope, light passes through an objective lens and an eyepiece lens. The objective lens forms an image and the eyepiece lens magnifies it.

The main purpose of a telescope is to gather as much light as possible from distant objects. Refractive telescopes use a large objective lens to gather light from distant objects. In a telescope, the larger the lens is, the more light it can gather. When more light enters the telescope, images of faraway objects look brighter, sharper, and more detailed. ✔

What is a reflecting telescope?

A **reflecting telescope** has a concave mirror instead of a concave objective lens to gather the light from distant objects. As you can see in the figure, the large concave mirror focuses light onto a secondary plane mirror. That mirror directs the light into the eyepiece. The eyepiece magnifies the image.

Because they are easier to make and keep clean, concave mirrors are less expensive than similar-sized objective lenses. Concave mirrors can be supported at their edges and on their backside. They can be made much larger than objective lenses. The largest telescopes in the world are reflecting telescopes.

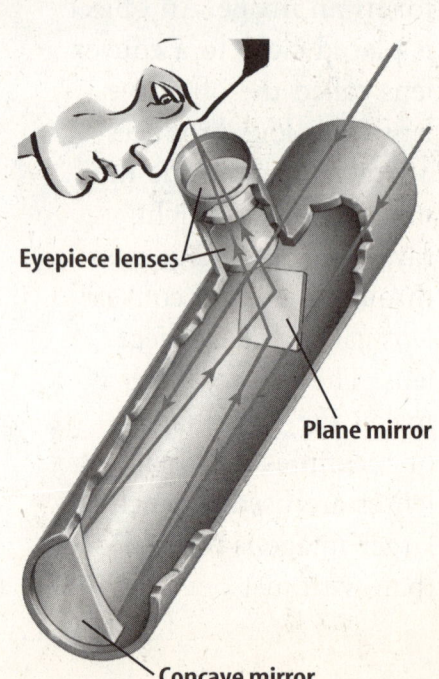

Eyepiece lenses

Plane mirror

Concave mirror

Reflecting Telescope

✔ **Reading Check**

1. **Explain** What is the purpose of an objective lens in a refracting telescope?

_____

_____

Picture This

2. **Identify** Circle the part of the reflecting telescope that gathers the light.

Cameras

You probably see photographs taken by cameras every day. A camera uses a convex lens to form an image on a piece of film. This is similar to the way the eye focuses an image on the retina. The convex lens has a short focal length. It forms an image that is smaller than the object and upside down, or inverted, on the film.

Look at the camera in the figure. When the shutter is open, the convex lens focuses an image on a piece of film that is sensitive to light. The film contains chemicals that undergo chemical reactions when light hits it. A device called a diaphragm controls how much light reaches the film. The brighter parts of the image affect the film more than the darker parts do.

Picture This

3. **Describe** the image that is formed on the film by the convex lens.

_____

_____

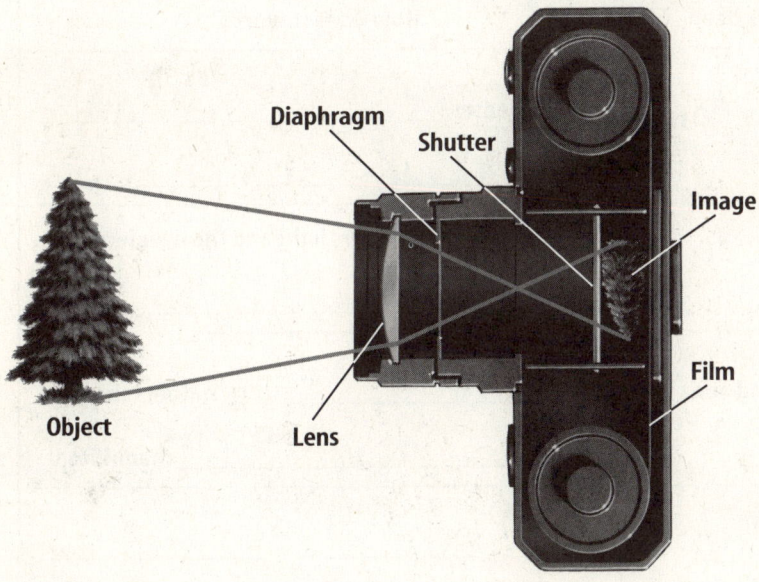

Diaphragm
Shutter
Image
Object
Lens
Film

Lasers

If you shine a flashlight on a wall in a darkened room, you see a spot of light. If you move back from the wall, the light spreads out and becomes less powerful. Laser light is different. The light waves in laser light are all the same wavelength. The crests and troughs of the light waves overlap, so the waves are in phase. Laser light does not spread out like regular light.

A laser can focus a large amount of energy on a very small area, so lasers are used for cutting and welding materials. They also can be used instead of scalpels in surgery. Less powerful lasers are used to read and write to CDs or to scan bar codes.

Think it Over

4. **Explain** Why do the waves of laser light have so much energy?

_____

_____

_____

● After You Read
Mini Glossary

reflecting telescope: uses a concave mirror to gather light from distant objects

refracting telescope: uses two convex lenses to gather light and form an image of a distant object

1. Review the terms and their definitions in the Mini Glossary. Explain the difference between a reflecting telescope and a refracting telescope.

_____

_____

2. Fill in the table below to describe how microscopes, telescopes, and cameras work.

Device	Lenses or Mirrors Used	How does it work?
Microscope	two convex mirrors	The two lenses _____ _____.
Refracting telescope		The objective lens gathers light and the eyepiece _____.
Reflecting telescope	one concave mirror and one convex lens	The _____ gathers the light and the _____ magnifies it.
Camera		The _____ focuses a smaller, upside down image on the film.

3. Which device with lenses described in this section do you use most often? Explain.

_____

_____

_____

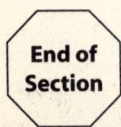
End of Section

 Visit **ips.msscience.com** to access your textbook, interactive games, and projects to help you learn more about using mirrors and lenses.

Electricity

section ❶ Electric Charge

● Before You Read

You use electricity every day. What would be different in your life if you didn't have electricity?

_____

_____

● Read to Learn

Electricity

To understand electricity, first you must think small—very small. Remember that all solids, liquids, and gases are made of tiny particles called atoms. Atoms are made of even smaller particles called protons, neutrons, and electrons. Look 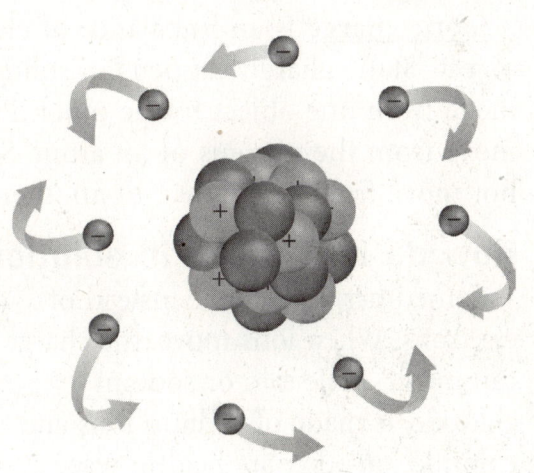 at the figure. Protons and neutrons are held together tightly in the nucleus at the center of an atom. Electrons swarm around the nucleus in all directions. Protons and electrons have electric charge. Neutrons have no electric charge.

What are positive and negative charges?

There are two kinds of electric charge—positive and negative. A proton has a positive charge. An electron has a negative charge. Atoms have an equal number of protons and electrons. So, atoms are electrically neutral. They have no overall electric charge.

What You'll Learn

- how objects become electrically charged
- about electric charges
- about conductors and insulators
- about electric discharge

▸ **Mark the Text**

Find the Main Idea As you read, highlight the main idea of each paragraph.

Picture This

1. **Label** Use three different colored highlighters, crayons, or pencils to mark the protons, neutrons, and electrons. Use a different color for each.

2. Explain Does an object that is positively charged have more electrons than protons or fewer electrons than protons?

_____

_____

🅐 **Compare and Contrast** Use two quarter sheets of notebook paper to compare and contrast information about gaining and losing electrons.

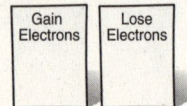

Gain Electrons | Lose Electrons

Picture This

3. Describe Look at the figure. How would you describe the chloride and sodium ions in the water? Circle your answer.

a. tightly held together
b. all the ions are negative
c. spread out evenly
d. all the ions are positive

Ions An atom can gain electrons. When it does, it becomes negatively charged. An atom also can lose electrons and become positively charged. A positively or negatively charged atom is an **ion** (I ahn).

How can electrons move in solids?

Electrons can move from atom to atom. They also can move from object to object. Rubbing is one way that electrons can move. Have you ever had clinging clothes when you took them out of the dryer? If so, you have seen what happens when electrons move from one object to another.

Imagine rubbing a balloon on your hair. The atoms in your hair hold their electrons more loosely than the atoms in the balloon. The electrons from the atoms in your hair move to the atoms on the surface of the balloon. So, your hair loses electrons and becomes positively charged. The balloon gains electrons and becomes negatively charged. Your hair and the balloon become attracted to one another. Your hair stands on end because of the static charge.

What is static charge?

Static charge is an imbalance of electric charge on an object. Static charge happens in solids because electrons move from one object to the other. Protons cannot easily move from the nucleus of an atom. So, protons usually do not move from one object to another.

How do ions move in solutions?

Sometimes, ions move instead of electrons. When ions move, the charge can move. Table salt, or sodium chloride, is made of sodium ions and chloride ions that are held in place and cannot move through the solid. Ions cannot move through solids, but they can move through solutions. Look at the figure. When salt is dissolved in water, the sodium and chloride ions break apart. The ions spread out evenly in the water and form a solution. In the solution, the positive and negative ions are free to move. Solutions that have ions make parts of your body able to communicate with each other. Nerve cells use ions to send signals to other cells. These signals move throughout your body so that you can see, touch, taste, smell, move, and even think.

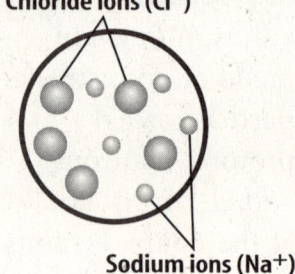

Chloride ions (Cl⁻)

Sodium ions (Na⁺)

Electric Forces

Remember that electrons in an atom swarm around the nucleus. What keeps the electrons close to the nucleus? The positively charged protons in the nucleus exert an attractive electric force on the negatively charged electrons. **Electric force** is the force between charged objects. All charged objects exert an electric force on each other. The electric force can attract or it can repel, or push away.

Look at the figure below. Objects with unlike charges, like positive protons and negative electrons, attract each other. Objects with like charges repel each other. Two positive objects repel each other. Two negative objects repel each other.

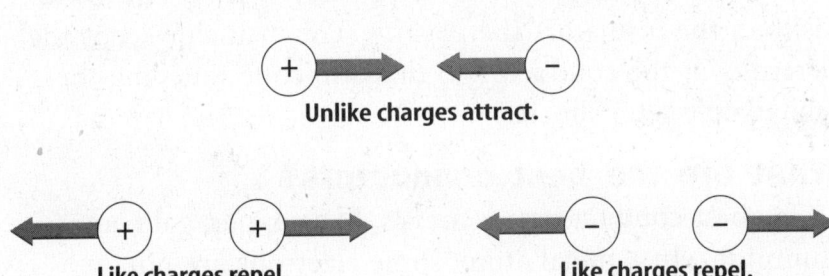

Unlike charges attract.

Like charges repel.   Like charges repel.

The electric force between two charged objects depends on the distance between them. Electric force also depends on the amount of charge on each object. The electric force between two charges gets stronger as the charges get closer together. As positive and negative charges come closer together, the attraction gets stronger. When two like charges come closer together, they repel each other more strongly. If the amount of charge on at least one object increases, then the electric force between the two objects increases. ☑

What are electric fields?

Charged objects don't have to touch each other to exert an electric force on each other. Imagine two charged balloons. They push each other apart even though they do not touch. Why does this happen?

Every electric charge has a space, or a field, around it. An **electric field** is the space in which charges exert a force on each other. If an object with a positive charge is placed in the electric field of another positive object, the objects repel each other. If an object with a negative charge is placed in the electric field of an object with a positive charge, the objects attract each other. Also the closer the objects are, the stronger the electric fields.

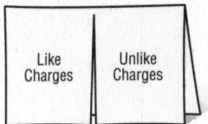

B Contrast Make the following Foldable to show the differences between like charges and unlike charges.

Like Charges   Unlike Charges

Picture This

4. **Apply** Circle the part of the figure that shows the force between two electrons.

☑ Reading Check

5. **Explain** What happens as a positively charged object comes closer to a negatively charged object?

_____

_____

_____

FOLDABLES™

C Contrast Make the following Foldable to show the differences between insulators and conductors.

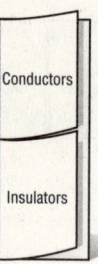

✓ Reading Check

6. **Determine** Give an example of each.

Conductor: _____

Insulator: _____

💡 Think it Over

7. **Explain** How do the extra electrons get on your hand?

_____

_____

_____

Insulators and Conductors

When you rub a balloon on your hair, the electrons from your hair move to the balloon. But, only the part of the balloon that is rubbed on your hair gains the electrons. Electrons cannot move easily through rubber. So, the electrons that move from your hair to the balloon stay in one place on the balloon. The balloon is an insulator. An **insulator** is a material in which electrons cannot move easily from place to place. Plastic, wood, glass, and rubber are examples of insulators.

A **conductor** is a material in which electrons can move easily from place to place. An electric cable is made from a conductor coated with an insulator, like plastic. The electrons move easily in the conductor (the wire), but do not move easily in the insulator (the plastic). The insulator keeps the electrons in the conductor so that someone touching the cable won't get a shock.

What are the best conductors?

The best conductors are metals, like copper, gold, and aluminum. In a metal atom, some electrons are not attracted as strongly to the nucleus as others. When metal atoms form a solid, the atoms cannot move far. But, the electrons inside the atoms that are not strongly attracted to each nucleus can move easily in a solid piece of metal. Insulators are different. In an insulator, the electrons of an atom are strongly attracted to the nucleus. The electrons in an insulator cannot move easily. ✓

Induced Charge

Have you ever walked on carpet and then touched a doorknob? Maybe you felt an electric shock or saw a spark. Look at the figure on the next page to see what happened.

As you walk, electrons rub off the carpet onto your shoes. The electrons spread over the surface of your skin. As your hand gets near the doorknob, the electric field around the extra electrons on your hand repels the electrons in the doorknob. The doorknob is metal, so it is a good conductor. The electrons on the doorknob move easily away from your hand. The part of the doorknob closest to your hand becomes positively charged. This separation of positive and negative charges because of an electric field is called an induced charge. The word induce means "to cause". You induced, or caused, a positive charge on the doorknob.

A As you walk across the floor, you rub electrons from the carpet onto the bottom of your shoes. These electrons then spread out all over your skin, including your hands.

B As you bring your hand close to the metal doorknob, electrons on the doorknob move as far away from your hand as possible. The part of the doorknob closest to your hand is left with a positive charge.

C The attraction between the electrons on your hand and the induced positive charge on the doorknob might be strong enough to pull electrons from your hand to the doorknob. You might see a spark or feel a mild electric shock.

If the electric field between your hand and the doorknob is strong enough, charge can be pulled quickly from your hand to the doorknob. The quick movement of extra charge from one place to another place is an **electric discharge**. Lightning is an example of an electric discharge. Imagine a storm cloud. The movement of air causes the bottom of the cloud to become negatively charged. The negative charge induces a positive charge on the ground below the cloud. Cloud-to-ground lightning strikes when electric charge moves between the cloud and the ground.

Grounding

Lightning is an electric discharge that can cause damage and hurt people. A lightning bolt releases a large amount of electric energy. Even electric discharges that release small amounts of electric energy can cause damage to electrical objects, like computers. One way to avoid damage caused by electric discharges is to make the extra charges flow into Earth's surface. Earth is a good conductor. Since it is so large, it can absorb, or take in, a large amount of extra charge. You may have seen a lightning rod at the top of a building. The rods are metal and are connected to metal cables. These cables conduct the electric charge into Earth if the rod is struck by lightning. So, the extra charge goes to Earth and the building is protected. ☑

Picture This

8. **Describe** Look at the figure. When do the electrons on the doorknob first begin to move away from your hand? Circle your answer.
   a. when you walk across the floor
   b. when your hand gets close to the doorknob
   c. when you induce a positive charge
   d. when you touch the doorknob

✔ Reading Check

9. **Determine** Is Earth an insulator or conductor?

After You Read

Mini Glossary

conductor: a material in which electrons can move easily from place to place

electric discharge: the quick movement of extra charge from one place to another place

electric field: the field, or space, in which charges exert a force on each other

electric force: the attraction or repulsion between charged objects

insulator: a material in which electrons cannot move easily from place to place.

ion: a positively or negatively charged atom

static charge: an imbalance of electric charge on an object

1. Read the key terms and definitions in the Mini Glossary above. What would cause the electric force between two objects to increase? Explain.

_____

_____

_____

2. The table below lists the charges of two objects. Use the words *attract* and *repel* to describe the electric force between the objects.

Charges of Two Objects	Electric Force
Positive and positive	
Positive and negative	
Negative and negative	

3. You were asked to highlight the main idea of each paragraph. Did this strategy help you learn about electric charge? Why or why not?

_____

_____

_____

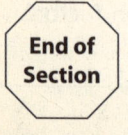
End of Section

Science Online Visit **ips.msscience.com** to access your textbook, interactive games, and projects to help you learn more about electric charge.

Electricity

section ⊇ Electric Current

● Before You Read

You can turn on the light in a room any time you wish. Where does the electricity come from?

_____

_____

● Read to Learn

Flow of Charge

Lights, refrigerators, TV's and other things need a steady source of electrical energy that can be controlled. Electric currents are used for steady and controlled electricity. An **electric current** is the flow of electric charge. In solids, the flowing charges are electrons. In liquids, the flowing charges are ions. Remember that ions can be positive or negative. Electric currents are measured in amperes (A). A model of an electric current is flowing water. Water flows downhill because a gravitational force acts on it. Electrons flow because an electric force acts on them.

What is a simple circuit?

The flow of water can create energy. Look at the figure. Water that is pumped high above the ground has potential energy because gravity acts on it. As water falls, it loses potential energy. When the water falls on a waterwheel and turns it, the waterwheel gains kinetic energy. The water flows through a continuous loop. A closed, conducting loop that electric charges flow continuously through is a **circuit**.

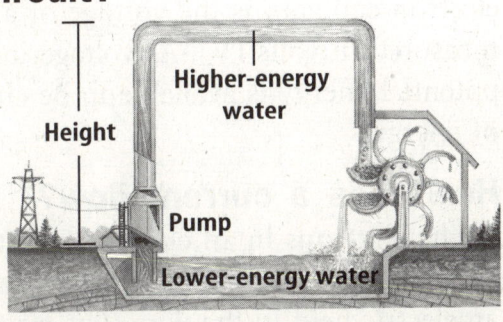

Height

Higher-energy water

Pump

Lower-energy water

What You'll Learn

- about electric currents and voltage
- how batteries make electric currents
- what electrical resistance is

Study Coach

Outline As you read the section, make an outline using each heading. Under each heading, write the main ideas that you read.

Picture This

1. **Highlight** Use a highlighter to mark the flow of water through the continuous loop.

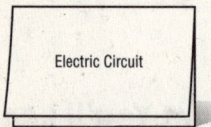

Picture This

2. Identify Circle the source of electric energy in the figure.

3. Identify In a circuit (like the one in the figure above), what increases the electrical potential energy of the electrons?

What are electric circuits?

The simplest electric circuit has a source of electric energy and an electric conductor. In the figure below, the source of electric energy is the battery. The electric conductor is the wire. The wires connect the lightbulb to the battery in a closed path. Electric current flows in the circuit as long as the wires, including the filament wire in the lightbulb, stay connected.

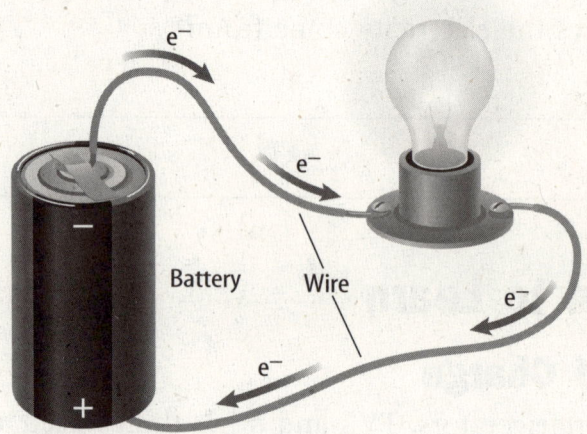

Battery   Wire

What is voltage?

Think of the example of the waterwheel. The pump increases the gravitational potential energy of the water by raising it from a lower level to a higher level. In an electric circuit, the battery increases the electrical potential energy of electrons. This electrical potential energy can be changed into other forms of energy. ☑

The measure of how much electrical potential energy each electron can gain is the **voltage** of a battery. Voltage is measured in volts (V). As voltage increases, more electrical potential energy is available to be changed into other forms of energy.

How does a current flow?

The electrons in an electric circuit move slowly. When the ends of a wire are connected to a battery, the battery makes an electric field in the wire. The electric field forces electrons to move toward the positive end of the battery.

Look at the plus sign on the battery in the figure on the previous page. As electrons move, they bump into other electric charges in the wire. Then, they bounce off in different directions. Electrons start to move again toward the positive end of the battery. An electron can have more than ten trillion of these bumps each second. So, it can take several minutes for an electron to travel even one centimeter.

How do batteries work?

Battery Terminals Batteries have a negative terminal, or end, and a positive terminal. You have learned that the battery in a circuit makes an electric field that forces electrons to move toward the positive terminal of the battery. When the positive and negative terminals of a battery are connected in a circuit, the electric potential energy of electrons in the circuit increases. As electrons move toward the positive terminal, the electric potential energy turns into other forms of energy. This also happens with water. The gravitational potential energy of the water turns into kinetic energy as the water turns the waterwheel.

Electrical Potential Energy The battery changes chemical energy into electric potential energy. In the figure on the previous page, the battery shown is an alkaline battery. Between the positive terminal and negative terminal is a moist paste. Chemical reactions in the moist paste move electrons from the positive terminal to the negative terminal. So, the negative terminal becomes negatively charged and the positive terminal becomes positively charged. This makes the electric field in the circuit that causes the electrons to move away from the negative terminal to the positive terminal. The chemical energy is now electrical potential energy.

How long can batteries last?

Batteries cannot supply energy forever. Do you know what happens if the lights on a car are left on for a long time? The car battery runs down and the car won't start. Why do batteries run down? Batteries have only a certain amount of chemicals in them that react to make chemical energy. As long as the battery is used, these chemical reactions happen. The chemicals change into other compounds. When the chemicals are used up, the chemical reactions stop. The battery is then "dead." ☑

Think it Over

4. Explain How does the negative terminal of a battery become negatively charged?

_____

_____

_____

Reading Check

5. Explain Why does a battery "die"?

_____

_____

_____

Resistance

Remember that electrons move more easily through conductors than through insulators. But, even in conductors, the flow of electrons can be slowed down. The measure of how difficult it is for electrons to flow through a material is **resistance**. The unit of resistance is the ohm (Ω). Insulators have a higher resistance than conductors.

You learned that, in a circuit, electrons bump into other electric charges. When this happens some of the electrical energy in the electrons turns into thermal energy in the form of heat or light. The amount of electrical energy that turns into heat and light depends on the resistance of the materials in the circuit.

Why are copper wires used in buildings?

The amount of electrical energy that turns into thermal energy increases when the resistance of the wire increases. Copper is one of the best conductors of electric energy. It also has a low resistance. So, less heat is made when an electric current flows through copper wire. Copper wire is used in houses and other buildings because copper wire usually will not become hot enough to cause fires.

How are length and thickness of a wire related to resistance?

A wire can have high or low electric resistance depending on what the wire is made of. The electric resistance of a wire also depends on the wire's length and thickness. The electric resistance of a wire increases as the wire becomes longer. The electric resistance also increases as the wire becomes narrower. ✔

How do lightbulbs work?

Lightbulbs have a tiny wire inside called a filament. The filament wire is so narrow that it has a high resistance. Remember that a material that has high resistance can turn electric energy into thermal energy in the form of heat or light. When electric current flows in the filament, the wire becomes hot enough to make light. Why doesn't the filament melt? The filament is made of tungsten metal. Tungsten has a much higher melting point than most other metals. So, the tungsten metal filament will not melt at the high temperature needed to make light.

Reading Check

6. **Recognize Cause and Effect** What increases when a wire is made longer or thinner?

_____

_____

💡 **Think it Over**

7. **Explain** Why is tungsten used for lightbulb filaments?

_____

_____

_____

● After You Read

Mini Glossary

circuit: a closed, conducting loop that electric charges flow continuously through

electric current: the flow of electric charge

resistance: the measure of how difficult it is for electrons to flow through a material

voltage: the measure of how much electrical potential energy each electron can gain.

1. Review the terms and their definitions in the Mini Glossary. Write one or two sentences to compare resistance in a conductor and an insulator.

   _____

   _____

2. Complete the graphic organizer to compare and contrast copper wire and tungsten wire using the information below.

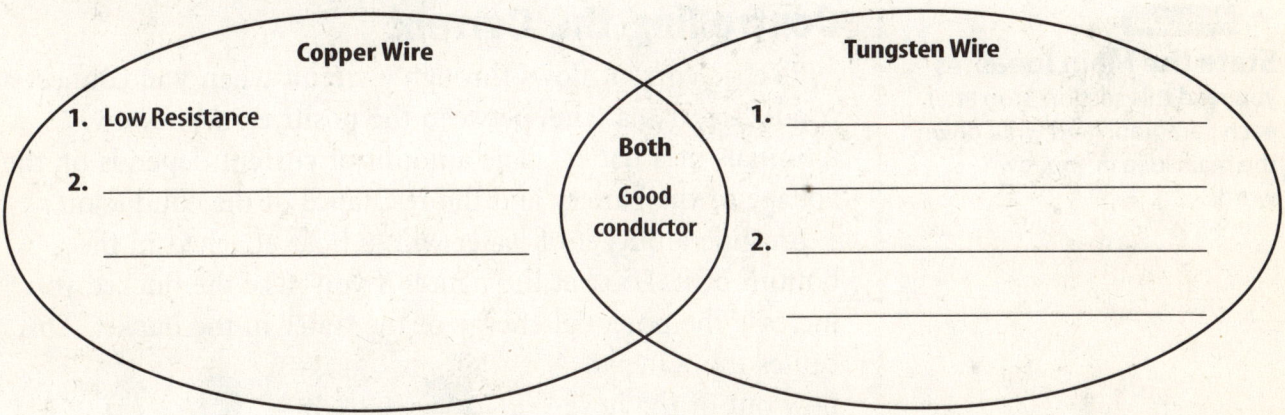

Copper Wire

1. Low Resistance

2. _____

   _____

Both

Good conductor

Tungsten Wire

1. _____

   _____

2. _____

   _____

3. At the beginning of the section, you were asked to create an outline of the section. How did the outline help you learn about electric current?

   _____

   _____

   _____

Science Online Visit **ips.msscience.com** to access your textbook, interactive games, and projects to help you learn more about electric current.

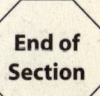

End of Section

Electricity

section ❸ Electric Circuits

What You'll Learn
- how voltage, current, and resistance are related
- about series and parallel circuits
- how to avoid dangerous electric shock

Study Coach

State the Main Ideas As you read this section, stop after each paragraph and write down the main idea in your own words.

Picture This
1. **Infer** Circle the bucket and hose that show greater resistance.

● Before You Read

You use circuits every day. Name some circuits you have used.

_____

_____

● Read to Learn

Controlling the Current

Electric current flows through a circuit when you connect a conductor, like a wire, between the positive and negative terminals of a battery. The amount of current depends on the voltage of the battery and the resistance of the conductor.

Imagine a bucket of water with a hose attached in the bottom of it. Look at the figure. If you raise the bucket, you increase the potential energy of the water in the bucket. This causes the water to flow out of the hose faster. This happens with electric current, too. If the amount of voltage increases, the amount of current flowing through a circuit will increase.

How do voltage and resistance affect current?

As the figure shows, the higher the bucket is raised, the more energy the water has. Increasing the voltage in a battery is like increasing the height of the water. The electric current in a circuit increases if the voltage increases. If the resistance in an electric circuit is greater, less current can flow through the circuit.

What is Ohm's law?

In the nineteenth century, a German scientist named Georg Simon Ohm measured how changing the voltage in a circuit affects the current. He found a relationship among voltage, current, and resistance in a circuit, know as Ohm's law. <u>Ohm's law</u> states that when voltage in a circuit increases, the current increases. The equation below shows this relationship.

Voltage (in volts) = **current** (in amperes) × **resistance** (in ohms)

$$V = IR$$

If the voltage in a circuit stays the same, but the resistance changes, the current will change, too. If the resistance increases, the current in the circuit will decrease.

Series and Parallel Circuits

Circuits control the movement of electric current by providing paths for electrons to follow. In order for a current to flow, the circuit must be an unbroken path. Imagine a string of lights with tiny light bulbs. In some strings of lights, if only one bulb is burned out, the whole string of lights won't work. This is an example of a series circuit. Some strings of lights will stay lit no matter how many bulbs burn out. This is an example of a parallel circuit.

What is a series circuit?

A <u>series circuit</u> is a circuit that has only one path for the electric current to follow. Look at the figure. If the path is broken, current cannot flow. The bulbs in the circuit will not light. The path could be broken if a wire comes off or if a bulb burns out. The filament in the lightbulb is also part of the circuit. So, if the filament breaks, then the flow of current stops.

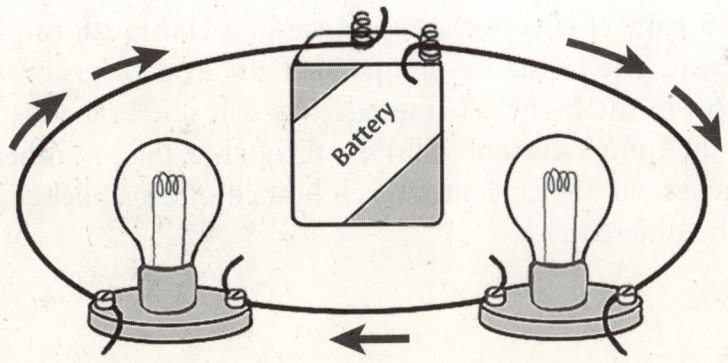

Applying Math

2. Calculate An iron is plugged in a wall socket. The current in the iron is 5 A. The resistance is 20 Ω. What is the voltage provided by the wall socket? Show your work.

_____

_____

_____

FOLDABLES™

E Organize Information
Use two half-sheets of notebook paper to write information about series and parallel circuits.

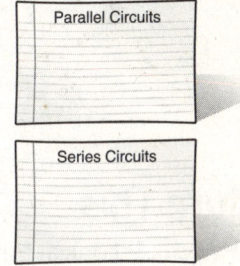

Picture This

3. Predict Look at the figure. What would happen if you remove a wire from one of the lightbulbs?

_____

_____

_____

What happens when resistance increases?

In a series circuit, electrical devices are connected along the same path. So, the current is the same through every device. But, if a new device is added to the circuit, the current will decrease throughout the circuit. Why? Each device has its own electrical resistance. In a series circuit, the total resistance increases as each new device is added. Ohm's law tells us that if resistance increases and the voltage doesn't change, the current will decrease. ☑

What is branched wiring?

What would it be like if all the electrical devices in your house were on a series circuit? You would have to turn on all the appliances in your house just so you could watch TV.

Parallel Circuits Your house, school, and other buildings are wired using parallel circuits. A **parallel circuit** is a circuit that has more than one path for the electric current to follow. The figure shows a parallel circuit. The circuit branches so that the electrons flow through each of the paths. If one of the paths is broken, electrons will still flow through the other paths. So, you can add or remove a device in one branch and the current will still flow.

Parallel Circuit

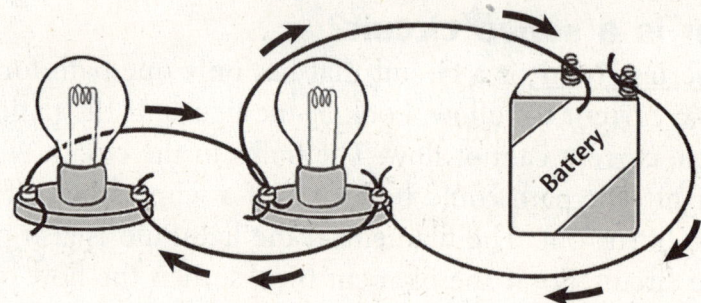

In a parallel circuit, the resistance in each branch can be different. The resistance in a parallel circuit depends on the devices in the branch. If the resistance in one branch is low, then more current will flow through it than in other branches. So, the current in each branch of a parallel circuit can be different.

Reading Check

4. Recognize Cause and Effect A series circuit has 2 lightbulbs on it. What happens to the resistance if you add another lightbulb to the circuit?

Picture This

5. Predict Look at the figure. What would happen to the lightbulb on the right if you remove the lightbulb on the left?

Protecting Electric Circuits

In a parallel circuit, electric current that flows out of a battery or electric outlet increases as more devices are added to the circuit. As the current through the circuit increases, the wires heat up.

What are fuses and circuit breakers?

If wires get too hot, they can cause a fire. To make sure that wires don't get too hot, the circuits in your house and other buildings have fuses or circuit breakers. Fuses and circuit breakers limit the amount of current in the wiring. If the current becomes greater than 15 A or 20 A, a piece of metal in the fuse melts or a switch in the circuit breaker opens, stopping the current. The device that caused the problem can be removed. Then, the fuse can be replaced or the circuit breaker can be reset.

Electric Power

When you use a toaster or a hair dryer, electrical energy changes into other kinds of energy. The rate, or speed, at which electrical energy is changed into other kinds of energy is **electric power**. In any electric device or electric circuit, the electric power that is used can be found by using the equation below.

Power (in watts) = **current** (in amperes) × **voltage** (in volts)

$$P = IV$$

The electric power is equal to voltage provided to the electrical device multiplied by the current that flows into the device. The SI unit of power is the watt. The table lists the electric power used by some common devices.

Power Used by Common Devices	
Device	Power (in watts)
Computer	350
Color TV	200
Stereo	250
Refrigerator	450
Microwave	700–1,500
Hair dryer	1,000

Think it Over

6. **Explain** What do fuses and circuit breakers do?

_____

_____

_____

Applying Math

7. **Calculate** A toaster is plugged into a wall outlet. The current in the toaster is 10 A. The voltage of the wall outlet is 110 V. How much power in watts does the toaster use? Show your work.

_____

_____

_____

Applying Math

8. **Interpret Data** How many more watts does a hair dryer use than a color TV?

_____

How do electric companies measure power?

Power is the amount of energy that is used per second. When you use a hair dryer, the amount of electrical energy you use depends on the power of the hair dryer. It also depends on how long you use it. Suppose you used the hair dryer for 10 minutes today and 5 minutes yesterday. You used twice as much energy today than you did yesterday.

How much does electrical energy cost?

Electric companies make electrical energy and sell it in units of kilowatt-hours. One kilowatt-hour (kWh) is equal to using one kilowatt of power continuously for one hour. This is about the amount of energy needed to light ten 100-W lightbulbs for one hour or just one 100-W lightbulb for 10 hours.

An electric company charges customers for the number of kilowatt-hours they use every month. An electric meter on the outside of each building measures the number of kilowatt-hours used in that building.

Electrical Safety

Electricity can be very dangerous. In 1997, electric shocks killed about 490 people in the United States. Here are some tips that will help prevent electrical accidents.

Preventing Electric Shock
Never use a device with frayed or damaged electric cords.
Unplug appliances before you work on them. For example, if a piece of toast gets stuck in a toaster, unplug the toaster before you take the toast out.
Never use an electric device near water.
Never touch power lines with anything, including a kite string or ladder.
Always pay attention to warning signs and labels.

How do electric shocks happen?

If an electric current enters your body, you feel an electric shock. Your body is like a piece of insulated wire. The fluids inside your body are good conductors of electric current. The electrical resistance of dry skin is much higher than the fluids in your body. Skin insulates the body in the same way that plastic insulates a copper wire. Remember that electrons cannot move easily in an insulator like plastic. Your skin works in the same way. ☑

Picture This

9. Infer Why should you not use an electric device near water?

_____

_____

_____

☑ **Reading Check**

10. Identify Is skin a conductor or an insulator?

_____

You actually become part of an electric circuit when current enters your body. The shock you feel can be mild or deadly, depending on the amount of current that flows into your body.

How much is too much?

The amount of current that can light a 60-W lightbulb is about 0.5 A. If this amount of current enters your body, it could be deadly. Even a current as low as 0.001 A can be painful. The table shows what you would feel when a certain amount of electric current flows through your body.

Current's Effects	
Amount of Current (in amperes)	**What You Feel**
0.0005 A	Tingle
0.001 A	Pain
0.01 A	Can't let go
0.025 A	
0.05 A	Difficult to breathe
0.10 A	
0.25 A	
0.50 A	Heart failure
1.00 A	

Picture This

11. Interpret Data Describe how you would feel if you were shocked by a current of 0.10 A.

_____

_____

_____

How do you keep safe from lightning?

Electricity in lightning can be very dangerous. Lightning can harm people, plants, and animals. In the United States, more people are killed every year by lightning than by hurricanes or tornadoes. Most of these lightning deaths happened outdoors. If you are outside and can see lightning or hear thunder, you need to go indoors right away. If you cannot go indoors, you need to take the following steps:

- Stay away from open fields and high places

- Stay away from tall objects like trees, flagpoles, or light towers

- Stay away from objects that conduct current such as water, metal fences, picnic shelters, and bleachers. ☑

✔ **Reading Check**

12. Explain Why should you stay away from metal fences when you see lightning or hear thunder?

_____

_____

_____

● After You Read

Mini Glossary

electric power: the rate, or speed, at which electrical energy is changed into other kinds of energy

Ohm's law: the relationship among voltage, current, and resistance; when the voltage in a circuit increases, the current increases

parallel circuit: a circuit that has more than one path for the electric current to follow

series circuit: a circuit that has only one path for the electric current to follow

1. Review the terms and their definitions in the Mini Glossary. Explain why it is better to have a parallel circuit in your home than a series circuit.

   _____

   _____

   _____

2. Explain the main ideas of Ohm's law in the cause-and-effect map below. Write *increases* or *decreases* in the blanks.

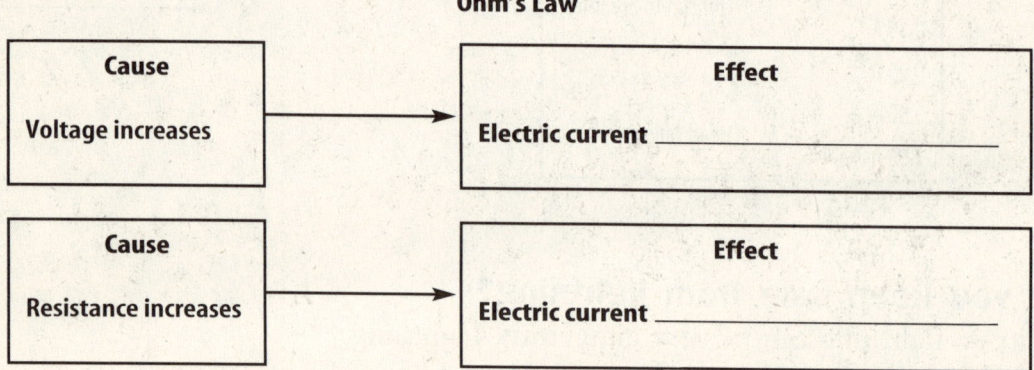

Ohm's Law

Cause		Effect
Voltage increases	→	Electric current _____
Resistance increases	→	Electric current _____

3. You were asked to write the main idea of each paragraph as you read this section. How did you decide which is the main idea for each paragraph?

   _____

   _____

   _____

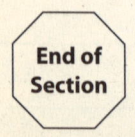

End of Section

Science Online Visit **ips.msscience.com** to access your textbook, interactive games, and projects to help you learn more about electric circuits.

Magnetism

section ❶ What is magnetism?

● Before You Read

Do you have magnets on your refrigerator? Why do magnets stick to a refrigerator and other things?

_____

_____

_____

● Read to Learn

Early Uses

Thousands of years ago, people found that a mineral called magnetite attracted other pieces of magnetite. It also attracted bits of iron. When they rubbed small pieces of iron with magnetite, the iron began to act like magnetite. If they let the pieces turn freely, one end pointed north. These might have been the first compasses. Compasses helped sailors and explorers know which direction they were going. Before compasses, sailors and explorers had to look at the Sun and stars to know which direction they were going.

Magnets

A piece of magnetite is a magnet. Magnets attract objects made of iron or steel, like nails and paper clips. Magnets also attract or repel other magnets. To repel means "to push away." Every magnet has two ends. The two ends are called poles. One end is called the north pole. The other end is called the south pole. The figure on the next page shows what happens when you put two magnetic poles together. Two north poles will repel each other. Two south poles also repel each other. But a north pole and a south pole are attracted to each other. ☑

What You'll Learn
■ how magnets behave
■ how the behavior of magnets and magnetic fields are related
■ why some materials are magnetic

Study Coach

Create an Outline Use the headings to make an outline of the information in this section.

✔ Reading Check

1. **Determine** Do the north poles of two magnets attract or repel each other?

_____

_____

Picture This

2. Highlight Use one color to circle the poles that are attracted to each other. Then use another color to circle the poles that repel each other.

Poles That Repel or Attract

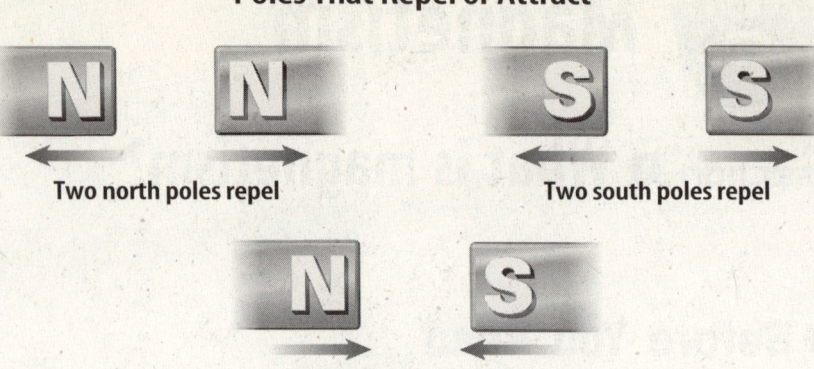

Two north poles repel   Two south poles repel

Opposite poles attract

What is a magnetic field?

Remember that a force is a push or a pull that can make an object move. Gravitational and electric forces can act on an object even when objects are not touching. Magnetic force also can act on objects when they are not touching. Notice that the magnets in the figure above are not touching and a magnetic force is acting on them. A magnet can even make an object move without touching it. The magnetic force gets weaker when the magnets move farther apart.

A **magnetic field** is the space around a magnet where the magnetic force is. Magnetic fields are around all magnets. If you sprinkle iron fillings near a magnet, the iron filings will show the magnetic field lines of the magnet. The figure below shows these curved lines. The lines start on one pole and end on the other.

Picture This

3. Highlight Trace the magnetic field lines as they leave the magnet. Which pole of the magnet did you always start at?

_____

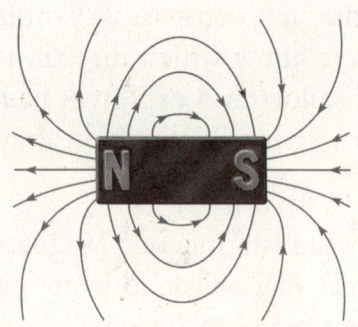

Magnetic field lines begin at a magnet's north pole and end at the south pole. The lines are close together where the field is strong. The lines get farther apart as the field gets weaker. The magnetic field is strongest close to the magnetic poles. It gets weaker farther away from the poles. Field lines that curve toward each other show attraction. Field lines that curve away from each other show repulsion.

How are magnetic fields made?

A moving electric charge produces a magnetic field. All atoms have negatively charged particles called electrons. These electrons spin around the nucleus of an atom. Each electron produces a magnetic field because of how it moves. The electrons in atoms that make up magnets are like even smaller magnets. The magnetic fields of many of the atoms in iron and other materials point in the same direction. A group of atoms with their magnetic fields pointing in the same direction is called a **magnetic domain**. ☑

How can some materials become magnetized?

A material, like iron or steel, that can become magnetized has many magnetic domains. When the material is not magnetized, these magnetic domains point in all directions as shown in the figure below. The material does not act like a magnet. This is because the magnetic fields made by the domains cancel each other out.

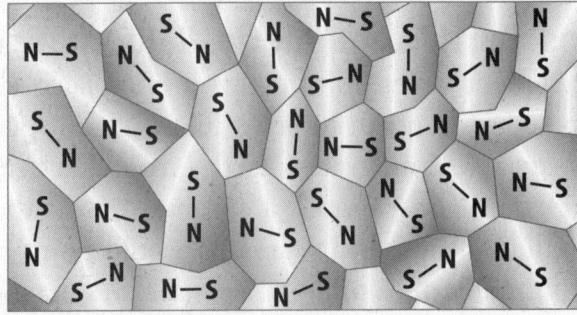

A magnet has a large number of magnetic domains that are lined up and pointing in the same direction. Suppose you hold a strong magnet next to a piece of iron. The magnet causes the magnetic field in many of the magnetic domains in the iron to line up with the magnet's field, as in the figure below. The magnetic fields of the iron's magnetic domains are added together. This magnetizes the iron.

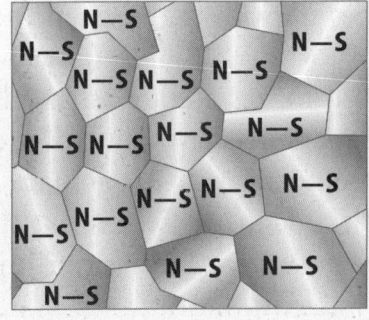 

✔ **Reading Check**

4. **Identify** Do the magnetic fields in a magnetic domain all point in the same direction or in different directions?

_____

_____

Picture This

5. **Explain** How do you know that the material in this picture is not magnetized?

_____

_____

_____

Picture This

6. **Label** Which pole of the bar magnet on the right should be closest to the figure on the left? Label this pole of the bar magnet N for north or S for south.

7. **Explain** How does Earth's magnetosphere protect Earth from charged particles from the Sun?

_____

_____

_____

Earth's Magnetic Field

Bar magnets are not the only objects that have magnetism. Earth has a magnetic field, too. The space affected by Earth's magnetic field is the **magnetosphere** (mag NEE tuh sfihr). The magnetosphere repels most of the charged particles from the Sun. Earth's magnetic field probably comes from deep within Earth's core. Moving melted iron in the outer core might produce the magnetic field. The shape of Earth's magnetic field is like the magnetic field of a huge bar magnet. ☑

What are magnets found in nature?

Some animals, including honeybees, rainbow trout, and homing pigeons, use magnetism to find their way. They have tiny pieces of magnetite in their bodies. These pieces are so small that they might contain only one magnetic domain. Scientists have shown that some animals use these natural magnets to find Earth's magnetic field. They use Earth's magnetic field and the position of the Sun or stars to help them find their way.

How does Earth's magnetic field change?

Earth's magnetic poles do not stay in one place. The magnetic pole in the north today is in a different place than it was 20 years ago. Sometimes, Earth's magnetic field also changes direction. For example, a compass needle that pointed south 700 thousand years ago would point north today. During the last 20 million years, Earth's magnetic field has changed direction more than 70 times. The magnetism of old rocks shows these changes in the magnetic field. When some kinds of molten rock cool, magnetic domains of iron in the rock line up with Earth's magnetic field. After the rock cools, the domains are frozen in place. So, the old rocks show the direction of Earth's magnetic field as it was long ago.

What is a compass needle?

A compass needle is a small bar magnet. It has a north and a south magnetic pole. When a compass is in a magnetic field, the needle turns until it lines up with the magnetic field line at its location.

Earth's magnetic field also makes a compass needle turn. The north pole of the compass needle points toward Earth's magnetic pole that is in the north which is actually a magnetic south pole. Earth's magnetic field is like that of a bar magnet with the south pole near Earth's north pole.

Think it Over

8. **Infer** If Earth's magnetic field is like a bar magnet, where is the north pole of the bar magnet?

_____

_____

● After You Read

Mini Glossary

magnetic domain: a group of atoms with their magnetic fields pointing in the same direction

magnetic field: the space around a magnet where the magnetic force is

magnetosphere: the space affected by Earth's magnetic field

1. Review the terms and their definitions in the Mini Glossary above. Circle two of the terms. On the lines below, tell how these two terms are related.

_____

_____

_____

2. Complete the flowchart below to describe how the magnetic domains of a paper clip change as it becomes magnetized.

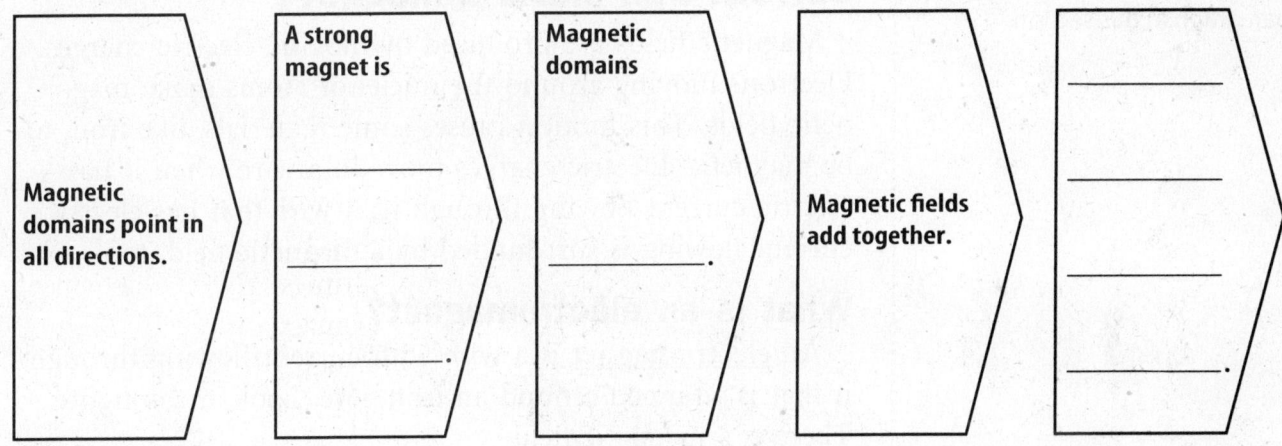

Magnetic domains point in all directions.

A strong magnet is _____ _____ _____.

Magnetic domains _____.

Magnetic fields add together.

_____ _____ _____ _____.

3. You were asked to make an outline of the section. How can you use the outline to help you study for a quiz?

_____

_____

_____

Science Online Visit **ips.msscience.com** to access your textbook, interactive games, and projects to help you learn more about magnetism.

End of Section

chapter 21

Magnetism

section ❷ Electricity and Magnetism

Mark the Text

Identify Main Ideas
Highlight the main idea of each paragraph in this section.

● Before You Read

Electricity makes your radio and other things work. Where does electricity come from?

_____

_____

● Read to Learn

Current Can Make a Magnet

Magnetic fields are produced by moving electric charges. Electrons moving around the nuclei of atoms make magnetic fields. This motion causes some materials, like iron, to be magnetic. Electric charges move in a wire when it has electric current flowing through it. A wire that has electric current flowing is surrounded by a magnetic field, too.

What is an electromagnet?

An **electromagnet** is a wire with current flowing through it that is wrapped around an iron core. Look at the figure. There is a magnetic field around each coil of wire. The magnetic fields add together to make a stronger magnetic field inside the coil. When the coils are wrapped around an iron core, the magnetic field of the coils makes the iron core magnetic. This makes the magnetic field inside the coils even stronger.

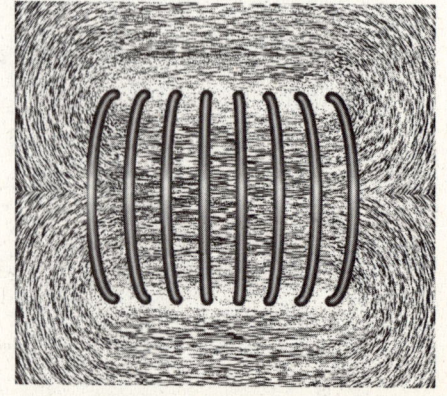

When a wire is wrapped in a coil, the field inside the coil is made stronger.

Picture This

1. **Draw** a line that shows the magnetic field inside the coils.

How do electromagnets work?

When an electric current is turned on, the magnetic field of an electromagnet is turned on. When the electric current is turned off, the magnetic field turns off, too. By changing the current, the strength and direction of the magnetic field of an electromagnet can be changed. This makes electromagnets useful.

How are electromagnets used?

The figure shows a doorbell that uses an electromagnet. When a button is pressed, a switch in a circuit that includes an electromagnet is closed. The magnet attracts an iron bar. There is a hammer attached to the iron bar. The hammer hits the bell. When it hits the bell, it has moved far enough to open the circuit again. The electromagnet loses its magnetic field. A spring pulls the iron bar and hammer back into place. This closes the circuit. This happens again and again as long as the button is pushed.

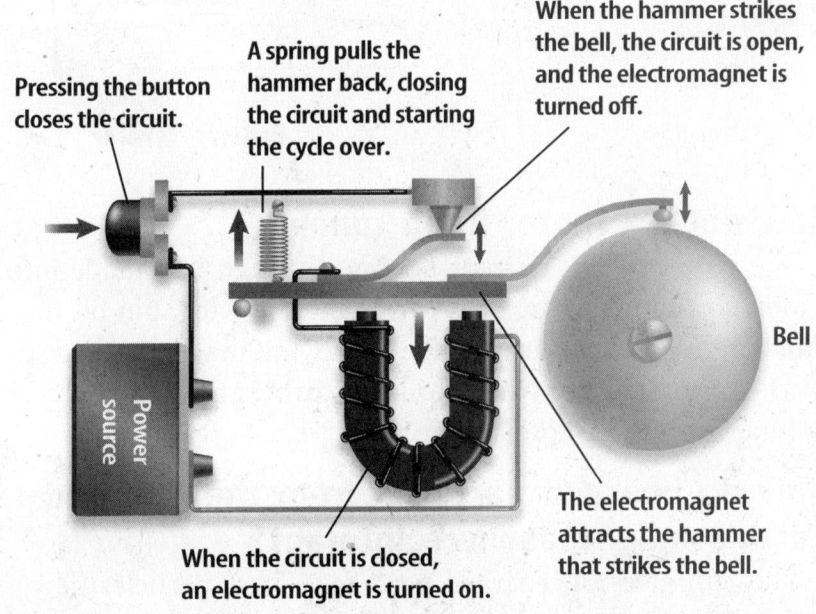

Pressing the button closes the circuit.

A spring pulls the hammer back, closing the circuit and starting the cycle over.

When the hammer strikes the bell, the circuit is open, and the electromagnet is turned off.

Power source

When the circuit is closed, an electromagnet is turned on.

The electromagnet attracts the hammer that strikes the bell.

Bell

Magnets Push and Pull Currents

Think of an electric device that produces motion, like a fan. How does the electric energy going into the fan change into kinetic energy? Remember that wires with electric current flowing through them produce a magnetic field. This magnetic field acts the same way as a magnetic field made by a magnet. Two wires that are carrying current in the same direction can attract each other as if they were two magnets.

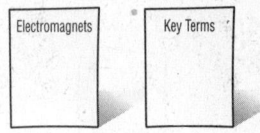
Picture This

2. **Identify** Circle the electromagnet in the figure.

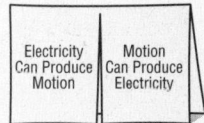

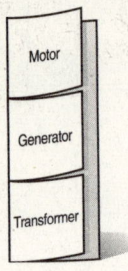

FOLDABLES™

C **Compare and Contrast**
Make the following Foldable to help you compare and contrast motors, generators, and transformers.

Motor

Generator

Transformer

Picture This

3. Analyze Look at the figure on the right. What would happen to the loop if there were no current running through it?

What is an electric motor?

Two magnets exert a force on each other. So do two wires that have current flowing through them. The magnetic field around a wire that has current flowing through it causes it to be pushed or pulled by a magnet.

Look at the first part of the figure below. The magnetic field will be pushed or pulled, depending on the direction the current is flowing through the wire. A magnetic field like the one shown will push a current-carrying wire upward. So, some of the electric energy carried by a current is changed into kinetic energy of the moving wire. Any machine that changes electric energy into kinetic energy is a **motor**.

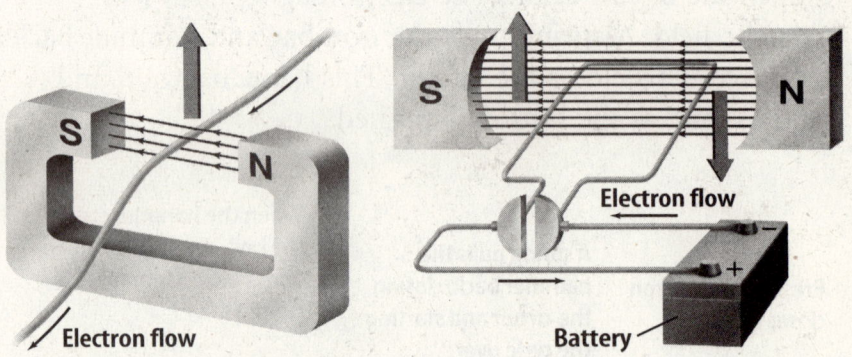

How does a motor keep running?

The wire that has current flowing through it is made into a loop so the magnetic field can make the wire spin all the time. In the second part of the figure, the magnetic field exerts a force on the wire loop. This causes the loop to spin as long as current flows in the loop.

How do charged particles from the Sun and Earth's magnetosphere interact?

The Sun gives off charged particles. These particles flow through the solar system like a huge electric current. Earth's magnetic field pushes and pulls on the electric current made by the Sun. This is just like how a magnetic field pushes and pulls on a wire that is carrying current. This pushing and pulling causes most of the charged particles from the Sun to be repelled. The charged particles do not hit Earth. This protects living things on Earth from damage that might be caused by the charged particles. The solar current also pushes on Earth's magnetosphere. It stretches the magnetosphere away from the Sun.

What is the aurora?

Sometimes the Sun gives off a great number of charged particles all at once. Earth's magnetosphere repels most of these charges. But some of the particles from the Sun produce other charged particles in Earth's outer atmosphere. These charged particles move along Earth's magnetic field lines. They move toward Earth's magnetic poles. At the poles, they crash into atoms in the atmosphere. These crashes cause the atoms to give off light. The light given off from the Sun's charged particles crashing into atoms in Earth's atmosphere is the **aurora** (uh ROR uh). In northern parts of the world, the aurora is called the northern lights.

Using Magnets to Create Current

In an electric motor, a magnetic field turns electricity into motion. A machine that uses a magnetic field to turn motion into electricity is a **generator.** In a motor, electric energy is changed into kinetic energy. In a generator, kinetic energy is changed into electric energy.

The figures below show how current can be produced in a wire that moves in a magnetic field. As the wire moves, the electrons in the wire also move in the same direction. If a wire is pulled downward through a magnetic field, the electrons in the wire also move downward. This is shown in the figure on the left.

The magnetic field exerts a force on the moving electrons. This force pushes the electrons along the wire in the figure on the right. This produces an electric current.

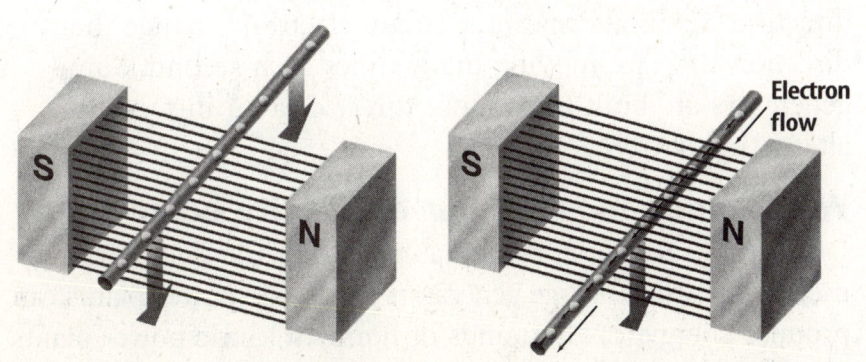

Electron flow

💡 **Think it Over**

4. **Explain** Would there be an aurora if Earth's magnetosphere repelled all of the electric charges from the Sun? Why or why not?

_____

_____

_____

Picture This

5. **Highlight Arrows** In the figure on the right, highlight the arrows that show the direction of electric current flow in the wire.

How does an electric generator work?

To produce electric current, the wire is made into a loop. Look at the figure below. A power source provides the kinetic energy to spin the wire loop. Every time the loop makes a half turn, the current in the loop changes direction. This makes the current change from positive to negative.

A current that changes direction is an **alternating current** (AC). To alternate means to switch back and forth. In the United States, electric currents change from positive to negative to positive 60 times each second.

Electric Generator

Power source turns loop

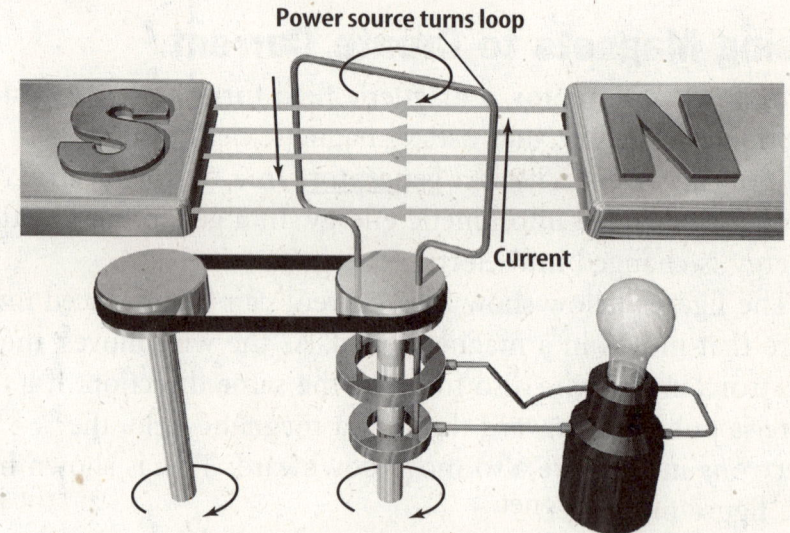

Current

What is a direct current?

A battery produces direct current instead of alternating current. In a **direct current** (DC), electrons flow in only one direction. In an alternating current, electrons change the direction they are moving many times each second. Some generators are built to produce direct current instead of alternating current.

Where does most of our electricity come from?

Electric generators produce almost all of the electric energy used in the world. Large generators in electric power plants can produce energy for thousands of homes. Electric power plants use different energy sources like gas, coal, and water to provide the kinetic energy needed to turn the coils of wire in a magnetic field. Coal-burning power plants are the most common. In the United States, coal-burning plants produce more than half of the electric energy made by power plants. ☑

What voltage do power plants transmit?

Electric energy made in power plants is carried to your home in wires. Remember that voltage is how much energy the electric charges in a current are carrying. Power lines from power plants send out electric energy at a high voltage of about 700,000 V. At a high voltage, less energy is changed into heat in the wires. But, high voltage is not safe to use in homes and businesses. So, the voltage must be reduced. ☑

Changing Voltage

A machine that changes the voltage of an alternating current without losing much energy is a **transformer**. Some transformers increase the voltage before sending out an electric current through the power lines. Other transformers decrease the voltage so the energy can be used in homes and businesses. Transformers also are used in power adaptors. Adaptors are used with devices that can be plugged into a wall outlet or can run on batteries. The adaptor changes the 120 V from the wall outlet to the same voltage that the batteries produce.

A transformer usually has two coils of wire wrapped around an iron core. One wire coil is connected to an alternating current source. The current produces a magnetic field in the iron core, just like in an electromagnet. The magnetic field it produces switches direction because the current is alternating. This alternating magnetic field causes an alternating current in the other wire coil.

How does a transformer change voltage?

A transformer increases or decreases the input voltage depending on the number of coils it has on each side. The number of coils on the output side divided by the number of coils on the input side equals the output voltage divided by the input voltage.

Look at the figure of a transformer. There are three coils on the input side. There are nine coils on the output side. 9 ÷ 3 = 3. The answer 3 can be used to find the output voltage of the transformer. If the voltage going into the transformer is 60 V, the output would be found by multiplying 60 × 3. The output voltage would be 180 V. If the input side has more coils, the transformer decreases the voltage. If the output side has more coils, the transformer increases the voltage.

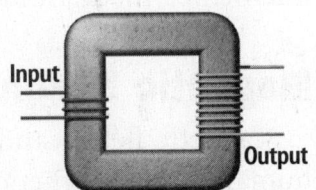

Input

Output

✔ **Reading Check**

8. **Explain** Why are transformers needed to decrease the voltage of electricity before it gets to homes and businesses?

_____

_____

_____

Picture This

9. **Analyze** Is the transformer in the figure increasing or decreasing voltage?

_____

Think it Over

10. **Compare and Contrast** How is a superconductor different from a conductor?

_____

_____

Superconductors

Electric current can flow easily through materials, like metals, that are electrical conductors. But even in conductors there is some resistance to the flow of current. Some of the electric current is changed into heat. This happens when electrons bump into atoms in the material.

A superconductor is a material that has no resistance to the flow of electrons. Superconductors are made when certain materials are cooled to low temperatures. For example, aluminum becomes a superconductor at about −272°C. No electric energy is changed into heat when electric current flows through a superconductor. So, no heat is made.

How does a superconductor affect a magnet?

Superconductors have other properties. For example, a superconductor repels a magnet. When a magnet gets close to a superconductor, the superconductor produces a magnetic field that is opposite to the field of the magnet. The field produced by a superconductor can cause a magnet to float above the superconductor.

How are superconductors used?

Superconductor materials can be used to produce very strong magnetic fields. If you make the wire of an electromagnet from superconductor material, the electromagnet will produce a very strong magnetic field. A particle accelerator is a machine that uses more than 1,000 superconducting electromagnets. A particle accelerator is used to speed up subatomic particles to nearly the speed of light.

Other uses for superconductors are being studied. If power lines were made from superconductors, they could carry electric current over long distances and not change electric energy into heat. Very fast computers could be built with microchips made from superconductor materials. ✓

Magnetic Resonance Imaging

Magnetic fields can be used to look at the inside of the human body. Magnetic resonance imaging, or MRI, uses magnetic fields to make images of the inside of a human body. MRI images can show if tissue is damaged. It also can show if there are tumors growing in the body.

Inside an MRI machine is an electromagnet made of a superconductor. The magnetic field is more than 20,000 times stronger than Earth's magnetic field.

How does an MRI make pictures?

About 63 percent of all the atoms in your body are hydrogen atoms. The nucleus of a hydrogen atom is a proton. The proton acts like a tiny magnet. The strong magnetic field inside the MRI tube makes all the hydrogen protons line up in the direction of the magnetic field. Then radio waves are applied to the part of the body being looked at. The protons absorb some of the energy in the radio waves. When this happens, the protons change the direction in which they are lined up.

When the radio waves are turned off, the protons go back to where they were. They line up with the magnetic field again and give off the energy they took in. How much energy they give off depends on the kind of tissue in the body. A computer uses the energy to make an image, like the one below. ✓

Copyright © Glencoe/McGraw-Hill, a division of The McGraw-Hill Companies, Inc.

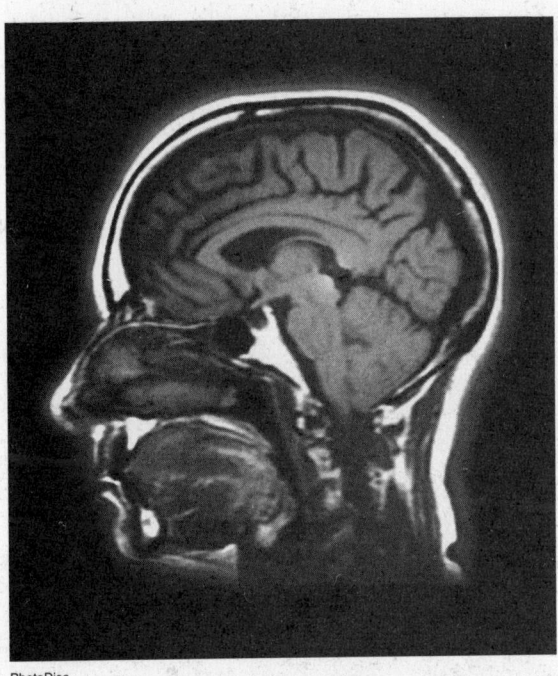

PhotoDisc

How are electric charges and magnets related?

Moving electric charges produce magnetic fields. Magnetic fields exert forces on moving electric charges. Together, these make electric motors and generators work.

✔ **Reading Check**

12. Explain What does the computer use to make an MRI image?

_____

_____

_____

Picture This

13. Identify What part of the body is shown in the upper part of the MRI image?

_____

● After You Read

Mini Glossary

alternating current: a current that changes direction

aurora: the light given off from the Sun's charged particles crashing into atoms in Earth's atmosphere

direct current: a current in which electrons flow in only one direction

electromagnet: a wire with current flowing through it that is wrapped around an iron core

generator: a machine that uses a magnetic field to turn motion into electricity

motor: any machine that changes electric energy into kinetic energy

transformer: a machine that changes the voltage of an alternating current without losing much energy

1. Review the terms and their definitions in the Mini Glossary. Describe how a generator and a motor can be used together to make kinetic energy from a magnetic field.

_____

_____

2. Match each machine with the description of what the machine does. Write the letter of each machine in Column 2 on the line in front of the description in Column 1.

Column 1	Column 2
_____ 1. turns motion into electricity	a. particle accelerator
_____ 2. uses magnetic fields to make images of the body	b. generator
_____ 3. moves electricity without making heat	c. motor
_____ 4. speeds up subatomic particles	d. MRI
_____ 5. changes the voltage of an alternating current	e. transformer
_____ 6. makes kinetic energy	f. superconductor

3. You were asked to highlight the main idea of each paragraph. How did you decide what the main ideas were?

_____

_____

_____

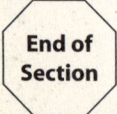
End of Section

Science Online Visit **ips.msscience.com** to access your textbook, interactive games, and projects to help you learn more about electricity and magnetism.

chapter 22 Electronics and Computers

section 1 Electronics

Before You Read

Describe a way you send information to your friends.

____

____

What You'll Learn
- how analog and digital signals compare
- how semiconductors are used

Read to Learn

Electronic Signals

You use electricity all the time. When you get ready to watch a movie, you turn on the TV, put a video in the VCR, and turn off the lamp. The TV, VCR, and lamp all use electricity. The TV and the VCR are electronic devices. An electronic device uses electricity to store, process, and transfer information.

The videotape has information recorded on it. As the videotape moves through the VCR, it produces a changing electric current. This current is the information the VCR uses to send signals to the TV. The TV uses the signals to make pictures and sounds.

A changing electric current that carries information is an **electronic signal**. The information is used to make sounds, pictures, printed words, or numbers. For example, a changing electric current causes a speaker to make sounds. If the current did not change, the speaker would not make any sounds. There are two types of electronic signals—analog and digital.

What is an analog signal?

Most TVs, VCRs, radios, and telephones use information that is in the form of analog electronic signals. An **analog signal** is a signal that changes smoothly over time. In an analog signal, the electric current increases or decreases smoothly over time.

Mark the Text

Locate Information Highlight in one color all the information about analog signals. Highlight in another color all the information about digital signals.

FOLDABLES™

A Construct a Venn Diagram Make the following Foldable to compare and contrast analog and digital signals.

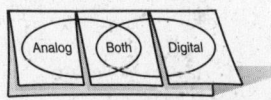

Other Analog Signals Electronic signals are not the only types of analog signals. An analog signal can be made by something that changes in a smooth, continuous way and contains information. A person's temperature changes smoothly and also contains information about a person's health.

What are examples of analog devices?

Look at the figure at the bottom of the page. Does your classroom have a clock like the first one in the figure? It is an analog device. The clock hands move smoothly from number to number all day long. A thermometer filled with a liquid is also an analog device. The liquid rises and falls smoothly as the temperature changes. ✔

A magnetic tape recorder is an analog device. When you record sounds, such as a song, on a tape recorder, it stores an analog signal of those sounds on a magnetic tape. When you play the tape, the tape recorder changes the analog signal to an electric current. This electric current changes smoothly over time and causes a speaker to vibrate. The vibrations make the sounds you hear.

What are digital signals?

Some electronic devices, such as CD players, use a type of electronic signal called a digital signal. A **digital signal** does not change smoothly, but changes in jumps or steps. If each jump in the signal is shown by a number, then a digital signal can be shown by a series of numbers.

Look at the digital clock in the figure below. It shows the time as numbers. The numbers change from 3:29 to 3:30 in one jump. The time on a digital clock does not change smoothly as on the analog clock. Digital thermometers also show temperature as a number. When the temperature changes, the number changes in one jump.

✔ **Reading Check**

1. **Explain** Why is a thermometer an analog device?

_____

_____

_____

Picture This

2. **Think Critically** Which of the clocks shown in the figure would be better to time something that takes less than one minute? Why?

_____

_____

_____

Can an analog signal be changed to a digital signal?

An analog signal can be changed to a digital signal. Suppose you take readings from an analog thermometer outside your house. Each hour you write down the temperature and the time you took the reading. Then, you record the time and temperature information on a graph. Your graph might look like the one below.

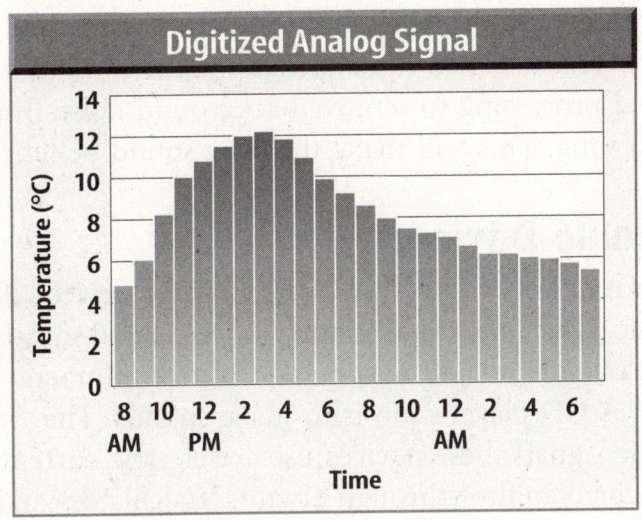

Picture This

3. Use Graphs What is the largest temperature-change step shown in the graph?

_____

The graph shows the temperature changing in steps, so this is now a digital signal. You have taken an analog signal and changed it to a digital signal.

How is an analog signal changed to a digital signal?

The temperature was recorded every hour to make the graph above. The analog signal was turned into a digital signal using a process called sampling. Sampling means the signal is repeatedly read and recorded at certain times, such as every minute or every hour. The smoothly changing analog signal is changed to a series of numbers. This series of numbers is a digital signal.

What is digitization?

The process of changing an analog signal to a digital signal is called digitization. You learned that a song recorded on a magnetic tape is an analog signal. The analog signal can be changed to a digital signal by sampling. In this way, a song can be shown as a series of numbers. ☑

☑ **Reading Check**

4. Explain What is one way an analog signal can be digitized?

_____

How are digital signals stored?

Why would you want to digitize an analog signal? After all, some information is lost in the process. Think about how analog and digital signals are stored. Suppose a song that is stored as an analog signal on a cassette tape is digitized. The digital signal is stored as a series of numbers. It might take millions of numbers to digitize one song. How can all these numbers be stored? A computer is an electronic device that can store all these numbers easily.

A computer stores a digital signal as a series of numbers. The computer uses mathematical formulas to change these numbers. This is called signal processing. A computer can use signal processing to remove background noise from a digitized song. This will make the song sound better.

Electronic Devices

Calculators and CD players are electronic devices. An electronic device uses information in electronic signals to do a job. A calculator's job may be to add two numbers together. A CD player's job is to make sounds. The electronic signals these devices use are electric currents. The electric currents flow through circuits. Calculators and VCRs may contain hundreds or thousands of electric circuits.

What are electronic components?

Electric circuits in an electronic device usually contain electronic components. Electronic components are small devices that use the information in electronic signals to control the electric current flow in the circuits. The electronic components in today's televisions and radios are made from semiconductors.

Semiconductors

On the periodic table, the elements found between the metals and nonmetals are called metalloids. Some metalloids, such as silicon, are semiconductors. A **semiconductor** is a metalloid that conducts electricity better than nonmetals, but not as well as metals.

Semiconductors have a special property that other elements do not have. Scientists can control how well a semiconductor conducts an electric current by adding small amounts of different elements to it.

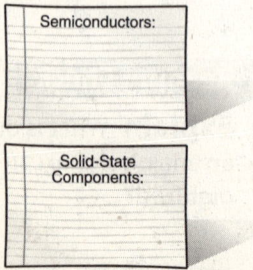

Are there different kinds of semiconductors?

Electrical conductivity measures how well a material conducts an electric current. The conductivity of a silicon crystal can be changed by adding a few atoms of an element like gallium or arsenic. Adding even a single atom of an element to a million silicon atoms greatly changes the conductivity. Atoms of a different element that are added to a semiconductor are called impurities. The process of adding impurities is called doping.

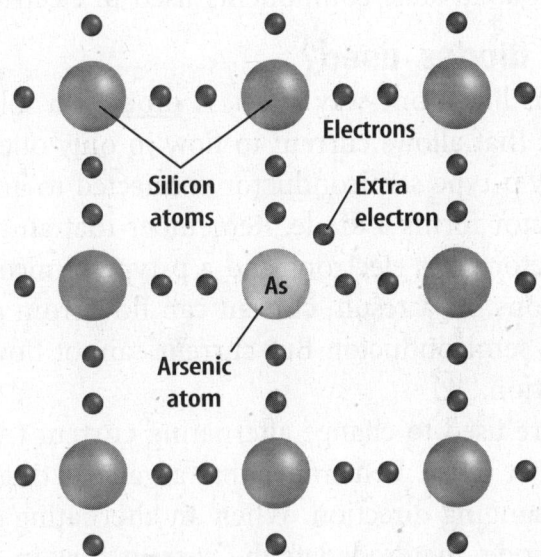

Silicon atoms

Electrons

Extra electron

As

Arsenic atom

Picture This

6. **Understand Scientific Illustrations** Highlight all the electrons in the n-type semiconductor.

N-type Semiconductor Doping can make two different kinds of semiconductors. The figure above shows what happens when atoms of arsenic are added to a silicon crystal. The arsenic atoms add extra electrons to the crystal. These extra electrons can move freely in the crystal. As a result, the electrical conductivity of the silicon crystal increases. A semiconductor with extra electrons is an n-type semiconductor. Because it has extra electrons, an n-type semiconductor can give, or donate, electrons.

P-type Semiconductor Atoms of gallium also can be added to a silicon crystal. When gallium atoms are added, the crystal has fewer electrons than it did before. This produces a p-type semiconductor. A p-type semiconductor can take, or accept, electrons.

Think it Over

7. **Explain** the difference between the two kinds of semiconductors.

_____

_____

_____

_____

Solid-State Components

P-type and n-type semiconductors can be put together to make electronic components that control the flow of electric current in a circuit. These components can be turned off and on like a switch.

The two types of semiconductors also can be put together to make components that increase the change in an electric current or voltage. Electronic components that are made by putting together different combinations of semiconductors are called solid-state components. Diodes and transistors are examples of solid-state components used in electric circuits.

How are diodes used?

A diode is like a one-way street. A **diode** is a solid-state component that allows current to flow in only one direction. A p-type semiconductor connected to an n-type semiconductor forms a diode. Remember that an n-type semiconductor gives electrons and a p-type semiconductor takes electrons. As a result, current can flow from an n-type to a p-type semiconductor. But current cannot flow in the other direction. ☑

Diodes are used to change alternating current (AC) to direct current (DC). Remember that an alternating current is always changing direction. When an alternating current reaches a diode, the diode lets the current flow in only one direction. This changes it to a direct current.

How are transistors used?

A **transistor** is a solid-state component that strengthens signals or acts as a switch in an electric circuit. An electronic signal tells a transistor to allow current to pass through it. Electronic signals can also tell a transistor to block the flow of current. ☑

What is an integrated circuit?

Personal computers and other electronic devices contain millions of transistors, diodes, and other components. All these components fit in computers and other devices because they use integrated circuits. An **integrated circuit** contains only one chip of semiconductive material and many solid-state components. An integrated circuit may be smaller than 1 mm on each side and still contain millions of transistors, diodes, and other components.

✔ **Reading Check**

8. Summarize How does a diode control current?

_____

_____

_____

✔ **Reading Check**

9. Explain how a transistor is used to control current.

_____

_____

_____

_____

● After You Read

Mini Glossary

analog signal: a signal that changes smoothly over time

digital signal: a signal that does not change smoothly, but changes in jumps or steps

diode: a solid-state component that allows current to flow only in one direction

electronic signal: a changing electric current that carries information

integrated circuit: made from only one chip of semiconductive material, containing many solid-state components

semiconductor: a metalloid that conducts electricity better than nonmetals, but not as well as metals

transistor: a solid-state component that strengthens signals or acts as a switch in an electric circuit

1. Review the terms and definitions in the Mini Glossary. Write a sentence that explains how integrated circuits make it possible for computers to contain all the components they need and still be small enough to fit on a desktop.

   _____

   _____

   _____

2. Use the outline below to help you review what you have read about electronic signals and electronic devices. Fill in the blanks where information is missing.

   Electronics

   I. Electronic Signals

      A. _____ signals

         An example of a device that uses these signals is a(n) _____.

      B. _____ signals

         An example of a device that uses these signals is a(n) _____

         _____.

   II. Electronic Devices

      A. Uses information in electronic signals to _____.

      B. An example of an electronic device is a(n) _____.

Science Online Visit **ips.msscience.com** to access your textbook, interactive games, and projects to help you learn more about electronics.

End of Section

Electronics and Computers

section ❷ Computers

What You'll Learn

- the different parts of a computer
- the difference between hardware and software
- the different types of computer memory
- the different types of computer storage

Mark the Text

Identify Specific Ideas As you read through this section, highlight the text each time you read about how information in a computer is stored.

✔ **Reading Check**

1. Explain Why were the first computers so large?

_____

_____

_____

_____

● Before You Read

You probably use computers every day. Write three things for which you use computers on the lines below.

_____

_____

_____

● Read to Learn

What are computers?

How many times have you used a computer this week? Computers are in libraries, grocery stores, banks, and even gas stations. A computer is an electronic device that carries out a set of instructions called a program. Different programs make computers do different tasks.

Some of the first computers were built in the United States between 1946 and 1951. Solid-state components and integrated circuits had not been invented yet, so the first computers were huge. One computer could take up an entire room and weigh more than 30 tons. The early computers were also very slow. ☑

How did computers improve?

Computers started using integrated circuits in the 1960s. Computers built with integrated circuits were much faster and smaller than earlier computers. Today, a hand-held game system can do many more things in one second than early computers could. For example, one of the first computers built could do 5,000 additions each second. Compare that to the millions of operations today's handheld game systems can do in one second.

Computer Information

How does a computer show words and pictures on a screen, do calculations, or play music? Every piece of information that a computer stores or uses must be changed into a series of numbers. Each word, number, picture, and sound is stored as numbers in a computer. Information stored as numbers is sometimes called digital information. ☑

What kind of numbers do computers use to store information?

What if you could only use the two words "on" and "off" to talk about everything that happens to you? Could you use these words to describe your favorite music? Could you read a book out loud using only "on" and "off"? Communication with just two words seems impossible, but that's exactly what a computer does.

All the digital information in a computer is changed to a type of number that uses only two digits—0 and 1. A number made of only two digits is called a binary (BI nuh ree) number. Each 0 or 1 is called a binary digit, or bit. A **binary system** is a number system that uses only two digits. A binary system is also called a base-2 number system.

How are binary digits combined?

You may think that only a small amount of information can be represented when you use only two digits. After all, it is almost impossible for you to describe your day using only two words. But a small number of binary digits can make a large number of combinations.

Look at the table below. It shows that one binary digit has only two possible combinations—0 or 1. However, there are four possible combinations for a group of two binary digits—00, 01, 10, and 11. Three binary digits will make eight possible combinations. The more binary digits used, the more possible ways there are to combine them. For example, 16 binary digits can be combined in 65,536 different ways.

Combinations of Binary Digits	
Number of Binary Digits	Possible Combinations
1	0 1
2	00 01 10 11
3	000 001 010 011 100 101 110 111

FOLDABLES

ⓒ Build Vocabulary
Make the following Foldables to help you learn the definitions of some of the terms in this section.

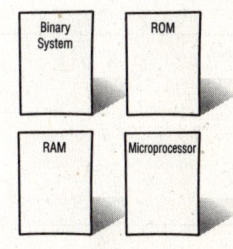

Picture This

3. **Use Tables** How many more possible combinations can be made with three binary digits than with two binary digits?

_____

_____

How do binary digits show information?

Combinations of binary digits can be used to represent information. The English alphabet has 26 letters. Each letter could be represented by one combination of binary digits. A total of 52 different combinations of binary digits would represent all lowercase and uppercase letters.

How do computers represent letters and numbers?

Computers use a common system to represent letters and numbers. A group of eight binary digits is used to represent each letter, number, or character. Eight binary digits equal one byte. There are 256 combinations possible for a group of eight binary digits. In this system, the letter "A" is represented by the byte 01000001. The lowercase letter "a" is represented by the byte 01100001.

Computer Memory

A binary number is a series of bits that can have only one of two values—0 or 1. Think of a light switch on a wall. The switch can have only two positions—on or off. A switch would be a good way to represent the two values of a bit. A switch in the "off" position could represent a 0. A switch in the "on" position could represent a 1. The table shows how switches could be used to represent different binary numbers.

Representing Binary Digits	
Binary Number	Switches
0000	
0001	
0010	
0011	
0100	
1010	

How does a computer store information?

The memory of a computer is made of integrated circuits. Each integrated circuit contains millions of tiny electronic circuits. These circuits are so tiny you cannot see them. In the most common kind of computer memory, each circuit can store an electric charge. The circuit can be either charged or uncharged. ☑

Reading Check

5. **Explain** What is a computer's memory made of?

_____

_____

Charged or Uncharged The circuit represents the bit 1 if it is charged and the bit 0 if it is uncharged. Because computer memory is made of millions of these circuits, it can store very large amounts of information using only the numbers 0 and 1.

What types of memory does a computer have?

When you remember something from long ago, you use your long-term memory. But when you work on a math problem, you may only keep the numbers in your head long enough to find the answer. You use your short-term memory. Computers also have a long-term memory and a short-term memory. Each is used for a different purpose.

What is a computer's short-term memory?

A computer's <u>random-access memory,</u> or RAM, is short-term memory that stores information and programs while the computer is using them. For example, when you use a word-processing program to write a report, your report is stored in RAM while you are working on it. Information stored in RAM is lost when you turn the computer off. Random-access memory cannot store anything that you want to use later.

Computers have different amounts of RAM. It is measured in megabytes. Remember that a byte is eight bits. A megabyte is more than one million bytes. A computer that has 128 megabytes of memory can store 128 million bytes of information in its RAM. That is nearly one billion bits.

What is a computer's long-term memory?

Information also can be stored in a computer's long-term memory. Information in the long-term memory cannot be changed in any way. It can only be read. <u>Read-only memory,</u> or ROM, is memory that cannot be changed and is permanently stored inside a computer. Information stored in ROM is not lost when the computer is turned off.

Computer Programs

A recipe has instructions that tell you how to make a certain food. A computer program is like a recipe. A program is many instructions that tell the computer how to do a job. Computer programs can contain millions of instructions that tell a computer how to do many different jobs.

Think it Over

6. **Identify** Give an example of something that might be stored in RAM.

_____

_____

Think it Over

7. **Summarize** How are RAM and ROM different?

_____

_____

_____

What does a computer program control?

Everything a computer does is controlled by programs. Instructions in programs tell the computer how to add two numbers, how to display a word, or even how to change a picture on the monitor when you move a joystick. A computer's memory can store many different programs.

What is computer software?

You use computer software when you type a report or play a video game on a computer. **Computer software** is any list of instructions that tell a computer what to do. The instructions tell the computer what to show on the monitor. When you move the mouse, the software instructions tell the computer how to respond to your action.

FOLDABLES™

D Identify Main Ideas Make the following Foldable to help you identify the main ideas about computer hardware and computer software.

Computer Programming

Computer programming is the process of writing computer software. A person who writes computer programs is called a computer programmer. Computer programmers use the following steps. First, they decide what they want the computer to do. Then, they plan the best way to organize the instructions they are going to write. Next, they write the instructions. Finally, they test the program to make sure it works. Computer programmers write software in computer languages, such as Basic, C++, and Java.

The figure shows part of a computer program. After the program is written in a computer language, it is changed into binary digits. Then the computer can store the digits in its memory and can follow the program's instructions.

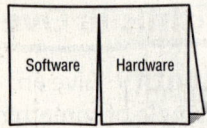

Picture This

8. Sequence What must happen to the program in the figure before the computer can carry out the program's instructions?

____

____

____

```
int request_dma(unsigned int dmanr, const char * device_id)
{
      if (dmanr > = MAX_DMA_CHANNELS)
            return -EINVAL;

      if (xchg(&dma_chan_busy[dmanr].lock, 1) != 0)
            return -EBUSY;

      dma_chan_busy[dmanr].device_id = device_id;

      /* old flag was 0, now contains 1 to indicate busy */
      return 0;
} /* request_dma */

void free_dma(unsigned int dmanr)
{
      if (dmanr > = MAX_DMA_CHANNELS) {
            printk("Trying to free DMA%d\n", dmanr);
            return;
      }
```

Mark Burnett

Computer Hardware

Press a key on a computer keyboard and a letter shows up on the monitor. This seems to happen all at once. But it takes three steps.

Step One First, the computer gets information from an input device. A keyboard or a mouse is an input device. When you press a key on the keyboard, the computer gets a signal from the keyboard and stores it.

Steps Two and Three In the second step, the computer changes the input signal from the keyboard into an electronic signal the monitor can understand. Instructions in the programs in the computer's memory tell the computer how to change the signal. In the third step, the computer sends the signal to the monitor. All three steps are carried out using a combination of hardware and software.

What is computer hardware?

Input devices, output devices, storage devices, and integrated circuits for storing information are examples of computer hardware. A keyboard and a mouse are input devices. A monitor, printer, and speakers are output devices. Floppy disks, hard disks, and CDs are storage devices. They are used to store information outside of the computer memory. A device called a microprocessor controls the computer hardware. ✔

What is a microprocessor?

A microprocessor is the brain of a computer. A **microprocessor** is an integrated circuit that controls the flow of information between the different parts of a computer. The microprocessor is also called the central processing unit, or CPU. A microprocessor can contain millions of transistors and other solid-state components. ✔

The microprocessor gets electronic signals from many different parts of the computer. It changes these signals and sends them to other parts of the computer. For example, the microprocessor might tell the monitor to change the picture on the screen. It does this by following instructions in computer programs that are stored in the computer's memory. Microprocessors can contain many millions of components on one chip and may be only 1 cm by 1 cm.

✔ Reading Check

9. Classify Which of the following is an input device? (Circle your answer.)

  a. CD
  b. keyboard
  c. floppy disk
  d. hard disk

✔ Reading Check

10. Explain What is the brain of a computer?

_____

Storing Information

Imagine you want to use a computer to type a report. The finished report is over ten pages long. You make changes to the report each time you read it. How can the computer let you store your report and also make changes to it?

You learned that information stored in RAM is lost when you turn off the computer. Information stored in ROM cannot be changed. It can only be read. To store information that you want to change but not lose when you turn off the computer, you need a storage device. Disks are storage devices. There are several different types of disks you can use. ✔

How does a hard disk store information?

A hard disk is usually located inside a computer. The figure below shows the inside of a hard disk. A hard disk is made of one or more metal disks that have magnetic particles on one surface.

Read/Write Head A device called a read/write head inside the disk drive saves the information onto the hard disk. The read/write head moves over the surface of the disk. As it moves, it changes the direction that the magnetic particles face on the surface of the disk. When the magnetic particles face one direction, it means a 0. When the particles face the opposite direction, it means a 1. When a computer reads the disk, the read/write head changes the digital information stored on the disk to an electric current.

11. Explain Why could you not use RAM to store a report you may want to change later?

_____

_____

Picture This
12. Infer Circle the read/write head in the figure.

Thomas Brummett/PhotoDisc

Information on a hard disk cannot be read as quickly as information stored on RAM and ROM. But it is stored magnetically instead of with electronic switches. This means the information stored on a hard disk can be changed and not lost when the computer is turned off.

How do floppy disks store information?

A floppy disk is a thin, bendable, plastic disk. You could use a floppy disk if you want to carry stored information with you. If you have ever held one, you may think that a floppy disk isn't very floppy at all. The bendable floppy disk is protected by the hard plastic case that surrounds it.

Floppy disks are coated with magnetic particles, just like hard disks. To store information on a floppy disk, a read/write head changes the direction the magnetic particles face. Floppy disks can store less information than hard disks can and are written to and read from more slowly.

What are optical storage disks?

CDs, DVDs, and laser disks are all optical storage disks. Information is stored digitally on an optical storage disk. An optical disk has a series of pits and flat spots on its surface. The pits and flat spots are arranged differently depending on the information stored on a disk. ☑

A tiny laser beam shines on the disk's surface to read it. The information is read by measuring the amount of energy that is reflected back from the surface of the disk. This depends on whether the laser beam hits a pit or a flat spot.

Some optical disks are read-only. The information stored on these disks only can be read. CD-RW (Read-Write) disks can be erased and rewritten several times. Information is written by a CD burner. It causes a metal alloy in the disk to change form when it is heated by a laser.

Computer Networks

A computer network lets people communicate using computers. A computer network is two or more computers that are connected to share information. These computers can be connected by cables, telephone lines, or radio signals.

The Internet is a collection of computer networks from all over the world. The Internet itself contains no information. You use the Internet to connect to other computers and find information. ☑

The World Wide Web is part of the Internet. The World Wide Web is the collection of information on computers all over the world. The computers that share this information are called servers. When your computer connects to a server through the Internet, you can view the information stored there. A collection of information stored in one place on the World Wide Web is called a Web site.

✔ Reading Check

13. Explain How is information stored on an optical disk?

_____

_____

_____

✔ Reading Check

14. Determine Which of the following does *not* contain information? (Circle your answer.)

   a. World Wide Web
   b. Internet
   c. servers
   d. a Web site

After You Read

Mini Glossary

binary system: a number system that uses only two digits

computer software: any list of instructions that tell a computer what to do

microprocessor: an integrated circuit that controls the flow of information between the different parts of a computer

random-access memory (RAM): a short-term memory that stores information and programs while the computer is using them

read-only memory (ROM): memory that cannot be changed and is permanently stored inside a computer

1. Review the terms and definitions in the Mini Glossary. Write two sentences that explain how a microprocessor uses computer software to control the flow of information in a computer.

   _____

   _____

   _____

2. Use the concept web below to complete the information you read about computer hardware.

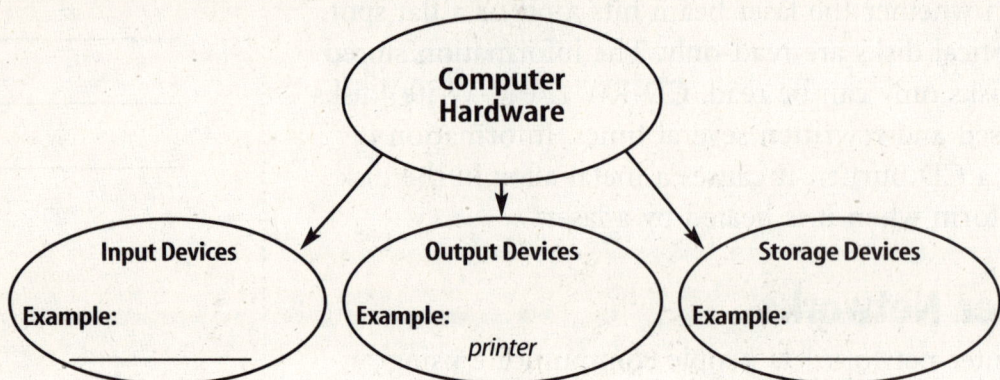

3. You were asked to highlight the text each time you read about how information in a computer is stored. How did this help you learn about storing computer information?

   _____

   _____

   _____

End of Section

Science **nline** Visit **ips.msscience.com** to access your textbook, interactive games, and projects to help you learn more about computers.

PERIODIC TABLE OF THE ELEMENTS

Columns of elements are called groups. Elements in the same group have similar chemical properties.

🎈 Gas
💧 Liquid
⬜ Solid
⊙ Synthetic

Element — Hydrogen
Atomic number — 1
Symbol — H
Atomic mass — 1.008
State of matter

The first three symbols tell you the state of matter of the element at room temperature. The fourth symbol identifies elements that are not present in significant amounts on Earth. Useful amounts are made synthetically.

	1	2	3	4	5	6	7	8	9
1	Hydrogen 1 H 1.008								
2	Lithium 3 Li 6.941	Beryllium 4 Be 9.012							
3	Sodium 11 Na 22.990	Magnesium 12 Mg 24.305							
4	Potassium 19 K 39.098	Calcium 20 Ca 40.078	Scandium 21 Sc 44.956	Titanium 22 Ti 47.867	Vanadium 23 V 50.942	Chromium 24 Cr 51.996	Manganese 25 Mn 54.938	Iron 26 Fe 55.845	Cobalt 27 Co 58.933
5	Rubidium 37 Rb 85.468	Strontium 38 Sr 87.62	Yttrium 39 Y 88.906	Zirconium 40 Zr 91.224	Niobium 41 Nb 92.906	Molybdenum 42 Mo 95.94	Technetium 43 Tc (98)	Ruthenium 44 Ru 101.07	Rhodium 45 Rh 102.906
6	Cesium 55 Cs 132.905	Barium 56 Ba 137.327	Lanthanum 57 La 138.906	Hafnium 72 Hf 178.49	Tantalum 73 Ta 180.948	Tungsten 74 W 183.84	Rhenium 75 Re 186.207	Osmium 76 Os 190.23	Iridium 77 Ir 192.217
7	Francium 87 Fr (223)	Radium 88 Ra (226)	Actinium 89 Ac (227)	Rutherfordium 104 Rf (261)	Dubnium 105 Db (262)	Seaborgium 106 Sg (266)	Bohrium 107 Bh (264)	Hassium 108 Hs (277)	Meitnerium 109 Mt (268)

The number in parentheses is the mass number of the longest-lived isotope for that element.

Rows of elements are called periods. Atomic number increases across a period.

The arrow shows where these elements would fit into the periodic table. They are moved to the bottom of the table to save space.

Lanthanide series	Cerium 58 Ce 140.116	Praseodymium 59 Pr 140.908	Neodymium 60 Nd 144.24	Promethium 61 Pm (145)	Samarium 62 Sm 150.36
Actinide series	Thorium 90 Th 232.038	Protactinium 91 Pa 231.036	Uranium 92 U 238.029	Neptunium 93 Np (237)	Plutonium 94 Pu (244)